The Phylogeny of

VERTEBRATA

The Phylogeny of
VERTEBRATA

SØREN LØVTRUP

*Department of Zoophysiology,
University of Umeå,
Sweden*

A Wiley–Interscience Publication

JOHN WILEY & SONS

London · New York · Sydney · Toronto

Library of Congress Cataloging in Publication Data:

Løvtrup, Søren:
  The phylogeny of vertebrata.

  'A Wiley–Interscience publication.'
  1. Vertebrates—Evolution. 2. Phylogeny. I. Title.
QL607.5.L63   596'.03'8   76–18707

ISBN 0 471 99412 X

Printed by William Clowes & Sons, Limited
London, Beccles and Colchester

A Ma Bien Aimée

'Why don't you, my dear peers, think instead of but compiling?'
(L. Croizat, 1962 [1964], p. 673).

'L'intelligence des phénomènes évolutifs exige une solide connaissance de la zoologie, de la paléontologie, de la génétique, de la biochimie, voire des mathématiques .... Un seul homme ne saurait posséder une aussi vaste erudition et pourtant, sans elle, comment traiter de l'évolution avec pertinence?'
(P.-P. Grassé, 1973, p. 10).

'From the soil of ignorance sprout the seeds of fresh ideas.'
(P. S. Prescott, 1974, p. 52)

'Any profession that does not supply its own criticism and iconoclasm will discover that someone else will do the job, and usually in a way it does not like.'
(N. Macbeth, 1971, pp. 149–150)

PREFACE

When, many years ago, I read *The Material Basis of Evolution*, I became thoroughly convinced of the correctness of Goldschmidt's ideas on large-scale evolutionary steps and, in particular, of the importance of epigenetics for phylogenetic innovation.

In my book *Epigenetics* I tried, in the last chapter, to outline some of the consequences of these ideas. My endeavours were incomplete in various respects, and I therefore continued to work on the problem. The result is the present book.

I have published, or tried to publish, sections of the book as separate papers. Frequently, the referees involved have shown little appreciation of my work, on occasions so little as to make me feel that I might be totally wrong. Fortunately, the papers which I succeeded in publishing have met with a response that has convinced me of the existence of a latent dissatisfaction with the ruling theoretical superstructure of present-day biology which has encouraged me to continue the work.

I have had the opportunity to present parts or the whole of the manuscript to various colleagues. Among these I may mention Dr. Gunnar Bertmar, Professor C. Devillers, Professor Bo Fernholm, Professor Arne Lindroth and Dr. Huguette Løvtrup-Rein. Their suggestions have entailed many improvements in the manuscript, for which I express my sincerest appreciation.

In particular I am indebted to Dr. Colin Patterson, who has been kind enough to subject the whole manuscript to a thorough revision.

I also wish to express my thanks to the personnel at the University Library in Umeå and at the other research libraries who have helped to satisfy my almost insatiable demands for literature.

The manuscript has been typed by Mrs. Ann Sofie Lindström, Mr. Ronald Grönlund and Mr. Nils Granberg. The original figures have been prepared by Mr. André Berglund. I should like to express here my appreciation of their cooperation.

CONTENTS

'Theory without fact is fantasy, but fact without theory is chaos. Divorced, both are useless; united, they are equally essential and fruitful.'
(C. O. Whitman, 1895, p. 205)

1

INTRODUCTION

With the establishment of the theory of phylogenetic evolution, notably through the work of Darwin, it followed that the ancestors of Vertebrata were to be sought among invertebrates living in the distant past.

The difficulties involved in the solution of the problem thus posed, *the phylogenetic ancestry of Vertebrata*, are obvious, for to all appearances the two groups, vertebrates and invertebrates, are separated by an almost unbridgeable abyss. Yet, nonetheless, Darwinian theory forces upon us the logical deduction that of the several invertebrate taxa one must be more closely related than all the others to Vertebrata, or to Chordata, because its members have ancestors in common with those of one of the above taxa which are not at the same time ancestors to members of any other invertebrate taxon. The disclosure of the invertebrate phylum in question constitutes a fascinating challenge to all biologists interested in phylogenetic problems.

Another question related to the origin of Vertebrata, arising as a consequence of Darwin's theory, is concerned with the phylogenesis of Vertebrata themselves, i.e. the succession of events through which the extant forms came into being once the realization of the vertebrate body plan had been effected by Nature. The taxon Vertebrata comprises a rather small number of taxa of high rank, each of which is distinguished by several unique characters. There is thus no problem about the initial subdivision of the taxon, and it might therefore *a priori* be anticipated that the answer to the second problem, *the phylogenetic divergence of Vertebrata*, would be easy to obtain compared to that concerning ancestry since, after all, it amounts to nothing more than establishing the kinship between these various high-ranking taxa.

Yet, curiously enough, a study of the literature shows that if concord may be taken to indicate that a problem is solved and discord that it is not, then the prevailing situation does not correspond to expectation. Rather, as far as the ancestry of Vertebrata is concerned, it appears that, but for some few timid voices of dissent, the overwhelming majority of biologists have, for the last half century, concurred in the belief that, outside the phylum Chordata, the closest relatives of Vertebrata among major invertebrate taxa are to be found in the phylum Echinodermata. In contrast, although agreement unquestionably obtains among phylogeneticists about the main outlines of the course of phylogenesis within the taxon Vertebrata, the pertinent literature, including the most recent, reveals that the opinions which have been voiced on this subject are many, widely diverging and, as a rule, mutually incompatible.

There is, of course, nothing wrong in scientists disagreeing, for in other empirical sciences it has often been found that through argument the issue between rival theories may be settled, with the subsequent general acclaim of the superior theory by all involved. This apparently is not the situation within phylogenetics; although it happens that certain propositions are abandoned, it rather seems that the number of rival hypotheses concerning the phylogeny of Vertebrata tends to multiply with time.

In the present book I have tried to find answers to these two problems. Chapter 2 outlines the theory of the method through which this goal has been approached. Since it deviates in some aspects from traditional phylogenetics, I have given a brief survey of the latter in the first subsection following. Chapters 3 and 4 present the answers I have reached, and Chapter 5 a theory of evolution. Against the background of the latter various consequences of my conclusions are tested through confrontation with empirical data. It is not the purpose of this book to give a detailed discussion of the mechanisms of phylogenetic evolution. Yet, since my theory of evolution asserts that large-scale mutations have been of utmost importance for phylogenetic evolution, I shall in the second subsection below advance various arguments in favour of this contention.

TRADITIONAL PHYLOGENETICS

Phylogenetics belongs to the large group of biological disciplines which do not readily lend themselves to formulation in terms of mathematics. This implies that thinking and argumentation in this field are deprived of the guidance for deductive reasoning which the resources of this science might provide.*

Another reason may be the overwhelming amount of empirical observations which have been accumulated through the assiduous work of many generations of biologists; no single person can survey all this knowledge, yet none of it can *a priori* be dismissed as irrelevant for the problem of phylogenetic evolution. One flaw in phylogenetic discourse, which probably may be referred to this cornucopia of empirical data, is to be found in the fact that phylogeneticists tend to base their views on the course of evolution on a single or, at most, a very limited number of characters, usually those which happen to be their speciality, often with high-handed disregard for other characters which may have led some of their colleagues to different, and contradictory, systems.

A further consequence of the wealth of existing information is that works on evolution are often so swamped with observational facts that the line of reasoning is obscured, a tradition which may be traced back to Darwin. And yet, of course, even in phylogenetics it is the case 'that only a clear statement of how the conclusions were arrived at can disclose the causes of divergent opinions' (Szarski, 1962, p. 195). The fact that this rule is seldom adhered to certainly appears to corroborate the assertion that 'the common biological methods of reasoning . . . are much less precise than those used by such sciences as physics and chemistry' (Szarski, 1962, p. 189). This, being a statement of fact, is at the same time an admission of failure, for there is, of course, no justification

---

*This statement refers to the traditional morphological approach. In numerical taxonomy and, particularly, in the studies on the phylogeny of cytochrome c and other proteins, mathematics has been employed with remarkable success.

whatsoever for the contention that the rules of logical thinking applicable in biology differ from those employed in other empirical sciences.

The basis for all phylogenetic work is the observation and listing of properties of all kinds possessed by living and extinct animals. The identity or close similarity between such characters, as found in two or more taxa, is usually interpreted as demonstrating phylogenetic kinship between the animals in question. The situation is more complex if distinct differences obtain, i.e. if the taxa have become disjoined through sizeable processes of evolution. Under such circumstances it becomes necessary to introduce rules of various kinds to account partly for the transformations which have occurred, allowing for tracing the affinity between the several taxa, and partly also perhaps for the agents that have been instrumental in those transformations. The generalizations, hypotheses or theories which have been invoked for this purpose may be subdivided into three groups: rules (or laws) of evolution; comparative–morphological interpretations; and teleological explanations.

The most important, in the sense of well-known, rules of evolution are Haeckel's 'biogenetic law', Cope's rule of the 'ascendancy of the unspecialized' and Dollo's rule of 'irreversibility in evolution'. The validity of the application of these hypotheses has been scrutinized by Szarski, whose final verdict is as follows: 'These rules indeed have their value, as they characterize in a concise way the course of evolution. But unlike the rules of physics and chemistry, they are not valid in every instance and . . . therefore one can never be sure of their validity in any particular case' (1962, p. 192). The reason for this state of affairs seems to be that, evolution being a chance event, the rules describing its course must be probabilistic ones.

Comparative morphology, or 'formanalysis' (Kuhlenbeck, 1967), the science concerned with the theoretical generalization and interpretation of observations of animal structure and form, is a branch of biology with proud traditions. Among the concepts which have been of particular usefulness in phylogenetics are those of 'homology' and 'coelom'.

Of these the first has been of special import in evolutionary studies, because the presence of homologous characters was taken as evidence of common ancestry. Much effort has been devoted to the purpose of establishing flawless criteria of homology (cf. Remane, 1956). There is no doubt that studies on homology have made many very significant contributions to phylogenetics, but it is obvious that biologists cannot always come to terms about what is homology and what is not; certainly, some evolutionists seem willing to stretch the application of this phenomenon beyond the limits of probability. As one example, among many, of the contributions of this 'school' may be mentioned Gegenbauer's hypothesis on the origin of paired limbs, according to which the latter are homologous with gill arches and gill rays in fish. By and large this interpretation is no longer in vogue, but, as pointed out by Szarski (1962), it is still occasionally encountered.

The concept of 'coelom' illustrates another working principle of comparative morphology, involving generalizations based on very abstract geometrical relationships. No doubt 'coelom' may in most cases be defined so as to be concerned with a circumscribed biological reality, but on closer scrutiny a coelom is a very simple morphological structure, namely, a cellular vesicle, arousing the suspicion that it may have arisen independently on more than one occasion. If this is the case it obviously loses in importance for phylogenetic studies. To make matters worse, a coelom may arise through two distinct morphogenetic mechanisms, as has been recognized by the comparative morphologists.

Originally, this difference was considered to be of phylogenetic significance, i.e. the two kinds of coelom were considered to be non-homologous, an apparently sound proposition. Subsequent discoveries have shown that the distribution of coelom-forming mechanisms does not correspond to initial suppositions, but, as we shall see in a later chapter, these observations have not led to any modifications in the classification based on the mechanism of coelom formation.

Stressing once more that comparative morphology has made very important contributions to phylogenetics, it must nevertheless be pointed out that the rules established in this discipline are not categorical; they may, but they need not, be followed, a fact which clearly impairs their general validity.

According to Darwinian and neo-Darwinian theories, the prime agent in evolution is natural selection, which, favouring the individuals that in a given situation are best adjusted to the environment, will ensure (1) a maximum adaptation to the latter and (2) a continuous modification of the form and function of animals in step with the environmental changes to which they are subjected. Through this principle, which apparently implies that animals are what they are because they live where they live, it is possible to explain the origination of all forms, past or present, on the assumption that either the environment changed or the animals changed their environment. Thus, fishes acquired lungs through exposure to air, limbs as a consequence of living in the vicinity of shores, etc.; the phylogenetic literature abounds with further examples.

It has been submitted that this teleological* theory, which certainly can account for evolution in a very simple way, is contradicted by diverse empirical observations. Generally, such objections are refuted through the introduction of *ad hoc* hypotheses of various kinds, the most usual of which is stated in the following quotation: 'Since Darwin's time, characters have been scrutinized with regard to their survival value. Where naturalists have failed to find such values, they have assumed that the data were merely incomplete' (Noble, 1954, p. 127). Thus once more we find that the adoption of a theory is at the discretion of the biologist: if observations are found to fit predictions, then they are accepted; if not, incompleteness is admitted. It is worthy of note, however, that this attribute is ascribed to the data, not to the theory.

It has often, and from many quarters (notably by Woodger, cf. 1929; 1952), been argued that we cannot expect to make any noteworthy headway in biology before we have succeeded in establishing a generally accepted theory, a framework of axioms and definitions, which forces all reasoning and argumentation to proceed along the lines of deductive logic. The truth of this contention is clearly demonstrated by the fact that those annexes of biology in which this has been accomplished, viz., genetics and molecular biology, have shown spectacular progress compared with other disciplines.

Before we discuss further the subject of biological theories, it may be opportune to deal with one fundamental aspect of empirical theories which has been so forcefully argued by Popper. I can present this point no better than by quoting from one of the works in which his views are expounded (Popper, 1969, p. 36): 'Every genuine *test* of a theory is an attempt to falsify it, or to refute it' and 'Confirming evidence should not count *except when it is the result of a genuine test of the theory*; and this means that it can be presented as a serious but unsuccessful attempt to falsify the theory'.

---

*Here, and elsewhere in the text, I use the word 'teleology' in the trivial sense, i.e. 'the use of design, purpose, or utility as an explanation of any natural phenomenon' (Webster's Third New International Dictionary).

It seems that among the shortcomings of phylogenetic reasoning the most serious is that Popper's principle of falsification is constantly violated. From the above discussion it appears that application of the three principal groups of theories employed in phylogenetics is a matter of individual preference and this, evidently, implies that they cannot be falsified. If this method is adopted as a standard approach, then the biological theory loses its predictive value at the same time as it becomes irrefutable. Instead of being a guide for logical deduction, it is a means of interpreting the relation between biological events which may be applied or ignored at the whim of the individual biologist, and thus it becomes a metaphysical, rather than an empirical, theory.

This reluctance to acknowledge falsifications is also displayed by some of those engaged in comparative biological sciences outside the morphological ones upon which, traditionally, phylogenetic classification is based. Obviously, results obtained in such studies are perfectly suited for the purpose of testing the established theories, and it would therefore appear that this should be one of their main purposes. This view stands in striking contrast to the following assertion: 'In the present situation, we must adopt the methodical rule to be led by knowledge of phylogeny in our search for biochemical evolution, rather than to be brought by biochemistry to the discovery of new aspects of phylogeny. The Ariadne's thread of comparative biochemistry can only be the knowledge of phylogeny, in which is integrated the knowledge about living organisms accumulated by generations of naturalists' (Florkin, 1966a, p. 13).

Acknowledging and admiring the veneration thus shown towards our predecessors, I think it needs to be pointed out that the attitude stated in this quotation implies that a monopoly is conceded to the morphologists as far as the solution of phylogenetic problems is concerned, the object of other disciplines of comparative biology being largely to confirm the results arrived at by systematists. Apart from the fact that, as we shall presently observe, in certain cases the use of characters other than the traditional ones is imperative, it appears to me that this attitude implies that all possibilities for the non-morphologist of making exciting discoveries are barred in advance. As a consequence most of the fun and the sense of the scientific vocation are lost.

The timidity thus expressed may in part be explained by the fact that the other comparative sciences are young relative to morphology, and that the amount of data compiled may appear too scanty to warrant its use in refutation of the established animal classification. There is some justification in this standpoint, but the main reason no doubt is that, as claimed by Kerkut (1960), academic teaching and learning in biology too uncritically accept established authorities.

This last point leads me to mention that, according to an anonymous critic of a manuscript of part of the present book, it is 'silly' to challenge prevailing phylogenetic theories as Kerkut and I do. And yet, so far as I can see, we do nothing but follow the procedure which Popper describes as the noblest pursuit in empirical sciences, and the only one that ensures progress, namely, to try to falsify the ruling theories. We may not be imaginative enough to put forward new and better theories, but if, by following logical lines of reasoning, we can show that the present theories lead to falsifiable predictions, then maybe those who are predestined to carry phylogenetics forward will be encouraged to make their contributions.

The attitude reflected by the quoted unknown critic appears to be complacency with the present state of affairs, and undoubtedly he has the majority of biologists behind him. Yet, uneasiness with established views on the agents and the course of phylogenetic evolution has been expressed repeatedly during the past Darwinian century. Some of these authorities have been quoted in my former book (Løvtrup, 1974), more are known, and many more, of course, unknown to me. Yet, unless the maxim 'The majority is always right' is made a premise in biological research, there is no reason to presume that these dissenters are necessarily wrong, and I, for one, find in their assertions, which to a large extent coincide with mine, the strong support which may be needed for venturing to publish the heretical views contained in the present book.

As stated above, there cannot exist a special biological way of reasoning. Biologists, like all others engaged in intellectual activities, must follow the only available expedient which will ensure a meaningful outcome, namely, deductive logic. It is, of course, true that we are not yet able, and may never be, to establish in biology a theoretical super-structure of the kind prevailing in physics, allowing rigorous and quantitative deduc-tions, primarily because we cannot make use of mathematics to any considerable extent. This circumstance, among others, implies that the beginning must be modest. Never-theless, it is my firm belief that even with a small number of premises it is possible to establish a framework which can guide our interpretation of the phylogenetic implica-tions of known biological data in a way that leaves little or, hopefully, no room for individual prejudices.

As I aim to show in the following chapter, it is particularly towards the problem of classification that this theoretical approach must be directed. With firm rules of classifi-cation it will be possible to establish a phylogenetic classification of all living animals, which can be tested and falsified through empirical observation.

In the literature one may often find statements to the effect that the solution of this or that phylogenetic problem must await further discoveries, for instance of some fossil 'missing link'. This obsession for facts, which so typically distinguishes present-day biology, reveals that induction still exerts a strong influence upon modern scientists. As is well known, the doctrine of induction implies that if we only have the relevant facts, then the theories will reveal themselves gratuitously, as it were, to the inquisitive mind.

This viewpoint should be contrasted with the epistemology of Popper (1968), accord-ing to which all theory construction is pure speculation and guesswork, inspired at its best. On the other hand, nothing is more inconsequential than the compilation of facts (protocol sentences); observations are relevant only to the extent that they form part and parcel of an empirical theory. In fact, it is to be suspected that heedless accumulation of empirical facts may have what is surely an undesired effect, namely, to deaden imagina-tive minds through the cornucopia of facts they must absorb.

It is my conviction that, for a long time now, the known biological facts have amply sufficed for the formulation of a testable and falsifiable theory on phylogenetic evolution. And it is in everybody's interest to strive towards this goal, for once it is reached we may replace unplanned research, i.e. the random collection of facts, with planned research, such that out of the infinitude of empirical observations which potentially we can make, we may choose to make only those which are relevant to scientific endeavour, i.e. those that will show whether or not it is possible to falsify the theory we cherish or distrust.

It may appear that the programme I have outlined here is one pertaining more to epistemology than to biology. In this I agree. The present work, at least the second chapter, and part of the fifth, should have been written by a philosopher of science, who, indubitably, could have done it much better than myself. Unfortunately, biologists have received little assistance from that quarter; with few exceptions, philosophers of science have vouchsafed all their acumen and industry to physics.

MECHANISMS OF EVOLUTION

In a different context (Løvtrup, 1974) I have recently discussed various mechanisms invoked to account for the occurrence of phylogenesis. The main conclusion of this survey was that up to now far too much emphasis has been placed upon genetic and molecular–biological, and far too little upon epigenetic, mechanisms. In particular, it was postulated that if this one-sided stand is abandoned, the number of mechanisms available for bringing about evolutionary change will be substantially increased, and many points which at the moment are difficult or impossible to explain by the theory of neo-Darwinism will find a simple resolution.

Since this question is of consequence for our understanding of the nature of evolutionary innovation, certain topics concerning evolutionary mechanisms, viz., mutations, inheritance, cell differentiation and epigenesis, will be dealt with in the following sections.

Mutation

A mutation is defined as 'any change in the genotype not due to gene recombination' (Dobzhansky, 1970, p. 43). It is generally claimed that all evolutionary changes are the outcome of either mutations or gene recombinations. By and large, this contention is no doubt correct; indeed, it is difficult to demonstrate that it is wrong. However, I believe that one such example may be found in the centriole. If, as is supported by some evidence, this organelle is engaged in its own replication, although possibly indirectly (cf. Fulton, 1971), then its origination cannot be referred solely to the properties of the protein subunits of which it is composed, and hence to the corresponding structural genes. Rather, the first centriole must have originated fortuitously, and its absence in the higher plants be due to its loss by chance.

Three kinds of mutations may be distinguished: genome mutations, chromosome mutations and gene mutations, listed in the probable order of increasing frequency. As discussed by Crow (1972), many observations suggest that mutations constitute a continuum with respect to the extent of their effects, that disadvantageous mutations are more frequent than advantageous ones and that their frequency is inversely related to their effect. If these premises are accepted, the distribution of mutations may be represented by the diagram in Figure 1.1. In disagreement with Crow, the distribution curve implies that highly advantageous mutations are possible, although highly improbable, events.

Using this figure we may outline the range of mutations that can be observed by researchers in various biological disciplines. First, geneticists can only work with observable mutations. This implies that they must occur at a frequency which is very high in evolutionary terms. It is suggested that these mutations lie above the horizontal line (4) in Figure 1.1. Further, they cannot observe the neutral and quasi-neutral mutations, lying between the two vertical lines (1 and 2) at some distance from the line of neutrality. What remain are two sections, a tiny one representing beneficial, and a large one representing deleterious, mutations. Being able to record even neutral and quasi-neutral mutations, molecular geneticists can work with one uninterrupted range of mutations.

All other biologists do not work with mutating genes, but study the effects of mutations on living and extinct organisms. Their working range is therefore cut off to the right by the line of viability (3), and is thus narrower here than the one available to genetic studies, which encompasses lethal mutations, but to the left there are no restrictions at all. Ordinary biologists have in common with geneticists the fact that they cannot observe neutral mutations; this limitation is not faced by molecular biologists (Figure 1.1).

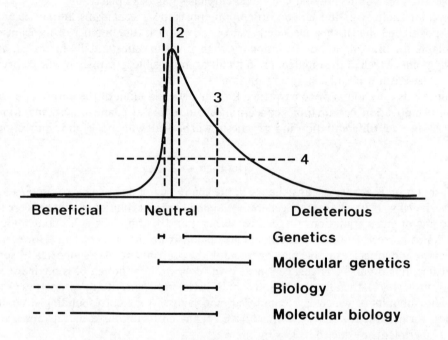

Figure 1.1 Conjectural correlation between the frequency and the effect of new mutations. Lines 1 and 2 delimit the range of neutral and quasi-neutral mutations. To the left of line 3 the mutations are viable, to the right they are lethal. Mutations above line 4 are frequent enough to be observable within a finite period of time; below are the rare mutations. Since probability is involved, it is clearly impossible to establish a clear-cut boundary. In this respect the figure is therefore slightly misleading. The labels 'Molecular biology' and 'Biology' are slightly ambiguous, but no misunderstanding seems possible. Based in part on Crow (1972)

Geneticists and evolutionists usually claim that beneficial mutations, particularly large-scale ones, do not exist, and that, indeed, most mutations are deleterious. From Figure 1.1 it appears that this viewpoint is correct as concerns the mutations observable in genetic experiments. But only an extreme positivist position can justify the inference that, if we cannot observe large-scale beneficial mutations, then their possible occurrence may be rejected. If it is admitted that mutations are chance events with widely varying frequencies, and allowance is made for the fact that Nature has had billions of years at her disposal, while genetic observations have been recorded at most for some thousand years, then this inference loses all credibility. I shall endeavour to present arguments for the occurrence of large-scale beneficial mutations in this and the next subsection.

In most cases mutations have been studied with reference to their visible effect on the phenotype, which implies that nothing is known about their immediate effect. With the advent of molecular biology it has been demonstrated that a certain, possibly rather small, fraction of the genome, the structural genes, specifies the amino acid sequences of all the polypeptides which can be synthesized by a certain organism, except for some proteins synthesized in the mitochondria. The information relevant for this purpose is represented by the base sequences in the DNA molecules which are the genes. After the ascertainment of the code which allows for the translation from base to amino acid sequences, it was found that this code has remained virtually unchanged during the course of evolution from bacteria to man. This point may be noted as one example of the remarkable conservatism displayed by evolution at the chemical level.

According to the concepts of molecular biology, any change in the sequence of bases in a DNA molecule is a mutation. Most common are point mutations, involving a base substitution, but mutations are known which entail different, and more extensive, changes in a gene (Jukes, 1966).

So far as the structural genes are concerned, it is possible to some extent to specify the consequences of mutations. Thus, confining ourselves to point mutations, it follows from the degeneracy of the genetic code that about 25 per cent of the mutations represent synonymous codons, and consequently do not lead to any changes in the amino acid coded (Sonneborn, 1965; Jukes, 1966; King and Jukes, 1969). Such mutations are mostly 'silent', their effect not being transmitted beyond the level of transcription. Hence, they do not correspond to the genetic notion of 'mutation', according to which a mutation must be operative, i.e. give rise to a phenotypic modification.

From comparative studies on the amino acid sequences in various proteins, it is possible to distinguish two extremes with respect to the effect of an amino acid substitution (Zuckerkandl and Pauling, 1965). Thus, it has been found that at certain 'invariant' sites no change has occurred in the course of evolution, while in other cases substitutions have occurred repeatedly, although to varying extents. It is assumed that in the former case the amino acid in question is of unique importance for the integrity of the polypeptide, any substitution leading to the loss of its function. If the latter is vital for the organism, every mutation will immediately be weeded out. The mutations affecting the sites where many changes have occurred may in some instances be envisaged as neutral, affecting the function imperceptibly and offering no appreciable selective advantage or disadvantage. Between these alternatives may be imagined a range of intermediate deleterious effects, and even beneficial ones should be possible. These mutations are supposed to be rejected or established as predicted by the neo-Darwinian theory of population genetics. It may be observed that according to King and Jukes (1969), Kimura and Ohta (1972) and King (1972) the amino acid composition of a number of proteins is indicative of random mutation.

According to Darwinian notions about the interaction between mutation and selection, the latter may ensure primarily the preservation, secondarily the improvement, of a given function. In order to account for the origin of evolutionary innovations on the chemical level, i.e. of proteins capable of carrying out hitherto non-existent functions, the phenomenon of gene duplication has been invoked (E. B. Lewis, 1951; Ohno, 1970; Britten and Davidson, 1971). The possible existence of redundant genes, exempt from selection towards preservation of a given function, evidently increases the likelihood that proteins with new functions may evolve. Yet it is obvious that in the creation of such novelties selection cannot be involved, since this agent does not enter into action before the molecule has assumed a function worth preserving. Hence, it appears that proteins with new properties can arise only fortuitously, without the supporting and directing influence of selection.

For this reason it may be anticipated that evolutionary progression on the chemical level is an extremely rare occurrence, and this prediction has been borne out by comparative biochemical studies which have shown that a number of fundamental metabolic processes, for example glycolysis, mitochondrial oxidations and amino acid syntheses, are almost identical in microorganisms and in the higher plants and animals. In point of fact, the survey presented by Britten and Davidson (1971) shows that, of the total of enzyme activities known from eukaryotes, 93 per cent are also present in prokaryotes.

Still, innovations have occurred on the chemical level. Numerous examples of this may be found in the chapter on the origin of the vertebrates. It seems justified that structural mutations giving rise to substances capable of exerting completely new functions are classed as large-scale mutations.

The function of the total of the DNA present in the metazoan genome is still a mystery, but the existence of regulatory genes may be taken for granted. The latter are purportedly involved in controlling the transcription of the structural genes. We know very little about the consequences of mutations in the regulatory genes, but it is reasonable to presume a range of effects from 'silence' to modifications of great significance. The latter point is supported by the fact, to be discussed in a later section, that morphogenetic processes are sustained through the participation of a very limited number of substances, the occurrence of which is ensured by structural genes. It is therefore at least probable that morphogenetic innovations in many cases are the outcome of large-scale mutations concerning regulatory genes.

Zuckerkandl (1975) has recently discussed the question of evolutionary innovation on the chemical level. His main conclusions are: (1) new structural genes can derive only from pre-existing structural genes, (2) new functions can derive only from the structural basis of pre-existing functions and (3) innovation is associated with the creation of new mechanisms of gene regulation, and thus with the formation of new differentiation patterns. Apart from a certain measure of preformation involved in Zuckerkandl's views, they are seen to agree with those submitted here.

In studies on amphibians and mammals it has been found that there is no correlation between protein evolution, the outcome of point mutations, and taxonomic divergence (Wilson, Maxson and Sarich, 1974; Wilson, Sarich and Maxson, 1974; Wilson *et al.*, 1975). On the other hand, a neat correspondence is found between the latter and the chromosome number. This observation has prompted the authors to propose that taxonomic divergence results from chromosome mutations, which may affect the expression of the structural genes. Should this turn out to be true, it amounts to a striking vindication of Goldschmidt's tenet (1940) that large evolutionary innovations are the result of chromosomal mutations.

A third kind of possible large-scale mutations includes those affecting the structural genes which code for polypeptide hormones, or for proteins involved in the synthesis of steroid hormones. It may be anticipated that even mutations of this kind cover a large range with respect to their effect. This point will be discussed in a later section.

Inheritance

With some few exceptions, genetic studies are of necessity confined to the inheritance of properties which are consequences of differences in the nuclear genome obtaining between members of the same species. The rules governing the inheritance of these nuclear elements, the genes, were first established by Mendel. This theory has, since its rediscovery in 1900, achieved a truly remarkable ascendancy; it is no exaggeration to assert that on the supra-molecular level Mendelism is the only biological theory which adequately meets the demands to be made on scientific theories. Owing to this success the notion has gradually become prevalent that all the information which through inheritance is transferred from generation to generation is located in the genes. This view has, of course, been further strengthened by the recent landslides in the field of molecular biology, including both genetic and biochemical aspects.

While, evidently, it is true that all the information whose inheritance can be studied in intraspecific crossings resides in the nucleus, one may ask whether the dogmatic assertion that this is the *only* information involved does not represent an unduly myopic view. Indeed, little reflection is needed to realize that the inherited material comprises more than the maternal and paternal genes, namely, the whole egg cell and certain extranuclear contributions made by the sperm, for instance a centriole in many species. These entities do indeed carry information important for ontogenesis, the former about the state of 'being a cell', and the latter about the process of cell division and certain other cellular functions (cf. Løvtrup, 1974). I do not believe that anybody will deny the indispensability of this information, nor do I think that even the most ardent advocate of molecular biology would maintain that it can be replaced by information residing in the genome, but it may possibly be argued that it is trivial insofar as it is common to all living organisms. This, incidentally, is not quite true, for, as mentioned above, centrioles are absent in higher plants.

Furthermore, the fact that cell differentiation takes place shows that transcription of the genome in the course of embryogenesis must be regulated by mechanisms of nucleo-cytoplasmic interaction. This circumstance apparently implies that even complete knowledge about the information coded in the genome, in structural as well as in regulatory genes, would not suffice to predict the course of the subsequent epigenetic process. Surely, if cytoplasmic factors can affect the transcription of the genes, then specification of the composition of the cytoplasm would be required with respect to the substances involved in nucleo-cytoplasmic interactions, since changes in the latter might upset the normal sequence of events.

This is not the place to discuss the influence of extranuclear factors on epigenesis; I have done that elsewhere (Løvtrup, 1974). The importance of this topic in the present context is that although these agents may be inherited according to the Mendelian laws, the possibility exists that they are not. Consequently, their contributions to the 'phenotype' may not be demonstrable in genetic experiments, and this could be the reason why so little is known about them. In spite of their elusiveness, it is possible that the extranuclear factors are of great ontogenetic consequence. Indeed, it is not impossible that in many instances they are involved in the establishment of the characters defining taxa of relatively high rank.

Another shortcoming of the genetic approach is that it is by necessity confined to characters which to a large extent concern slight modifications of the phenotype. This very fact implies that the mutations in question must become effective at a late onto-genetic stage. It is generally suggested that mutations of this kind, through accumulation, have been responsible for all the changes which have taken place in the course of evolution. It is a remarkable fact, however, that the basic body plan is laid down in various embryos before the transcription of the genome has begun to any significant extent. This seems to leave little room for the kind of mutational effects envisaged by geneticists, at least as far as the initial phases of morphogenesis are concerned. In fact, the results of comparative embryology rather suggest that taxonomic divergence has occurred through modifications at successive ontogenetic stages, a course of evolution which is also implied by the phylogenetic classification of the kingdom Animalia.

The reservations made here are not intended to detract from the importance of genetics. The subject of this discipline, the inheritance of nuclear genes and their modification through mutation, is a phenomenon which has been of the utmost consequence for phylogenetic evolution. Nevertheless, it appears that we cannot expect to solve all questions about the importance of heredity for phylogenesis exclusively on the basis of the approach adopted in classical genetics.

Cell differentiation

It is a distinctive trait of metazoans that their body is composed of a number of different cell types, the members of each of which synthesize only a limited part of the whole repertoire of polypeptides encoded in the genome. The genes representing the proteins not synthesized by a given cell must therefore exist in a repressed state. It may thus be inferred that genes are usually repressed rather than expressed. In point of fact, it appears that a very important aspect of epigenesis consists in the ordered derepression of different parts of the genome, thereby giving rise to the various patterns of synthetic activity represented by the several kinds of cells found in the body. I have proposed (Løvtrup, 1974) that it is possible to distinguish two separate steps in this process of 'cell diversification'. The first of these, called 'cell transformation', leads from the original cell type, as represented by the egg and the early blastomeres, to eight different types, or orders, of cells. Subsequently, the members of a cell order may undergo 'cell differentiation', assuming one of a number of distinct differentiation patterns, each characterized by the elaboration of a particular group of polypeptides among those represented in the genome. It seems likely that the process of cell diversification is a sequential process, involving a number of steps of branching, probably dichotomies, through which new patterns arise.

Very little is known about this process of derepression which controls the compart-mentalization of the transcription of the structural genes. It seems certain that some part of the mechanism resides in the nucleus, and, as mentioned above, it is likely that DNA molecules, in the form of regulatory genes, are involved, even if their mode of action may well be very different from that observed in bacteria, as described by Jacob and Monod (1961; 1963). It is evident, however, that in each dichotomy, the causal postulate requires the participation of an 'external' factor to account for at least one of two differentiation patterns represented by the branches. Since this agent can affect the nucleus only by way of the cytoplasm, it seems justified, for this reason alone, to postulate the existence and involvement of nucleo-cytoplasmic interaction in the process of cell diversification.

Although a differentiation pattern may comprise many structural genes, it appears that one mechanism of nucleo-cytoplasmic interaction may suffice to initiate the activity of each new differentiation pattern. This does not necessarily mean that all the genes are united into one 'operon'; it is also possible to imagine that a differentiation pattern encompasses several 'operons', among which one is set working through the influence of an 'external' agent, whereas the remaining ones are activated through 'inductors' arising as metabolic products in the differential activity of the cell.

Considering that differentiation patterns represent the concerted activity of several structural genes, it is natural to presume that the appearance of new differentiation patterns will be an extremely rare occurrence, more uncommon even than the origination of proteins with new functions. It is not possible to establish the relative frequency of these two kinds of evolutionary events, since their total numbers are unknown, but it is a matter of observational fact that metazoan phylogenesis has been associated with the origin of many new differentiation patterns, a remarkable accomplishment when contrasted with the paucity of innovation on the chemical level. It must be emphasized, however, that the probability of the origination of a new differentiation pattern is enhanced by the fact that, irrespective of the number of structural genes comprised by the differentiation pattern, only one gene, a regulatory one, may be necessary for its instigation. On the basis of empirical observations it therefore seems plausible, as submitted by Britten and Davidson (1971), that evolutionary novelties in metazoan phylogenesis may have been confined largely to the establishment of new regulatory genes, responsible, among other things, for selection of those DNA sequences in the genome which are to be transcribed in the several types of differentiated cells.

One feature which must increase the likelihood of this proposition is that, in combination with gene duplication, the allotment of various genes to disparate differentiation patterns actually paves the way for selectional diversification in a given function, as indicated by the demonstration of Adelson (1971) that the amino acid sequences in a number of digestive enzymes (rennin, pepsin, chymotrypsinogen, trypsin and elastase) exhibit similarities suggestive of a common origin. Examples of this kind, already quite numerous, show that cell differentiation constitutes a mechanism through which specialization may be achieved without the concomitant establishment of new kinds of function. That a number of hormones, for example growth hormone, secretin, glucagon, cholecystokinin, gastrin and insulin, appear to be related to the listed proteolytic enzymes suggests that new types of activity may also be ensured through gene duplication and that this has occurred, presumably, at one time in the course of the evolution of all these compounds.

As revealed by isozymes, gene duplication need not lead to the adoption of new catalytic functions, but only to isomers with different catalytic properties. The first isozymes discovered were those of lactate dehydrogenase, but subsequently a variety of isozymes has been found (cf. Markert, 1975).

The truly momentous strides which have been made in the field of molecular biology in the last decade have, understandably enough, led the practitioners of this discipline to believe, if one may trust the occasional statement which is made, that with the discovery of the genetic code, etc., we are rapidly approaching a solution of the 'problem of life'. And indeed, if by this concept we understand the metabolic aspects of the existence of cells, notably prokaryotes, on which most of the results have been obtained, this belief may be true. However, when we consider the problem of the ontogenesis and phylogenesis of the metazoans, there may still be a long way to go. For one thing, the really outstanding feature of this group of animals is the division of function between separate types of cells, as represented by the disparate diversification patterns. And, as we have seen, the latter may owe their existence primarily to the action of regulatory genes, about whose function we know next to nothing. Furthermore, it should be stressed that, even if we had solved this question, we would still face a formidable problem, namely, the mechanisms through which these various diversified cells, and the several extracellular products elaborated by them, contrive to cooperate in the creation and the functioning of the metazoan body. This topic is, however, a problem of epigenetics and will be dealt with in the following section.

Epigenesis

Epigenetics is the study of the mechanisms through which the inherited information, during the course of ontogenesis, is exploited in the elaboration of the metazoan body. Since, as emphasized by Garstang (1922, p. 98), 'the phyletic succession of adults is the product of successive ontogenies', it follows that epigenetics is also of great importance for phylogenetics, for, without understanding how the epigenetic processes contrive to accomplish individual cases of ontogenesis, we obviously cannot hope to account for the ways and means through which phylogenetic modifications are brought about. The particular contribution of epigenetics to phylogenetics comprises three aspects.

First, as discussed above, biological inheritance comprises elements in addition to those located in the genome.

Second, the construction of the metazoan body is, to all intents and purposes, a physical process. The latter is carried out by the elements of the ontogenetic substrate, consisting partly of cells and partly of extracellular substances, elaborated and secreted by the various embryonic cells. It follows that the physical properties of those elements must be responsible for the essential aspects of morphogenesis.

Third, epigenesis is a series of processes which are causally dependent. This means that any particular event, morphological, physiological or chemical, may exert a regulatory effect on a number of subsequent steps. This implies that any modification, small by itself, may be magnified through *epigenetic amplification.*

These three points demonstrate that ontogenesis, and hence phylogenesis, comprise events which lie beyond the scope of molecular biology and genetics. Since the various effects concern epigenetic mechanisms, they are naturally included in the domain of epigenetics. In the following two subsections we shall discuss the second and third points outlined above.

SUBSTRATE ELEMENTS

Innovations in the construction of the metazoan body, and these are by far the most frequent evolutionary occurrences, may arise either from the inclusion of new substrate elements, or by putting the ones available to new uses.

As an illustration, it may be mentioned that phylogenetic evolution has been associated with an increase in the variety of glycosaminoglycans present in the animal body, although losses have also occurred at times (cf. Hunt, 1970; Mathews, 1975; Rahemtulla and Løvtrup, 1975; Rahemtulla, Höglund and Løvtrup, 1976). It is certain that the appearance or disappearance of this kind of material must be one-step events; if the enzymes required for a certain type of glycosaminoglycan are present, the latter will be synthesized, otherwise not. And since the particular physical properties of the individual glycosaminoglycans make possible the occurrence of specific kinds of morphogenetic processes, it would seem that evolutionary innovations on the morphological plane have occurred repeatedly in association with changes in the repertoire of glycosaminoglycans produced in the animal body.

The most important morphogenetic entity in the vertebrate embryo is the notochord. The latter may be defined as a linear aggregate of chordocytes, a special kind of glycosaminoglycan-producing cells (myxocytes), surrounded by a collagenous sheath, produced by collagenocytes. The physical properties of the notochord, which are so important for vertebrate morphogenesis, may be accounted for completely by the physical properties of these differentiation products.

The latter are quite common in invertebrates, and yet no notochord is ever formed. This shows clearly that the presence of the appropriate structural genes does not ensure a given morphological outcome. A further requirement appears to be the simultaneous occurrence of the two differentiation patterns in a circumscribed region of the embryo at a particular stage of development, and this amounts to a prescription of the particular spatio–temporal pattern which distinguishes vertebrate epigenesis. It is difficult to specify the mechanism responsible for the latter, but it may be safely asserted that regulatory genes are involved in the control of events of this kind. From this it follows that on some occasions mutations in this type of gene, or in the cytoplasmic factors interacting with them, may result in large-scale innovations.

The view that the origination of new biological forms is primarily a matter of changing design, within a given framework of constructional materials and principles, has been advocated by several authors, among whom are Pantin (1951) and Novikoff (1952), and many of the geometrical rules to which these principles conform have been revealed by Thompson (1942). It is a curious and regrettable fact that this work has been largely neglected, and it becomes particularly distressing when it is realized that the main reason for this state of affairs is surely that it cannot be accommodated within established Darwinian and neo-Darwinian orthodoxy, although some of the advocates of the latter have realized the importance of epigenesis for the creation of variations in form (e.g. Haldane, 1932).

Another point which is worth making is that the substances required for carrying through the decisive morphogenetic events must be very few indeed. This assertion may be corroborated by considering a specific instance of embryogenesis, for instance that of the amphibian embryo. In the latter it is possible to distinguish three separate phases, pregastrulation and larva formation and, between these two, a period comprising gastrulation, neurulation and tail-bud formation. It will be noticed that at the end of this intermediate phase the vertebrate body plan has been established in all essentials, including the formation of most organs. Yet tissue differentiation has barely commenced, for histological observations show that this process only operates at full scale during the larval stages. These findings may be complemented and corroborated by chemical observations, for instance by analyses of the patterns of change in various enzyme activities. Thus, very few enzymes have been found to change before gastrulation, some do so during the middle, morphogenetic, phase, while all enzymes investigated so far increase during larval development (cf. Løvtrup, 1974).

Clearly, a number of proteins must be synthesized during the morphogenetic phase, but they can represent only a small part of the genome. The substances which are known to be elaborated during this phase comprise notably some glycosaminoglycans, hyaluronate, chondroitin, chondroitin sulphate and heparan sulphate (Höglund and Løvtrup, 1976), and collagen. These compounds are substrate elements which to a large extent may account for the various morphogenetic activities occurring in the early embryo.

This epigenetic interpretation thus claims that the realization of the total genome, as evidenced through the appearance of the various differentiation patterns, is a gradual process that has not gone very far at the early stages of embryogenesis during which the primordia of many important organs originate. Hence, it seems a misunderstanding to presume that one or several *specific* genes exist, responsible for executing the process of gastrulation or for the formation of notochord, neural tube, eyes, limbs, etc. Rather, these happenings, being physical events, are realized through the concerted action of the substrate elements, cells as well as extracellular substances, and may be referred to the particular physical properties possessed by each of these entities.

EPIGENETIC AMPLIFICATION

Ontogenesis is the outcome of a diversity of causally related epigenetic events. This has the consequence that a small change at one stage may entail far-reaching modifications—deleterious or beneficial, as the case may be—in subsequent ontogenesis.

In the preceding section we discussed two kinds of change which may have important morphogenetic consequences, namely, the acquisition of a new type of glycosaminoglycan, and alterations in the spatio–temporal pattern of the interaction between the substrate elements. If mutations of this kind were effective only at the stage of embryogenesis when they are first expressed phenotypically, then their effects might still be of limited scope. But if they happen to give rise to new structural entities, then these may in turn elicit whole series of other modifications. This phenomenon of epigenetic amplification may be illustrated by the notochord. In normal vertebrate embryos, this structure is responsible for the inception of a series of epigenetic events, neural tube formation, somite segmentation, etc. (cf. Løvtrup, 1974). One important inference to be made from this fact is that the notochord must have had the same result the first time it ever occurred. Certainly the sequence may have become modified in various respects in the course of time, but there is good reason to believe that the initial steps occurred also in the first individuals possessing a notochord, even if this structure happened to arise in different phylogenetic lines. The statement of Gregory (1951, p. 89): 'It is remarkable that this essentially early Palaeozoic organ [the notochord] persists in the human embryo no less than in the embryos of man's humbler relatives', is typical of the frequent lack of understanding of epigenetic mechanisms among biologists. The formation of the notochord is clearly a recapitulatory phenomenon which is an essential part of a chain of causally connected epigenetic events leading to the creation of the vertebrate body plan.

As observed by von Baer one and a half centuries ago, the embryos of the several vertebrate classes exhibit striking similarities during the early stages of development. Within each of the classes the likeness continues to later developmental stages. Indeed, one might claim with little exaggeration that the taxonomic divergence of, say, the mammals is nothing but a variation on a particular theme, namely, the mammalian body plan or, still better, the mammalian archetype (see D 2.13, p. 27).

We know that many variations of this kind are under hormonal control. It may therefore be inferred that evolution within the class Mammalia must to a significant extent have been a matter of changes in the activities of hormones controlling growth and differentiation. As I have suggested above, it is not unlikely that point mutations may affect the activities of various hormones. The fact that hormones control the rate of processes involved in morphogenesis means that, although the mutational changes are small by themselves, they may have far-reaching effects on the ontogenetic end-product. This kind of epigenetic amplification of mutational changes was proposed as an agent in evolutionary innovation by Goldschmidt (1940).

The three kinds of large-scale mutation listed above have been represented by the examples discussed here and in the preceding section. I believe that there is good reason to suspect that just this type of mutation has been responsible for most of the morphological innovations realized in the course of phylogenetic evolution. Therefore, if we wish to understand the mechanisms responsible for phylogenetic evolution we must understand how such mutations affect ontogenesis. As I have submitted, this study is the subject of epigenetics, a discipline which spans all levels of organization, from those of molecular biology and genetics to those of morphology.

'The wealth of their consequences has to be unfolded deductively; for as a rule, a theory cannot be tested except by testing, one by one, some of its more remote consequences; consequences, that is, which cannot immediately be seen upon inspecting it intuitively.'
(K. R. Popper, 1969, p. 221)

2

THE LOGIC OF PHYLOGENETICS

The discussion in the Introduction may wrongly have given the impression that, in my opinion, the reasoning applied in biology in general, and in phylogenetics in particular, is not, or does not intend to be, deductive. This is not my view, but I do submit that the premises employed are not usually stated explicitly and that, even when they are, the argument is based on far too few of them to warrant that correct conclusions are obtained.

In the present chapter we shall outline the logic of phylogenetic analysis. Nothing is consequently more appropriate than to begin with two premises which form the basis for all phylogenetic research, as they justify the feasibility of establishing phylogenetic relationships between diverse animal taxa.

The first premise asserts:

(i) *All animals, living and extinct, have a common phylogenetic origin.*

The general acceptance of this premise, implying that all members of the animal kingdom have originated in the course of a unique evolutionary process, may be considered the principal contribution made by Darwin.

The second premise, which presumes the validity of the first, may be formulated in the following way:

(ii) *Barring convergence, the varying degrees of similarity obtaining with respect to the properties possessed by various animals are expressions of phylogenetic kinship.**

These premises, as well as the following theoretical discussion of phylogenetic classification, are phrased with respect to animals. This does not necessarily imply that the various theorems do not apply to the members of the other kingdoms; indeed, I hope they do, but I shall refrain from making any explicit assertions on this point.

Clearly, if (ii) is correct, then it must be possible to use the characters through which the similarities are displayed for the purpose of establishing the phylogenetic relationship between the animals, an achievement which clearly amounts to the erection of a phylogenetic classification. These two premises, asserting that phylogenetic evolution is a reality and phylogenetic classification a possibility, are endorsed by most, if not all, biologists.

Owing to the fact that various characters are manifested by different numbers of

---

*As will be shown later, 'convergence' here ought to be replaced by 'non-apotypy'.

animals, classification was found to lead spontaneously to hierarchical systems. This indicates that some kind of relationship exists between diverse animals, the degree of which is demonstrated by their place in the hierarchy. To a believer in Special Creation, the revelation of affinities of this kind must be a mirror of the creative inspiration of the mind of the Almighty, the workings of which might be disclosed through the classification. Modern biologists, who possess evidence that life has originated through an evolutionary process, feel entitled to infer that the hierarchical arrangement outlines genetic kinships and thus, to some extent, the course of phylogenesis.

It is perfectly well known that systematists and evolutionists are anything but unanimous about the way the members of the animal kingdom should properly be classified. Hence, if, through phylogenetic classification, we want to establish the ancestry and divergence of Vertebrata, it is clearly necessary to lay down much firmer rules for classification than those usually applied today. Certain aspects of this problem have been dealt with in two recently published theories on phylogenetic classification, called 'cladism' and 'numerical taxonomy', respectively. We shall deal with these in the following two subsections; in the third the specific difficulties encountered in superphyletic classification are discussed, and in the last are stated three methodological premises for phylogenetic classification which have been adopted in the present book.

CLADISM

In the Introduction I have criticized evolutionists for want of logical stringency. This situation is changing as regards the subject of the present section, the principles of phylogenetic classification. This is primarily due to the advancement of a theory of cladism by Hennig (1966), which is a most outstanding contribution to the logical foundation of phylogenetics (cf. Brundin, 1968).

Hennig's theory has met with enthusiasm in some quarters, with coolness and rejection in others, but of late the trend seems to be towards the former alternative (Nelson, 1972). Unfortunately, if I understand him correctly, Hennig has not drawn the ultimate conclusions of his own premises. This, at least, is the opinion I have reached when trying to axiomatize the theory (Løvtrup, 1973; 1974). In the present context I shall first present my version of Hennig's cladistic theory. Subsequently, I shall discuss certain qualities of the cladistic 'phylogenetic tree', the dichotomous dendrogram, and a classificatory expedient called 'basic classification'. In the following three subsections we shall deal with the utilization of empirical data for the purpose of classification, partly 'ordinary'[4] taxonomic characters and partly information pertaining to the relative age of taxa. Subsequently, certain aspects of the phylogeneticist's vocabulary will be briefly discussed. Finally, since phylogenetic classification according to cladistic principles deviates in important aspects from common systematics, an outline of the differences between phylogenetic and Linnean classification is presented at the end of the section. It should be mentioned that some of the conclusions derived below have been stated by Wiley (1975), and also that the most important aspects of my theory have been published elsewhere (Løvtrup, 1975c).

Cladistic theorems

I have mentioned above that a theory of cladism was put forward by Hennig (1966). Obviously, many aspects of this work are well known and generally endorsed. It is not too easy to delimit Hennig's own contribution, but I trust that I commit no injustice in stating that it consists primarily in introducing the very important concepts of 'sister group', 'symplesiomorphy' and 'synapomorphy'. In any case, it seems fair to say that Hennig has given few directions for the practical application of his theory. In my experience, the simplest way to reach clarity on this point consists in the axiomatization of the theory of phylogenetic classification. The attempt towards this goal presented here embodies a number of axioms and definitions besides those advocated by Hennig (1966), but let me once more stress that the latter are undoubtedly the cornerstones of my axiomatization. The presentation constitutes a recapitulation, extended in certain respects, of work published earlier on this subject (Løvtrup, 1973; 1974); for this reason it has been possible to reduce the discussion to a minimum.

We may begin with a theorem derived from premise (i), namely:

(T2.1)* The various properties possessed by each and every animal have all been acquired in the course of phylogenetic evolution.

The first axiom states:

(A2.1) Animal classification involves the distinction and naming of classes of animals, defined by the possession of certain properties in common.

The classification referred to in A2.1 need not be a phylogenetic classification, but the subsequent discussion will deal only with this alternative.

(D2.1) A taxon is a concept defined by a set of properties distinguishing a particular class of animal. All individuals, past, present or future, possessing the whole set of properties are usually said to be members of the taxon.

I have been criticized for defining a taxon as a concept, rather than as a class of animals. For some time I was impressed by the arguments raised against me. However, after reading Popper's recent book (1972), I came to the conclusion that animals belong to Popper's World 1, and taxa to his World 3. Hence taxa cannot be groups of animals.

(D2.2) Taxonomic characters are properties which have been, or may be, used for defining a taxon.

We now encounter the problem of whether it is possible *a priori* to decide which of the various properties possessed by animals should be used as taxonomic characters and which not. According to Simpson (1961, p. 71), classification requires that 'the *whole* organism shall be considered in all its parts and aspects'. Acceptance of this claim would imply that all properties possessed by an animal are either actual or potential taxonomic characters.

However, to avoid that parents and offspring belong to different taxa, it is necessary that the properties in question are possessed not only by all extant members of a given taxon, but also by all members of succeeding generations. One might contrive to make sets of Mendel's pea plants according to the colour of their flowers, but since this trait may change in the offspring it should not be used as a taxonomic character. Clearly, the features which are the subject of genetic studies cannot be used as taxonomic characters.

As an approximation to the ideal envisaged by Simpson the following axiom may be submitted:

(A2.2) All properties possessed by an animal, and which are transferred unchanged from generation to generation, are either actual or potential taxonomic characters.

---

*In what follows, T, A and D stand for 'theorem', 'axiom' and 'definition', respectively.

One logical difficulty is encountered with respect to sexual characters. Either these must be excluded from the characters referred to in A2.2, or else sexual subtaxa must be erected for all bisexual animals. The former alternative is obviously not acceptable to practical systematists, because sexual characters are often very valuable for taxonomic distinctions. It must be emphasized that this is a matter of expediency, but I shall venture to postulate that it may be possible, in each and every instance, to find some characters, perhaps chemical ones, which are shared by both males and females of a given taxon.

As I have emphasized before (Løvtrup, 1973), A2.2 should not be interpreted to mean that all the taxonomic characters known to distinguish a certain taxon must be used as diagnostic characters. I do not think it is useful here to make the commonly encountered distinction between characters and character states; all properties, except for those discussed above, are potential taxonomic characters. Yet it may be assumed that in the classification of higher taxonomic categories, as in the present book, the properties involved will be characters rather than character states.

We shall now introduce a further axiom about which no doubt can be raised:
(A2.3) All actual and potential taxonomic characters possessed by an animal can be known only for extant ones.

From A2.1, A2.2 and A2.3 we may derive the following theorem:
(T2.2) Only extant animals can be classified.

In order to appreciate the implication of this theorem, which may well be regarded as a provocation by some palaeozoologists, it should be recalled that a phylogenetic classification is an empirical theory which must be exposed to continuous Popperian attempts at falsification (cf. Løvtrup, 1973). If we happen to have classified some living and extinct animals on the basis of osteological characters, and subsequently discover that other properties, say chemical parameters, dictate changes in the classification of the living animals, then we shall never be able to decide whether or not corresponding modifications are required for the extinct forms. The fact that, in contrast to what holds for living animals, the classification of fossils is testable only within narrowly circumscribed limits, is the intended meaning of T2.2.

I think that the case of *Latimeria* may serve to corroborate this theorem. Until recently the taxon Actinistia was known only through fossils. On the basis of knowledge of the hard parts of these extinct animals, they were included in the taxon Crossopterygii, together with the extinct Rhipidistia. Since the latter are known as the ancestors of the Tetrapoda, they are often presumed to have been rather 'advanced' fishes. The recently acquired insight into the soft anatomy of *Latimeria* has shown the latter to be so 'primitive' that it can only be very distantly related to the Rhipidistia. In fact, the kinship is so remote that suppression of the taxon Crossopterygii has been suggested (Jarvik, 1968a).

I hope that I may appease the palaeozoologists by stating that fossils, like every other kind of object, can be classified, and also, as I shall try to demonstrate in a subsequent chapter, that most fossil groups may be allocated to definite sites in the phylogenetic classification of living animals. This approach seems better than the independent classification of fossils proposed by Crowson (1970).

In any case, it seems possible to infer from T2.2:
(T2.3) The discovery of a new fossil has no impact on classification.

This conclusion is of great significance for several reasons. For one thing, one often encounters statements in the literature on phylogenetic evolution to the effect that the solution of some problem must await the emergence of new fossil data. Secondly, and of more specific import in the present context, it has been pointed out to me that Calci-chordata (Jefferies, 1968), which purportedly represent one of the missing links between Echinodermata and Vertebrata, are of the utmost consequence for the phylogenetic classification of Vertebrata. Hence, according to these critics, I commit a serious error in neglecting this taxon in my work. I do not ignore the significance, for the present enterprise, of fossils in general, and of Calcichordata in particular, even if I may not rate it as all-important as it is usually made out to be. I claim, however, that if, with the wealth of information which it is possible to extract from living animals, we cannot arrange the latter in a correct phylogenetic system, then no amount of fossil data will ever help us to approach this goal, and therefore I feel justified in postponing discussion of the fossil record until the phylogenetic classification of the living taxa has been established.

It has already been mentioned that the various taxonomic characters are found in numerically different assemblies of animals. This observation may be presented by the following axiom:

(A2.4) Taxonomic characters differ with respect to the number of individuals in which they are found. Some occur in all animals, some only in subsets among the latter, some only in subsets within the subsets, etc.

A 2.4, together with D2.1 and D2.2, lead us to the conclusion:

(T2.4) Taxa can be arranged hierarchically.

The hierarchy, of course, must be constructed by placing at the apex the taxon comprising all animals and at the base those taxa whose members have no taxonomic characters in common other than those defining taxa at higher levels in the hierarchy.

Since the taxa are based upon taxonomic characters possessed by individual animals, we may infer:

(T2.5) Each animal is the bearer of the taxonomic characters defining all the taxa to which it belongs.

The establishment of a hierarchical system allows for the ranking of the various levels. The rank of a taxon may be defined in the following way:

(D2.3) A taxon T_j, of rank j, is a set of n_{j+1} taxa T_{j+1}, of rank j+1 ($j \geq 1$).

It appears that according to D2.3 the numbering begins at the apex of the hierarchy, an expedient which is contrary to the custom adopted in Linnean classification, where the most comprehensive taxa, class, phylum and kingdom, are characterized as 'high-ranking'. It will be shown below that the approach adopted here is a necessity in phylogenetic classification, where inconsistencies are encountered unless the numbering begins at the hierarchical apex. Lest misunderstandings arise, the epithets 'superior' and 'inferior' will be used when reference is made to the position of a taxon in the hierarchy, a superior taxon being of lower numerical rank than an inferior one.

Hence:

(D2.4) For $i < j$, the taxon T_i is superior to the taxon T_j, and T_j inferior to T_i.

At this stage it may be appropriate to recall that axioms (i) and (ii), stating that all animals have originated through a unique evolutionary process and that their various properties, i.e. taxonomic characters, are expressions of phylogenetic affinities, are premises for phylogenetic classification.

From (ii) and A2.1 it follows:

(T2.6) Classification of animals will bear some relation to their phylogenetic kinship.

Apparently, no matter how we classify, some phylogenetic affinities will be reflected.

We shall, however, specify the demands to be made on a phylogenetic classification:
(D2.5) A phylogenetic classification of Animalia is a classification which correctly sets out the phylogenetic kinship between the various taxa in the kingdom.

From (i), T2.4 and D2.5 it follows:
(T2.7) The phylogenetic classification of Animalia is a unique hierarchy.

The phenomenon of convergence, for which reservation was made in (ii), involves, as will be discussed later, the origin of the same character on more than one occasion. Under these circumstances it cannot reflect a genetic relation, and from (ii) and D2.5 it therefore follows:
(T2.8) In phylogenetic classification a property can be used as a taxonomic character for a taxon T_j only if it is non-convergent, i.e. if its presence in all the members of the purported taxon can be referred to its acquisition at a unique occasion in the course of evolution.

A taxon which obeys the criterion established in T2.8 is called a 'monophyletic taxon', a concept that may be defined as follows:
(D2.6) A monophyletic taxon T_j is a concept defining a class of animals comprising all those individuals which are descendants of the first animals endowed with all the taxonomic characters of T_j.

From T2.7, T2.8 and D2.6 we may deduce:
(T2.9) The hierarchy of phylogenetic classification comprises only monophyletic taxa.

Henceforth, when the word 'taxon' is used in the present context, it may therefore be regarded as synonymous with 'monophyletic taxon', unless otherwise specified.

In order to establish the correspondence between phylogenetic classification and the process of evolution, it is necessary to introduce a postulate concerning the mechanism through which members of new taxa arise.* Here we may refer to the concept of 'sister group', defined by Hennig (1966, p. 139) as follows: 'Species groups that arose from stem species of monophyletic group by one and the same splitting process may be called "sister groups"' . . . and 'every monophyletic group, together with its sister group (or groups), forms—and forms only with them—a monophyletic group of higher rank'.

From these statements it appears that Hennig allows the possibility that a 'species group' may have two or more sisters. However, from Hennig's book (1966), and elsewhere in the literature, it appears that sister groups are usually twins. I shall adopt this notion in the following axiom:
 (A2.5) The taxa T_{j+1} have arisen through dichotomy from the taxon T_j in which they are included.

It appears that this axiom, which I propose to call *Hennig's axiom*, is regarded with considerable aversion by systematists. I acknowledge that the existence of trichotomies, tetrachotomies, etc., cannot be excluded, but I submit that their occurrence is inversely proportional to the purported number of branches and, furthermore, that among the superior taxa they are probably non-existent. On the genus–species level this phenomenon is possibly more common, maybe primarily because we do not possess the knowledge required to accomplish further subdivision. But this admission does not imply that a genus with hundreds of species correctly reflects a phylogenetic reality;

---

*Clearly taxa, being concepts, do not arise, they are invented. However, for the sake of simplifying the following discourse I shall write of the 'origination', 'creation', 'extinction', 'discovery', etc. of taxa. If it is kept in mind that the discussion concerns the biological reality behind the taxa, i.e. their members, then this phrasing should not lead to misunderstandings.

rather, this marvel is a consequence of the straitjacket which the Linnean categories impose upon classification, as evidenced by the fact that various auxiliary categories are often invoked to subdivide such genera.

Thus, although bifurcation must be the most usual type of branching, the categorical formulation of A2.5 is not theoretically justified. But this expedient, adopted for the sake of simplicity, ought to be acceptable when it is realized that, after proper modifications, most, if not all, of the various theorems will be valid for trichotomous, tetrachotomous, etc., dendrograms.

From A2.5 we may now conclude:

(T2.10) In phylogenetic classification there are two taxa T_{j+1} included in every taxon T_j, except for the terminal ones.

I propose to replace 'sister groups' by 'twin taxa', 'triplet taxa', 'quadruplet taxa', etc. The first of these concepts may be thus defined:

(D2.7) The taxa T_{j+1} included in the taxon T_j are twin taxa.

From the principle adopted in the construction of phylogenetic hierarchies, and from D2.7, we may derive the following theorem:

(T2.11) If x is a taxonomic character in the taxon T_j, then $\bar{x}$, i.e. not-x or absence of x, must be a taxonomic character in its twin taxon, or x will distinguish a superior taxon.

It is easy to see that if the phenomenon dealt with in T2.11 is extended to comprise all T_j characters, then the taxon is undefinable. Hence we may introduce the following definition:

(D2.8) The taxon T_{j+1}, purportedly distinguished by the characters x, y, z, etc., is undefinable unless at least one member of T_j possesses the characters $\bar{x}$, $\bar{y}$, $\bar{z}$, etc.

From T2.10 also follows:

(T2.12) The hierarchy of phylogenetic classification can be represented by a dichotomous dendrogram.

Evidently, a dendrogram need not be comprehensive in the sense that it represents the whole of the animal kingdom; the latter, incidentally, represents only a part of the phylogenetic hierarchy of living beings. Rather, it is possible to select a part of the animal kingdom for special attention without regard to the remainder of it. This, of course, represents a fragment of the hierarchy, itself constituting a hierarchy, about which we may conclude:

(T2.13) Any part of the phylogenetic hierarchy of the animal kingdom comprising more than one taxon is a dichotomous dendrogram.

From T2.9 and T2.13 we get:

(T2.14) A phylogenetic dendrogram always represents the classification of a monophyletic taxon, provided the classification is inclusive.

The concept 'inclusive classification' is defined in D2.16 (p. 33).

From A2.5 further follows:

(T2.15) The taxon T_j is older than the taxa T_{j+1} which it includes.

A property that has originated at a unique occasion must have been present before the taxon defined by it gave rise to new taxa through dichotomy, and we may therefore infer from T2.8 and A2.5:

(T2.16) No taxonomic character defining the taxon T_j can have originated later than any of those defining the taxa T_{j+1} included in T_j.

And conversely:

(T2.17) No taxonomic character defining the taxa T_{j+1} included in T_j can have originated earlier than any of those defining T_j.

From A2.5, T2.11, T2.16 and T2.17 we may also deduce:

(T2.18) The initial step in the dichotomy leading from the taxon T_j to the taxa T_{j+1} must have been that at least one member of the taxon T_j acquired a T_{j+1} character x, whereby the remaining members of T_j came to belong to the T_{j+1} twin taxon distinguished by $\bar{x}$.

From the statements embodied in T2.16 and T2.17 it is possible to delimit the period during which a taxon T_j comes into existence,* namely:

(T2.19) The creation of the taxon T_{j+1}, i.e. the origination of all its taxonomic characters, has occurred during a period of time lasting from the subdivision of the taxon T_j, in which it is included, until it itself became subdivided into two taxa T_{j+2}.

The decisive event in the creation of a pair of twin taxa is that some members of the ancestral taxon, distinguished by a particular taxonomic character, become isolated. Since the process of isolation and/or the origination of a new taxonomic character must be fortuitous, we may state as an axiom which is empirically supported:

(A2.6) The origination of a pair of twin taxa T_{j+1} from the taxon T_j, in which they are included, is a chance event.

From this it follows that it is impossible to predict when each of a pair of twin taxa undergoes dichotomy, and hence one cannot make any predictions about their relative age, since this parameter refers to the time when each of them gave rise to a new set of twins. When it undergoes bifurcation, the taxon itself, and not the taxa T_{j+1}, becomes defined. Besides, of course, the taxonomic level $j+1$ is established. Therefore it would be more correct in T2.19 to write 'two incipient taxa T_{j+2}'. We may thus infer:

(T2.20) Nothing can be predicted about the relative age of a pair of twin taxa.

This conclusion stands in apparent contradiction to the assertion made by Hennig (1966) that 'sister groups' are of equal age, the reason being that Hennig measures the age of a taxon from the first, and I from the last, bifurcation involved in the history of a taxon.

From A2.5, T2.12 and T2.19 we may conclude:

(T2.21) The horizontal lines in a phylogenetic dendrogram represent the process of dichotomy through which new taxa arise, while the vertical lines represent periods of time during which taxonomic characters have been acquired.

Since, in T2.15, the taxon symbolized by T_j is nearer the apex than are the taxa T_{j+1}, it follows:

(T2.22) The direction of the course of phylogenetic evolution corresponds to a movement from the apex to the base in the phylogenetic hierarchy, or from taxa of low, to taxa of high, numerical rank.

The idea that evolution has passed in the direction from the high to the low ranks in the Linnean hierarchy has been put forward by many biologists, notably by Schindewolf (1950). I therefore suggest that T2.22 should be called *Schindewolf's first theorem*.

From this theorem, together with D2.5, we may derive:

(T2.23) Phylogenetic classification must begin at the apex of the hierarchy.

From A2.6 we may also conclude:

(T2.24) Nothing can be predicted about the frequency of dichotomy in the various regions of a phylogenetic hierarchy.

---

*The expression 'comes into existence', used in association with taxa, should not be taken too literally. The implication of these words is only that a taxon becomes fully definable. This situation prevails when a subdivision of the incipient taxon occurs. It will be shown later that according to this usage the terminal taxa in a completely resolved classification have not yet 'come into existence'.

The implication of T2.24 is, of course, that the frequency may be expected to vary from one region to another. In order to elucidate the consequences of this axiom, we shall introduce a concept called 'terminal taxon':

(D2.9) A terminal taxon is a taxon located at the base of the hierarchy, irrespective of the degree of resolution of the classification.

The concept 'resolution' is defined in D2.10. It is seen that if, in two twin taxa near the base of the hierarchy, one or more branchings have occurred in one of them subsequent to their separation, but none in the other taxon, then the terminal taxa cannot be of equal rank. Hence we may conclude from A2.6, T2.24 and D2.9:

(T2.25) The terminal taxa in the hierarchy need not all have the same rank.

This theorem bears upon the problem of 'undefinable taxa'. According to taxonomic convention the original seven Linnean categories must always be applied. If a high-ranking taxon, say, the order Tubulidentata, contains only one kind of animal, then it is still necessary to introduce names for the family, the genus and the species to which these animals purportedly belong. As shown by Gregg (1954), this convention introduces certain logical difficulties, because it implies that the taxon at each successive level must be included in, and at the same time be identical with, the closest taxon of higher rank. As I have tried to show elsewhere (Løvtrup, 1973; 1974), this 'Gregg's paradox' is a deception; the taxa below the level where further resolution is possible are undefinable (D2.8), and the names given to them are synonyms, alternative names given to the lowest definable taxon, the rank of which may vary from case to case.

From T2.23 and T2.25 we may deduce:

(T2.26) The numbering of the taxonomic levels in the hierarchy is consistent only when it begins at the apex.

This conclusion thus justifies the decision made above concerning the numbering of the taxonomic levels.

We may now introduce a concept called 'resolution':

(D2.10) A phylogenetic classification has been brought to the point of complete resolution when it is impossible to subdivide the terminal taxa any further.

Clearly, complete resolution is the ultimate goal which is but seldom reached; most classifications must be regarded as representing 'incomplete' or 'partial' resolutions. It is important to note, however, that, depending upon the degree of resolution, the terminal taxa may vary considerably with respect to their rank in the Linnean classification, if they happen to occur in the latter. They may represent phyla, classes and orders, but also genera, species and subspecies, and nothing prevents the degree of resolution from varying in different regions of a dendrogram. In this connection it must also be stressed that in a completely resolved classification the terminal taxa need in no way coincide with the lowest recognized Linnean taxa, species and subspecies. Rather, if it is possible, on the basis of variations within a taxon of this kind, to establish subgroups that have arisen through dichotomy as envisaged in A2.5, then it is clearly not only permissible, but mandatory, to introduce these as terminal taxa in the system. This does not necessarily imply, however, that they must be given Latin names.

Since it is possible to assign a number to the rank of a taxon, it follows:

(T2.27) Within any monophyletic taxon the absolute rank of the included taxa may be objectively evaluated, irrespective of the degree of resolution, provided the classification is inclusive.

It is evident that the 'absolute' rank thus determined will be relative if the classified taxon comprises only part of the living world. It is seen that the ranks established in an incompletely resolved dendrogram are immutable, for since classification must begin at the apex it follows that further resolution can be concerned only with taxa inferior to those already under consideration. However, as stated, this theorem is valid only if the classification is inclusive, as defined in D2.16.

It was observed in T2.19 that a taxon comes into existence, in the sense of being fully defined, only at the moment it becomes subdivided into two inferior taxa. From this, together with D2.9 and D2.10, we may conclude:

(T2.28) The terminal taxa in a completely resolved classification are taxa under creation and, as such, not fully definable.

Before we proceed with the argument, we shall define the concept of 'phylogenetic line' as follows:

(D2.11) A phylogenetic line is a sequence of taxa in the hierarchy, $T_1, T_2 \ldots T_n$, each of which is included in the preceding one.

It is readily grasped that a phylogenetic line will be step-like, the horizontal and the vertical parts having the implications stated in T2.21.

In T2.15 and T2.20 we have deduced relations between taxonomic rank and age. These theorems may be generalized as follows:

(T2.29) In a phylogenetic line T_j is older than T_{j+1}, which in turn is older than T_{j+2}, etc. Nothing can be predicted about the relation between age and rank of taxa in disparate phylogenetic lines.

For the purpose of the following discussion it is necessary to introduce one more concept, viz., 'independent' taxonomic character, through the following definition.

(D2.12) If two or more taxonomic characters, for reasons other than chance, always occur together, then only one of them is independent; the other(s) are dependent characters.

We need not explain the phenomenon of dependence here except to point out that if, as discussed elsewhere in this book and previously (Løvtrup, 1974), epigenesis is a chain of causally related events, then it is reasonable to presume that the independent taxonomic character initiates the process through which the dependent one(s) arise.

It is a well-known fact that many taxa, particularly, of course, those of superior rank, differ from their twin taxon in a number of taxonomic characters. Even if some of these may not be independent, it is apparent in many instances that these characters have not all been acquired at one occasion, and I shall therefore submit the following postulate:

(A2.7) The several independent taxonomic characters defining certain taxa have been acquired at separate occasions.

From T2.18, T2.19 and A2.7 we conclude:

(T2.30) If the taxon T_j is distinguished by n independent taxonomic characters, then the creation of the taxon has involved n steps of dichotomy.

I have recently (Løvtrup, 1974) tried to revitalize the concept of 'archetype', specifically in connection with the body plan which distinguishes each of the animal phyla. It seems, however, that with reference to T2.30 it may be possible to introduce a much more general definition of this concept in the following way:

(D2.13) The archetype of the taxon T_j is the sum of the taxonomic characters, independent as well as dependent, which were acquired during the creation of the taxa in the phylogenetic line $T_1, T_2 \ldots T_j$, to which the latter belongs.

If T2.30 is granted, then apparently the only objection which can be raised against this definition is the name, which to some people seems to invoke notions of idealistic philosophy. This is unfortunate for, as appears from D2.13, such ideas are not implied by the concept.

Since, clearly, only one of the branches which arose through the repeated dichotomies implied by T2.30 exists today, we may conclude:

(T2.31) In the steps of dichotomy involved in the creation of the taxon T_j, all the branches that did not acquire T_j characters became extinct.

These extinct groups may be represented in the dendrogram by side branches to the vertical line which represents the phase of evolution during which they arose. To the extent they are known as fossils we may therefore conclude:

(T2.32) All fossil groups represent extinct side branches to existing taxa, and more than one fossil group may be thus affiliated with the same taxon T_j.

We shall finish the present section by considering the consequences of two very likely events, namely, that a contemporaneous taxon becomes extinct, or that a new taxon is discovered. From the rules involved in the construction of the dendrograms it follows that:

(T2.33) The extinction of a taxon T_{j+1} requires that its twin taxon is raised to the level j, its $j+1$ characters becoming j characters, and that the numerical value of the ranks of the taxa it includes is lowered by one; and:

(T2.34) The discovery of a taxon T_{j+1} requires that, from the taxon T_j which includes the new taxon, a twin taxon to the latter is erected, defined by some of the previous T_j characters, and that the numerical value of the ranks of the taxa included in the twin taxon is raised by one.

The dichotomous dendrogram

According to Darwinian notions, evolution involves the transformation of species, the direction of change thus being from the lowest to the highest taxonomic ranks. This mode of thought has important consequences for how the course of evolution is conceived, since it is supposed that the common ancestors of any two high-ranking groups must always be looked for at a stage before either of the taxa existed, or, in other words, 'that evolutionary change has affected chiefly generalized types' (Neal and Rand, 1936, p. 643). One is left to wonder about the nature of these 'generalized types' which do not belong to any high-ranking taxa. The only possible conclusion is that certain kinds of lowly organisms, perhaps larvae, remained as a reserve for mobilization whenever evolution of new forms took place.

This reason has had certain implications for 'phylogenetic botany' which were pungently described by Jepsen in the following quotation (1944, p. 83):

'Paleontologists, in their anxiety, founded the sceptical school of phylogenetic tree surgery, sharpened their erasers, and went to work. They did a thorough job. When they had finished pruning they reassembled the faggots into two general kinds of sketches. One kept the old oak pattern but consisted entirely of dashed lines [Figure 2.1] and the other looked like the candlewood ocotillo with its twigs ascending almost parallel from the ground [Figure 2.2]. No one cared or dared to draw a solid line from any one fossil group to any other. Paleontologists who worked with abundant material, like that provided by the rodents, composed the chant "Parallelism, parallelism, and still more parallelism"'.

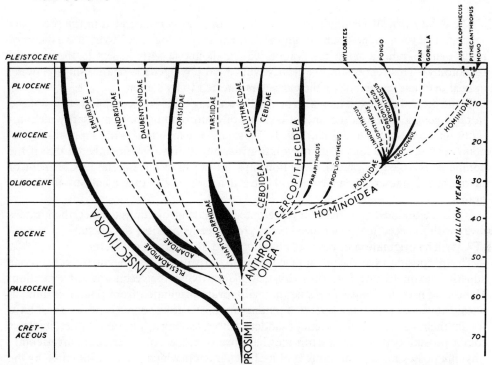

Figure 2.1. Phylogenetic classification of Primates. (Reproduced from J. Z. Young (1962), *The Life of Vertebrates* (2nd ed.), by permission of The Oxford University Press)

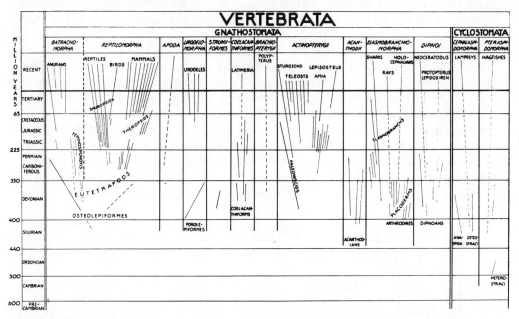

Figure 2.2 Phylogenetic classification of Vertebrata. (Reproduced from E. Jarvik (1968b), in *Current Problems of Lower Vertebrate Phylogeny*, by permission of the author and Almquist & Wiksell)

The great merit of Hennig's theory (1966) is that, as demonstrated in the preceding section, it leaves us in no doubt about the shape of the phylogenetic 'tree': it is a dichotomous dendrogram (a slight reservation was made above with respect to the point of dichotomy). This representation clearly will reveal phylogenetic relationships between animal taxa undreamed of by the gardeners referred to by Jepsen.

In the present context it may also be worth recalling that when computers are engaged in phylogenetic research the output consists of dendrograms, not necessarily dichotomous ones (cf. Sneath and Sokal, 1973). This may be explained on the supposition that computers are more logical than evolutionists, but it must be remembered that if the computer is asked to establish relationships on the basis of a certain assortment of taxonomic characters, then it can hardly do it in a simpler way than by means of dendrograms.

If the hierarchical classification represented by a dendrogram is correct, then it does away with the need for 'generalized types', for it seems possible to conclude:
(T2.35) The origination of new taxa has occurred at all taxonomic levels.

This proposition has the advantage that many expedients may be inherited from one superior taxon to another, thus drastically improving the 'economy' of evolution. Hence, we need no longer presume that Vertebrata originated from a humble animal, hardly advanced above the level of a pluteus larva, the descendants of which must have fought their way up the phylogenetic ladder without leaving any traces. Rather, it allows for the possibility that the ancestors may have been members of an advanced invertebrate phylum, possessing a great number of useful characters which could be inherited by the incipient Vertebrata. It is, of course, possible to imagine that some of the characters possessed by these ancestors could not be exploited or otherwise realized epigenetically in the first vertebrates, but, on the other hand, it may be anticipated that to be 'hopeful monsters' they had to take over a maximum of the advantageous characters of their predecessors.

The dichotomous dendrogram may become of great importance in future phylogenetic studies. It is therefore of some interest to know the number of different phylogenetic dendrograms that can be established in a classification with n terminal taxa. In order to arrive at an answer to this question we may first note that in a dendrogram with three terminal taxa, A, B and C, three different classifications are possible, since replacing A in Figure 2.3 with either B or C will give different alternatives. From the figure it appears that I have turned the hierarchy upside down, a measure implying that the time-scale is represented by the ordinate. This method of illustration is a convention in phylogenetic literature which unquestionably facilitates the visualization of the course of evolution. I have not, however, changed the terminology correspondingly, thus, it

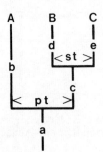

Figure 2.3. Phylogenetic classification of the three terminal taxa, A, B and C, in a dichotomous dendrogram, in the present context called a 'basic classification'. The vertical lines a–e represent the five separate taxa comprised by the dendrogram; the first three of these coincide with the terminal taxa. pt, the set of primary twins; st, the set of secondary twins

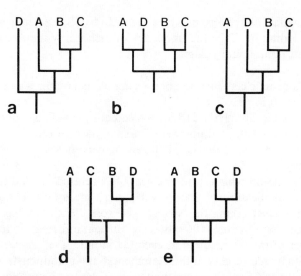

Figure 2.4. The five possible alternatives for the classification of a fourth taxon, D, relative to the basic classification represented by Figure 2.3

must be recalled, the base of the hierarchy and the terminal taxa are to be found at the top of the diagrams.

Figure 2.3 is seen to embody five vertical lines, each of which, as we know, represents a separate taxon. If we have to incorporate a new taxon, D, in this dendrogram, it may obviously be a twin to any one of the five taxa. These five possibilities are shown in Figure 2.4. Since three different classifications are possible for n = 3, it follows that for n = 4 the number of classifications, C_4, becomes $3 \times 5 = 15$. This example is a special application of the general equation for the number of different dichotomous dendrograms,

$$C_n = \sum_{i=1}^{n-1} (2i-1),$$

which can be constructed with n separate taxa (Fitch and Margoliash, 1969).

The first ten values of C_n are listed in Table 2.1. It is apparent that the value of C_n increases very rapidly. Obviously, if no special method exists for systematizing phylogenetic classification, then our only hope would seem to be computer assistance. However, as will be shown in the next section, it is possible to simplify the procedure so that even the human brain can engage in this endeavour.

It is also possible to estimate the minimum number of taxonomic levels which are required to accommodate n terminal taxa in a dichotomous dendrogram.

Thus, if n is a power of 2, and the frequency of dichotomy uniform, then the number of taxonomic levels, j, will be $j = 1 + \log n/\log 2$. The extra taxon added represents the mono-

Table 2.1
The number of possible phylogenetic dendrograms for n taxa

n	1	2	3	4	5	6	7	8	9	10
C_n	1	1	3	15	105	945	10 395	135 135	2 297 295	43 648 605

phyletic taxon in which all the terminal taxa are included. Since n may not be a power of 2, and since, according to T2.24 the frequency of dichotomy may vary, the calculated number represents a minimum value.

Hence:

(T2.36) The phylogenetic classification of n terminal taxa requires a number of taxonomic levels $j \geq 1 + \log n/\log 2$.

It is quite clear that although the 21 Linnean categories available for animal classification may suffice for a substantial number of taxa in a dichotomous dendrogram, their number is far too small to allow the phylogenetic classification of, at least, the most populous superior taxa.

In the lively discussion which in recent years has centred around the consequences for phylogenetic classification of adopting Hennig's ideas (1966), various arguments have been raised against taking this step (cf. Nelson, 1972). Among these objections may be mentioned Mayr's claim (1969) that for efficient information retrieval a reasonably large number of taxa is required at each taxonomic level. As a non-taxonomist I am unable to estimate the validity of this statement, but if taxonomists have to rely upon computers in their work, then the programmers, in order to obtain 'efficient information retrieval', must introduce a binary classification of the various taxa. It requires little imagination to realize the enormous advantage of phylogenetic classification under these circumstances for, owing to the classification in dichotomous dendrograms, each taxon has in principle a binary code number at the outset. Admittedly, some of these would be so large that some kind of shorthand would be required for non-computer use, but I do not think that the difficulties involved in creating such an expedient would be insuperable.

It may be mentioned that it would be easy to incorporate the present Linnean system in the type of classification required by cladism; thus, for instance, a genus including n species would be classified in a dichotomous dendrogram the minimum number of levels of which is stated in T2.36. The only case that might cause embarrassment is represented by the undefinable taxa, but this may be an indication of the soundness of cladism.

The basic classification

We now face the problem of how to apply in practice the theory outlined above. In order to approach this project we may first define a concept called 'basic classification':

(D2.14) A basic classification is a classification comprising three terminal taxa.

With reference to the discussion in the preceding section, we may conclude:

(T2.37) A basic classification contains five taxa, representing the ranks j, $j+1$ and $j+2$.

This is easily seen from Figure 2.3, where the vertical line e represents the level j, the lines a and d the level $j+1$ and the lines b and c the level $j+2$. For convenience, we may introduce here the concepts 'primary' and 'secondary' twin taxa:

(D2.15) In a basic classification the taxa of rank $j+1$ are primary, the taxa of rank $j+2$ secondary, twin taxa.

We have seen above (T2.33) that the extinction of a taxon entails certain changes in rank number, but otherwise the classification of the various taxa is not upset. Since

extinction may be replaced by the mental operation of omission, we may conclude:
(T2.38) Omission of one or more taxa in a dendrogram does not upset the relative positions of the taxa remaining.

From this in turn we infer:
(T2.39) Barring trichotomy, any three taxa can be classed in a basic classification in such a way that their proper phylogenetic kinship is evident.

Clearly, then, a basic classification can be carried out completely without regard to the rank of, or degree of affiliation between, the three taxa. We shall presently discuss some practical examples of this stratagem, but it is necessary first to evaluate the consequences of the process of omission sanctioned by T2.38. The distinction between a classification comprising all relevant taxa and one in which one or more taxa are excluded may be made by the application of the following terminology:
(D2.16) In an inclusive classification of a taxon T_j the included taxa represent all members of T_j; in an exclusive classification they do not.

The introduction into an exclusive basic classification of an omitted taxon will upset the twin relationships between the various taxa. In Figure 2.4 it is seen that the introduction of D in three cases, a, b and c, preserves the secondary twin relationship between B and C, while other relationships are changed. In d and e the secondary twin relationship between the two taxa is altered to a primary one. The possibility of changing the various twin relationships does not obtain in an inclusive classification, because here the only way new taxa can be accommodated involves resolution of one of the terminal taxa. To distinguish these two cases we shall introduce the concepts 'absolute' and 'relative' twin relationships:
(D2.17) An absolute twin relationship is immutable; a relative twin relationship is mutable.

From D2.16 and D2.17 we derive:
(T2.40) All twin relationships in an inclusive classification are absolute; some twin relationships in an exclusive classification are relative.

It may be appropriate at this stage to illustrate the principles outlined here with an example. From T2.39 it follows that it is possible to classify the taxa Insecta, *Myxine glutinosa* and *Petromyzon marinus* in a basic classification, and nobody will presumably contest that in this instance A stands for Insecta, B and C for the two species of Cyclostomata (Figure 2.3). This classification is, however, exclusive, and the possibility therefore obtains that the introduction of further taxa may alter the twin relationship between these taxa. If the taxon D in Figure 2.4 stands for Gnathostomata or any taxon included in the latter, for example Dipnoi, Muridae or *Homo sapiens*, then clearly only three classifications are possible, namely, those represented by Figure 2.4c, d and e. We shall discuss later which of these alternatives is the most probable one.

It is clear that the successful ordering of three taxa in a basic classification will permit a very important assertion concerning the phylogenetic kinship between the three taxa, namely:
(T2.41) The members of the secondary twin taxa in a basic classification have ancestors in common that were not ancestors to the members of the primary twin taxon.

This statement I propose to call *Hennig's first theorem*. It may be noted that it might be more correct to write 'the other primary twin taxon' in T2.41, but since no possibilities of mistakes seem to exist, I adopt the simplification implied by the wording above.

Plesiotypy, apotypy and teleotypy

On the basis of the concepts of 'plesiomorphy' and 'apomorphy' a series of further concepts has been introduced by Hennig (1966) to distinguish separate classes of taxonomic characters and their distribution. It appears that the chosen suffix may refer to properties other than morphological ones. However, in modern taxonomy, and particularly in phylogenetic classification, it has become more and more usual to employ non-morphological characters, and under these circumstances Hennig's terminology may be confusing, at least to persons not versed in classical philology. I therefore propose, in the present context, to adopt Tuomikoski's suggestion (1967), whereby 'morphy' is replaced by 'typy'. As will appear from the present discussion one further concept, 'teleotypy', is required in a basic classification. These three notions may be defined as follows:

(D2.18) In a basic classification, the taxonomic characters distinguishing the taxon T_j, and those superior taxa in which the latter is included, are plesiotypic, those of the non-terminal taxon T_{j+1} are apotypic, and those of the three terminal taxa are teleotypic.

Using these concepts we can now derive another important rule of phylogenetic classification, *Hennig's second theorem*:

(T2.42) Barring non-apotypy, the only taxonomic characters that are common to two out of the three taxa in a basic classification are apotypic ones.

I have chosen to introduce the concept of 'non-apotypy' in T2.42, even though the phenomenon is not dealt with until the following section. However, non-apotypy is so important for phylogenetic classification that it would have been a serious shortcoming if the theorem had been stated without this reservation.

Generalizing T2.42 beyond basic classifications we may conclude:

(T2.43) Barring non-apotypy, a taxonomic character x is, in an inclusive classification, apotypic only for one taxon, T_j. For all taxa of rank $r \geq j+1$, belonging to the phylogenetic lines including T_j, x is plesiotypic, and for the taxa of rank $r \leq j-1$, belonging to the one phylogenetic line including T_j, x is teleotypic. When, in an exclusive classification, new taxa are introduced, the status of a taxonomic character may change according to rules which can be derived from the preceding paragraph.

Thus, when facing a basic classification involving three arbitrary taxa, the task of the phylogeneticist amounts to determining which of the taxa are the secondary twins, or, as Hennig would phrase it, the sister groups. And this problem may in principle be resolved by determining which two taxa have apotypic characters in common.

If subsequently a fourth taxon D is to be collocated, the first point to decide is whether D possesses none, some or all of the apotypic characters shared by the secondary twins, B and C in Figure 2.3. In the first case D is either the primary twin, and the taxa A and B+C the secondary twins (Figure 2.4a), or else D is the secondary twin to A, the taxon B+C being the primary twin (Figure 2.4b). The choice between these alternatives depends upon whether or not it is possible to establish apotypic characters in common between A and D. The second possibility implies that D is the secondary twin taxon to B+C (Figure 2.4c). If all of the apotypic characters for the taxa B and C are found in D, then the classification will be a matter of determining which of the two taxa B and C is the secondary twin to D, the outcome being represented by the alternatives d and e in Figure 2.4. This approach can be continued *ad infinitum*, and it is therefore possible to state this working programme as a theorem:

(T2.44) Phylogenetic classification can be carried out through a succession of basic classifications.

At this stage it may be appropriate to recall the reservation made above with respect to the phenomenon of dichotomy. Should ardent search for apotypic characters between two out of three taxa be in vain, then, of course, no other expedient remains than to class them in a trichotomous dendrogram. This means that we renounce establishing the phylogenetic kinship between the taxa concerned (cf. T2.39). From this stage on the usual cladistic approach may be resumed. It must be remembered, however, that a classification must be subjected to continuous testing and, should new data permit the establishment of a more detailed relationship between the taxa in a plurichotomous dendrogram, then obviously the possibility obtains that we may gradually approach the cladistic ideal, dichotomy.

Convergence and non-apotypy

Before we investigate the possibilities of circumventing the impediment to phylogenetic classification constituted by non-apotypy, we shall deal with another phenomenon which may also complicate phylogenetic studies. It is often observed that 'expected' similarities do not materialize, while 'unexpected' similarities do. The direction of anticipation must, of course, be based on preconceived notions about the phylogenetic affinities between various animal taxa. In order to account for such unexpected events a special terminology has been developed. The several concepts are not very well defined and it is therefore possible that the following renderings do not cover all implications, but by and large I trust that they represent common usage. Thus, the first case mentioned above is referred to as 'specialization' or 'degeneration', the second as 'convergence' or 'parallelism'. In general, 'degeneration' stands for loss of some anticipated character, 'specialization' for its replacement by a new one. These concepts are discussed in a subsequent section. 'Parallelism' is used when similar or identical properties are presumed to have been independently acquired in closely affiliated lines, 'convergence' when the taxa involved are supposed to be distantly related. Since this distinction implies acceptance of an established classification, an attitude which is hardly compatible with an unprejudiced approach towards phylogenetic classification, I suggest to use 'convergence' to cover all cases of this kind. Since this phenomenon is clearly of importance for phylogenetic classification, I propose to incorporate the concept 'convergence' in the theory, defined as follows:

(D2.19) Convergence consists of the independent gain or loss of the same taxonomic character in two separate taxa in a basic classification.

A certain ambiguity resides in the word 'same'; the essential implication is, of course, that the character in question is regarded by taxonomists as the same. It may be noted that the definition does not exclude a certain character having been acquired independently in three or more taxa. The consequence of this situation is only that the character in question will be convergent in more than one basic classification. There exists an inverse relation between convergence arising through loss and through gain, respectively, in such a way that any authentic case of convergent non-apotypy can be interpreted in terms of either one of these two phenomena. This 'substitution principle' is most easily explained by means of a diagram. In Figure 2.5 it is presumed that a certain taxonomic character is found in the taxon A and in any one of the taxa C, D and F. The first case may be accounted for by the independent gain of the character in A and C, or through its being a plesiotypic character that has been lost in B, thus without invoking convergence. The second case may be explained as convergent gains as before, or, if the character is plesiotypic to all the taxa A–G, through its convergent loss in B, C and in the twin taxon to D, thus three times. Finally, if the character is present in A and F, its distribution will be accounted for through losses in B, C, D, E and G, or five times.

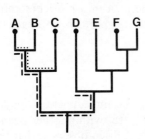

Figure 2.5. Diagram of the substitution principle of convergence. The presence of the same character, symbolized by a filled circle, in A and in either one of the taxa C, D and F may be accounted for through the number of convergent gains and losses shown in the following table:

	Gains	Losses
A–C	2	1
A–D	2	3
A–F	2	5

For the first two instances the lines used for the calculation of the number of losses have been indicated in the figure

This example may be generalized in the following theorem:

(T2.45) The presence of the same taxonomic character in two terminal taxa which are not secondary twin taxa can be explained through its convergent gain in the two taxa, or through its convergent loss in certain taxa affiliated with the two possessing the character. The number of losses required equals the number of taxa (i.e. vertical lines) in the phylogenetic lines uniting the two taxa.

This conclusion may be of interest, for instance, if it can be presumed that the probability of losses is substantially larger than that of gains.

If non-apotypy never occurred, then it follows from T2.42 that, among the three pairs which can be formed from the taxa in a basic classification, only one pair should have characters in common. The characters in question would be apotypic, and the particular pair of taxa the secondary twins. One single attribute of this kind would suffice to establish a basic classification. Under these circumstances the yoke of the phylogeneticist would be light. However, in reality matters are not so simple. The reason for this is that frequently it is found that two or three of the pairs of taxa in a basic classification share some characters. This situation corresponds to the phenomenon of non-apotypy, which may be defined as follows:

(D2.20) Non-apotypy is said to obtain when, in a basic classification, a property is found to be common to the primary, and to one secondary, twin taxon.

It is clear that a non-apotypic character will be apotypic to the twin taxa into which the primary, and the one secondary, twin taxon may be resolvable. Hence:

(T2.46) A non-apotypic taxonomic character is an apotypic character in two basic classifications.

Further, since non-apotypic taxonomic characters unite taxa that are not secondary twins, the discovery of a character which does not conform to a certain classification cannot *a priori* be regarded as falsifying evidence, for it may be a non-apotypic character.

Therefore:

(T2.47) Allowing for non-apotypy, no classification can be falsified by the discovery of a character whose distribution runs counter to prediction.

The reason for the appearance of this ostensibly anti-Popperian theorem is, of course, that the course of evolution is a stochastic process, and theories about evolution are consequently probabilistic theories. It is true that a probability statement cannot be falsified, but this does not imply that we must give up working with probabilistic theories. For, as Popper (1968) has shown, it is possible to formulate a methodological rule which forbids the occurrence of certain empirical events, and hence implies that even this kind of theory can be falsified. In Popper's words (1968, p. 205): 'What this rule forbids is the predictable and reproducible occurrence of systematic deviations; such as deviations in a particular direction. . . . Thus, it requires not a mere rough agreement, but the best possible one *for everything that is reproducible and testable*; in short, for all *reproducible effects*'. Therefore, I believe, T2.47 should not be regarded as frustrating, for in spite of it there is still hope that we may succeed in establishing phylogenetic classifications.

The easiest way to appreciate the implications of non-apotypy may be to survey the various ways in which two characters may end up to be common to two taxa. This is done in Table 2.2, from which it appears that altogether 10 different ways are imaginable. The first five concern plesiotypic characters; out of these only one (5) represents true apotypy, but of the remainder (1) and (3) will appear to be apotypic in the sense that they unite the set of secondary twins, while the other two are to be rated as non-apotypic. Case (6) is also truly apotypic, representing the gain of a character in the taxon comprising the two secondary twins. The last four alternatives deal with teleotypic characters, and

Table 2.2
The different ways in which a taxonomic character may end up being common to two out of three taxa in a basic classification

| | A | B | C | Number of taxonomic characters | | |
				AB	AC	BC
(1) Plesio-apotypy	ⓟ	p ⟷	p			Ft_1
(2) Plesio-non-apotypy	p ⟷	p	ⓟ	Ft_2		
(3) Convergent plesio-apotypy	p	ⓟ	ⓟ			$F^2 t_2^2 / N$
(4) Convergent plesio-non-apotypy	ⓟ	p	ⓟ	$F^2 t_1 t_2 / N$		
(5) Apotypy	p	ⓟ ↔	ⓟ			$F(t_1 - t_2)$
(6) Apotypy	$\bar{a}$	a ⟷	a			$F(t_1 - t_2)$
(7) Teleo-apotypy	t	$\bar{t}$ ⟷	$\bar{t}$			Ft_1
(8) Teleo-non-apotypy	$\bar{t}$ ⟷	$\bar{t}$	t	Ft_2		
(9) Convergent teleo-apotypy	$\bar{t}$	t	t			$F^2 t_2^2 / N$
(10) Convergent teleo-non-apotypy	t	t	$\bar{t}$	$F^2 t_1 t_2 / N$		

A is the primary, B and C the secondary, twin taxa. An arrow between two characters indicates non-convergence, a bar above a letter means absence of the character in question. Uncircled this absence implies 'non-gain', circled 'loss'. The small letters refer to the origin of the characters; a, apotypic; p, plesiotypic; t, teleotypic. N is the number of potential mutations. To designate the various phenomena some names have been proposed the derivation of which appears to be straight-forward.

again it is seen that two, (7) and (9), will appear to be apotypic, the other two non-apotypic. Thus, although only two cases represent true apotypy, it will be impossible to distinguish these from the other four cases unless we know about the phylogenetic history of the characters in question. Similarly, it may be impossible to decide which of the four instances of nonapotypy is represented by a particular type of character. It will be noticed that each of the cases of loss of a plesiotypic character is matched by one representing a gain of an apotypic or teleotypic character.

The concepts of 'parallelism' and 'convergence' almost always imply what is here called 'non-apotypic convergence'. However, as seen from Table 2.2, convergence may be apotypic as well.

Before we discuss the difficulties which the phenomenon of non-apotypy creates for the phylogeneticist it should be mentioned that there are a number of characters, for instance body form in Salientia, feathers in Aves and embryonic membranes in Amniota, which are so unique that we can with great confidence rate the taxa defined by these characters as monophyletic. It must be admitted that but for these taxa, which range to levels inferior to those indicated by the quoted examples, phylogenetic classification would probably be unachievable. The real problem facing the phylogeneticist is rather to establish the kinship between these undisputable monophyletic taxa.

However this may be, it follows from Table 2.2 that whenever we attempt to establish a basic classification by searching for characters in common between the taxa, we shall usually find that all three pairs share some characters. Under these conditions we face the problem of how to decide which of three sets of characters is apotypic and which is not. Two possibilities obtain: (1) the nature of the characters are such that the decision is obvious or (2) we have no means to settle this issue offhand.

We shall first discuss the second alternative, and our first approach to this problem must obviously be to consult Hennig (1966). Here we find various suggestions which purportedly ensure that a classification is phylogenetically correct. Among these are the rule of deviation, which states that one of two twin taxa generally has preserved many more plesiotypic characters than the other. Although it is undoubtedly true in most cases, this knowledge is of little value in the establishment of a basic classification. Thus, if Cyclostomata is a monophyletic taxon, then it is the twin taxon to Gnathostomata, and in this case there is no doubt which of them has preserved most plesiotypic characters. But if we were facing the task of classifying Hyperotreta, Hyperoartii and Gnathostomata in a basic classification, then the rule of deviation would apparently require that the two first taxa are made the secondary twins. But if the similarities between them are plesiotypic characters, then the outcome is not necessarily correct. The clue to the classification must be to determine whether either of the two taxa has characters in common with Gnathostomata, and, so far as I can see, the rule of deviation does not give any directives about how this may be done.

In order to solve the dilemma facing the phenetic approach, Hennig (1966) has suggested recourse to other sources of information. Thus, it is proposed to use the chronological data stored in the fossil record to distinguish apotypic characters. It is unquestionably true that characters found in fossils from earlier geological strata are older than some found in later deposits, and it is likely that in some classifications they are plesiotypic. But again, this knowledge cannot determine whether or not one of two 'plesiotypic' taxa have apotypic characters in common with a 'derived', or advanced, taxon. Furthermore, the number of important classifications which can be settled by means of fossil data is modest. Returning to the example quoted above, the modern cyclostomes do not become fossilized. Arguments have been advanced in support of affinity between the extant forms and separate groups of extinct ostracoderms, but even if these ideas are correct they do not seem to be of particular value for the basic classification discussed above.

Second, Hennig (1966) proposes using chorological data, as revealed by the geographic distribution of animals. Yet, this suggestion presupposes that the process of dispersal is associated with taxonomic divergence, and that consequently the 'plesiotypic' forms are found at the centre of origin. I shall contend in Chapter 5 that 'divergence through dispersal' is a common phenomenon, but nevertheless I am convinced that the criticism of Croizat, Nelson and Rosen (1974) of the notion of 'centres of origin' is justified. In Nature so many different things can happen, for instance that a derived form gains dominance over the original one and disperses retrogradely, extinguishing its more or less immediate ancestors. Furthermore, the application of this approach requires absolutely correct information about the history of our planet's surface. We should not too quickly forget the many stratagems used to explain the striking similarities between animals and plants on both sides of the South Atlantic before the theory of continental drift was accepted. Yet, even if the value of biogeographical observations for phylogenetics may be more limited than Hennig thought, it has to be admitted that such data may be employed with benefit, as demonstrated for instance by the work of Brundin (1966).

Third, referring to Haeckel's law of recapitulation, Hennig (1966) suggests the use of embryological data in phylogenetic classification. This approach may occasionally be of advantage; Vertebrata and Amniota are well-known examples. But these are instances where ordinary phenetic characters might suffice. At the superior and inferior taxonomic levels, where auxiliary data would be of greater value, this approach fails. As an illustration of this, I propose to show in Chapter 3 that it is possible to falsify the Protostomia–Deuterostomia theory, a classification which is based primarily on an embryological character.

It thus appears that the various non-phenetic aids to phylogenetic classification outlined by Hennig are of limited validity at best. But if we are to succeed in this pursuit we must rely on a general method, and one, it seems, which is based on phenetics. What remains for this purpose is the last of Hennig's suggestions, namely, combining the phenetic approach with the principle of parsimony. As we shall discuss later, this method has been adopted by the various molecular biologists working on phylogenetic problems.

The implication of the principle of parsimony is that a choice is made among the three alternatives in a basic classification on the basis of their respective probabilities. The guiding principle may thus be stated in the following axiom:
(A2.8) Among the possible alternatives in a basic classification, the one with the highest probability should be preferred.

This probability may be estimated on the basis of the frequency of origination of taxonomic characters. In relation to this point we shall introduce the following axiom:
(A2.9) The frequency with which taxonomic characters arise is distinguished by a certain mean value F.

Here no distinction is made between losses and gains of a particular property although, as mentioned above, losses are probably much more common than gains.

The number of taxonomic characters defining a certain taxon must be Ft, where t is the time during which the taxon has been under creation. Since this period of time is generally unknown, we shall here let t represent the time from the separation of the taxon from its twin taxon to the present day. This means that when we are dealing with superior taxa, the taxonomic characters should not be those defining only these taxa but, rather, those representing a phylogenetic line leading from the bifurcation in the dendrogram to some terminal taxon (species) which is included in each of the taxa under investigation.

If these stipulations are accepted, and the ages of the various taxa in the basic classification are assumed to be t_1 for the terminal primary twin taxon, t_2 for the secondary twin taxa, and thus $t_1 - t_2$ for the non-terminal primary twin taxon (Figure 2.6), then the characters in common in the various cases in Table 2.2 can be calculated, as shown in the last column of the table.

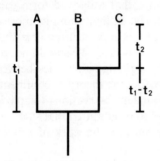

Figure 2.6. A basic classification in which the time t_1 has passed since the splitting of the primary twin taxa and the time t_2 since the splitting between the secondary twin taxa

The frequency of convergence may be calculated in the following way. Let the total number of potential mutations in the phylogenetic lines originating in a pair of twin taxa be N. Let further the number of mutations which actually have occurred be n $(=Ft)$. Then the number of identical mutations, i.e. number of convergent characters, will have a hypergeometrical distribution with the mean value n^2/N. Thus we have:

(T2.48)* The frequency of apotypic convergence in a basic classification is $F^2 t_2^2/N$. And:

(T2.49) The frequency of non-apotypic convergence in a basic classification is $F^2 t_1 t_2/N$.

It may be presumed that $N \gg Ft$, and therefore we may infer:

(T2.50) The probability of a phylogenetic classification is inversely proportional to the number of cases of convergence it implies.

With reference to A2.8, it is possible to derive the following rule from this theorem: (T2.51) In phylogenetic classification the number of cases of convergence should be minimized.

From this discussion on convergence we may make some important conclusions. First, in view of T2.51 it is rather disquieting to register the excessive use which has been made of convergence as an explanatory device in traditional phylogenetics.

Second, mathematically non-apotypic convergence is more frequent than apotypic convergence. But, in fact, if the possession of certain characters in common enhances the likelihood that a further property is gained convergently, we may then expect apotypic convergence to be the more common of the two phenomena. Therefore there

---

*The reader probably wonders why I have chosen to present some simple calculations in the form of this and the following theorem. The reason is that at a very late stage in the preparation of the manuscript I was forced to delete two other theorems in the present context, and in order to avoid changing the numbers of all the following ones I introduced these two rather trivial theorems.

is no reason to suspect that convergence will in general be a serious encumbrance in phylogenetic classification.

Third, this last conclusion is further corroborated by the fact that, as we have seen, convergence must be a relatively rare occurrence, rare enough, I believe, .to warrant its being neglected in the following discussion.

From the values listed we may calculate that the difference between the number of non-convergent characters uniting the pair of secondary twins and those common to any of the other pairs is

$$D = 2Ft_1 + 2F(t_1 - t_2) - 2Ft_2 = 4F(t_1 - t_2).$$

The requirement for the dendrogram being a basic classification rather than a trichotomy is that $t_1 - t_2 > 1/4F$, the time necessary for a character to arise in either of the pair of primary twin taxa. For $t_1 - t_2 \to 1/4F$, we have $D \to 1$. The requirement for the dendrogram being a basic classification rather than a dichotomy is that $t_2 > 1/4F$, the time required for a character to arise in either of the pair of secondary twin taxa. For $t_2 \to 1/4F$, we have $D \to 4Ft_1 - 1$.

The quotient between the two sets of characters is

$$Q = \frac{2Ft_1 + 2F(t_1 - t_2)}{2Ft_2} = \frac{2t_1 - t_2}{t_2} = 2t_1/t_2 - 1.$$

For $t_2 \to t_1$, we have $Q \to 1$, so that the numbers of apotypic and non-apotypic characters are the same, as expected. For $t_2 \to 1/4F$, we have $Q \to 8Ft_1 - 1$, which represents the maximum possible.

On the basis of these very crude assumptions we thus arrive at the following conclusion:
(T2.52) In a basic classification the pair of secondary twins have more characters in common than any of the other pairs of taxa.

This theorem formulates the *rule of the quantitative approach to phylogenetic classification*, according to which the relative number of characters in common decides the relationship between the taxa. It seems possible that Hennig had this approach in mind when he suggested that the use of many characters may increase the probability of a given classification. But I cannot agree with the following statement (1966, p. 121): 'The more characters certainly interpretable as apomorphous (not characters in general) that there are present in a number of species, the better founded is the assumption that these species form a monophyletic group'. Surely, if only *one* character is *certainly* interpretable as apomorphous (apotypic), then we need not assume, we know, that we are dealing with a monophyletic taxon. Rather, it is in all those cases where we are not absolutely sure about the nature of the characters that we must adopt the numerical approach implied by the above quotation.

Sneath and Sokal (1973) stress that 'invariant' characters should be excluded and, in a basic classification at least, invariant characters coincide with those, common to all three taxa, which with confidence can be rated as plesiotypic. If this interpretation is correct then the theorem also represents the basic methodological principle of numerical taxonomy. Anyhow, since the number of plesiotypic characters will be the same for each set of characters in a basic classification, their inclusion should not influence the outcome, but only reduce the precision with which decisions between alternative classifications are made.

In spite of its wording I therefore believe that T2.52 is an endorsement of the view 'that numerical phenetics will in general give monophyletic taxa, because we believe that phenetic groups are usually monophyletic' (Sneath and Sokal, 1973, pp. 46–47). I therefore propose to call it *Sneath and Sokal's theorem*.

It is possible to calculate the total number of characters, apotypic as well as non-apotypic, which must be sampled to make a correct classification. This value is a function of Q, and hence of t_1/t_2. A few examples are shown in Table 2.3, from which it appears that even when the secondary twins have been united into one taxon for 15/17 of the time since the separation of the primary taxa, one character is not enough to establish a correct classification; the chances are still one to two that it is non-apotypic. It is further seen that if the separation of the secondary twin taxa is of long standing, a large number of characters is required to ensure a reliable result.

Table 2.3

The minimum number of randomly sampled taxonomic characters required for the correct establishment of a basic classification

t_1/t_2	Q	n	P
1·5	2	43	0·95
2·5	4	10	0·95
4·5	8	5	0·96
8·5	16	3	0·97

In this table n stands for the number of samples and P for the probability of a correct choice. These calculations, based on the procedure published by Bechhofer, Elmaghraby and Morse (1959), have been made by Dr. Göran Broström, to whom I express my sincere gratitude.

T2.51 thus suggests as an approach to the establishment of a basic classification that a search is made for taxonomic characters which unite two out of the three taxa, and accepting those two taxa that have most characters in common as the secondary twins. Thus, the number of cases of apotypy is being maximized, those of non-apotypy minimized.

The preceding calculations are based on the assumption that it is possible to operate with a mean frequency for the origination of new taxonomic characters. I do not believe that this stratagem is inadmissible from a theoretical point of view, but it may tend to obscure the fact that certain characters, notably those which distinguish a number of superior taxa, have remained unchanged for hundreds of millions of years. These include the features mentioned above, viz., salientian body form, feathers and amniote embryonic membranes. The preservation of these characters, which, evidently, must contribute significantly to the survival of animals possessing them, may to a large extent be responsible for the fact that many Linnean taxa are monophyletic.

However, the establishment of the phylogenetic relationships between various groups of these several distinct taxa often meets with great difficulties, because it is next to impossible to find characters which unite some, but not all, of the taxa. Here, and also in many instances of classification at the inferior taxonomic levels, I believe that the quantitative method outlined here constitutes the only objective approach to phylogenetic classification.

Yet one cannot rid oneself of the nagging doubt that even this approach does not always provide the correct answer. This may happen if the characters employed fail to comply with the specifications stipulated above, namely, (1) that they should represent not individual taxa, but the phylogenetic lines leading from the bifurcations in the dendrogram, (2) that new characters should arise with a reasonably constant frequency and (3) that they should be independent.

The very fact that Hennig (1966) has been able to state his rule of deviation shows that the second requirement is not fulfilled by the morphological characters usually employed in classification. This appears to represent the most serious source of error. There is also reason to suspect that morphological characters are very often interdependent, partly because their appearance during the course of embryogenesis is due to a series of causally connected epigenetic events, and partly because the acquisition of one property may pave the way for others arising through separate mutations. However, these phenomena should rather increase the likelihood that a phenetic classification is phylogenetically correct.

It will be shown in the following section that the characters studied by molecular biologists meet in principle with the three demands outlined above. Unfortunately, data of this kind are scarce so far, and unless we are willing to wait for new ones to appear, we must use the information available. I therefore submit that certain non-morphological characters assume a position intermediate between morphological and molecular–biological ones as regards their reliability in phylogenetic classification. Characters of this kind will be used in Chapters 3 and 4.

The numerical approach outlined above is based on A2.9, stating that it is possible to ascribe a certain mean value to the frequency at which taxonomic characters arise. I have admitted that this premise is not generally valid, and some phylogeneticists will be inclined to reject it. But in that case phylogenetic classification becomes impossible, unless we are prepared to adopt another premise, namely, that *it is feasible to distinguish between apotypic and non-apotypic characters*. It appears that Hennigian cladists in general are convinced that they are able to make such evaluations correctly.

I shall not contend this point, but only observe that if they are right, then phylogenetic classification should be very easy, one single character would suffice to establish a basic classification. Yet, even if there is a measure of truth in the claim made by the cladists, I believe that it will be well advised to use as many characters as possible even in this case.

The particular cladistic insight will be of great value in cases where a classification must be based primarily on morphological features, because these characters, as mentioned above, occasionally may be very conservative. In this situation the two instances of non-convergent non-apotypy in Table 2.2 may impede a quantitative approach to the establishment of a basic classification. These cases, involving either (2) that the primary and one secondary twin taxon have plesiotypic characters in common which have been lost in the second twin taxon or (8) that one of the secondary twin taxa has gained characters through teleotypy, the other two taxa therefore being united through the absence of these characters. If this situation obtains it may be found that one of the taxa has many characters in common with both of the other taxa, making a numerical decision difficult or impossible.

But if we have the faculty required to decide which characters are apotypic and which are not, no difficulties are encountered in settling the issue. The situation outlined here may be represented in the following theorem:

(T2.53) If, in a basic classification, one taxon, which must be a secondary twin, has several characters in common with both of the other taxa, then one of the two sets of characters will represent the presence of plesiotypic characters absent, and/or the absence of teleotypic characters present, in the third taxon. The latter is the second secondary twin taxon.

This theorem states the *rule of the qualitative approach to phylogenetic classification*. In the decision required to establish the classification Hennig's rule of deviation may be of use, since it may help to determine which out of the three taxa is a 'derived' one, and hence one of the secondary twin taxa. Once this point has been established, it follows from T2.53 that the taxon which has characters in common with both of the others must be the second secondary twin taxon. Clearly, this method is much less laborious than the numerical approach but, although it is certainly an advance compared to the classical Linnean approach, it is less foolproof, and the risk of mistakes cannot be excluded.

An example of the evaluative or qualitative approach is found in the situation where it is required to classify two fish taxa, for instance Selachii and Dipnoi, and a taxon among Tetrapoda, say, Caudata. Here the first two taxa have a large number of characters in common, for example the absence of limbs, of keratinized epidermis and of a middle ear, but also the presence of certain features like scales, fins, lateral line system, etc. Yet, as it turns out, Dipnoi and Caudata share some properties which, by some evolutionists are least, are considered to reveal a kinship between the two taxa, i.e. they contend that the characters in question are apotypic, and that these taxa, in the given situation, must be the secondary twins, even though they may not use this terminology.

The discussion so far may give the impression that the taxon representing phylogenetic progress must always be one of the secondary twin taxa. That this is not so may be demonstrated by the basic classification of Selachii, Holocephali and Caudata. Compared to the latter the first two taxa have a large number of characters in common, among which some may unquestionably be regarded as absence of advanced characters. However, neither of them share, to any noticeable extent, characters other than plesiotypic ones with Caudata, and hence they must be classed as secondary twin taxa, and the amphibian order as the primary twin taxon.

It is sometimes asserted that taxonomic characters should be weighted, and, indeed, if two sets of characters support different classifications in a situation where it is impossible to use the qualitative approach, it might be valuable to be able to evaluate the relative weight of the characters in the two sets. Since, according to A2.8, our goal must be to establish the most probable classification, we may state:

(T2.54) The weight of a taxonomic character is inversely related to the probability of its origination.

This approach to character weighting was proposed by Farris (1966).

As has been discussed above, form seems to be the most labile of the various properties exhibited by the members of kingdom Animalia. Much more constancy is displayed by various chemical features, physiological mechanisms and histological differentiation patterns. Hence, we may postulate:

(A2.10) The origination of new non-morphological characters is much less probable than is the origination of new morphological ones.

From T2.54 and A2.10:

(T2.55) In phylogenetic classification non-morphological characters carry much more weight than morphological ones.

While it must be emphasized that chance plays a great role in evolution and that, consequently, strict adherence to T2.55 may lead to wrong conclusions, this theorem is still a reminder that the existing monopoly of the morphologists regarding phylogenetic classification cannot be sustained. If non-morphological and morphological data diverge as regards a certain classification, then the chances are reasonable that we are dealing with non-apotypy and/or convergence on the morphological, rather than on the non-morphological, plane, and the result obtained with non-morphological characters should therefore have priority.

Relative age of taxa

The establishment of a basic classification implies an estimation of the relative age of the taxa. Therefore, an evolutionary clock is required, and it is for this reason that it is demanded of the characters used for phylogenetic classification that they originate with a reasonably constant frequency.

This is a premise in the analyses on protein evolution made by amino acid sequence analyses and serological methods. In some cases it has been possible to demonstrate that changes occur at almost the same rate in disparate taxa, for instance albumin in frogs and mammals, but in birds these proteins have evolved much more slowly (Prager and Wilson, 1975). This is perhaps what might be expected when we are dealing with phenomena that are the outcome of chance events. It is therefore justifiable to criticize the assumption that mutations occur with constant frequency (cf. Jukes and Holmquist, 1972; Penny, 1974). It also appears that the methods used for evaluation of the results may need further improvement (Holmquist, 1972; Sneath and Sokal, 1973), and even sampling is an important source of error (Kimura, 1969). However, by and large the results obtained seem rather encouraging, not least because they usually fit quite well to the current classification, even though in some cases there are noteworthy differences. I have stated above that data of this kind should have priority over those obtained by means of morphological characters, but also that no single character can falsify a given classification. One such event can therefore at most throw doubt on the ruling views. Yet proteins are an almost inexhaustible source of taxonomic characters, and should subsequent work corroborate such heretical results, then there might be reason to contemplate revising the classification.

It should also be stressed that characters of this kind fulfil the requirements to estimate the changes which have occurred in a phylogenetic line rather than in a taxon, and to be independent. And they furthermore yield a large number of characters at one time, a valuable feature in numerical taxonomy. If therefore the molecular–biological approach is far from flawless, its potential as a source of information of relevance for phylogenetics is nevertheless vast. Thus, there is no reason to doubt that most, if not all, of the classificatory problems which cannot be settled satisfactorily by other means can be decided by molecular–biological methods.

On the phylogeneticist's vocabulary

Since it may throw some light on the background of the present theoretical approach, we shall in the present section discuss certain adjectives which are of common use in the phylogenetic literature, but which, it seems, vary in meaning from one author to another.

It is common to find statements to the effect that this or that taxon is 'diphyletic' or, possibly, 'polyphyletic'. It may not be without interest to try to establish the exact meaning of these words. First, we must emphasize that, as we have seen above, phylogenetic classification comprises only monophyletic taxa, so that the taxa in question, if their characterization is correct, must be rejected by the phylogeneticist. This does not, of course, prevent an individual taxonomist, or even the whole guild of taxonomists, choosing to work with taxa of this kind, for reasons of convenience or tradition, as long as they do not claim or believe that their classification is phylogenetic.

However, the negation of 'monophyletic' is, presumably, 'non-monophyletic', so the possibility exists that the words quoted above may have a significance of their own apart from that, evidently, the taxa are not monophyletic. This appears to be the case, for from the postulate that the class Aves sprang from some reptilian ancestor, implying that Aves and Reptilia + Aves are monophyletic taxa, whereas Reptilia is a non-monophyletic taxon, it is not customarily concluded that the latter is a diphyletic or polyphyletic taxon.

It may be simplest to survey the implications of these various concepts by means of a practical example, say, Cyclostomata. The assertion that this taxon is monophyletic would imply the classification shown in Figure 2.7a. If the Cyclostomata are not monophyletic, the most likely classification of the taxon must be the one represented by Figure 2.7b. This dendrogram shows that each of the agnathous taxa are monophyletic taxa, as are Vertebrata, Cyclostomata 2 + Gnathostomata and, of course, the latter by themselves.

With reference to the definition of monophyly, it is seen that in order to classify the two orders of cyclostomes as diphyletic it is required that the characters which these two taxa have in common, and which are otherwise either apotypic agnathous characters (Figure 2.7a) or plesiotypic vertebrate characters (Figure 2.7b), have arisen through convergence. Two alternative classifications may account for this proposition. The first of these implies that the two cyclostome orders have arisen independently from gnathostomous ancestors, or that two groups of gnathostomes have arisen independently from agnathous forebears (Figure 2.7c). The other alternative is that the two taxa of

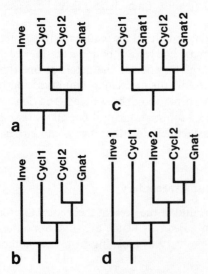

Figure 2.7. Classification of Cyclostomata. (a), as a monophyletic taxon; (b), as two monophyletic taxa; (c) and (d), as two 'diphyletic' taxa

cyclostomes have originated independently from invertebrate ancestors (Figure 2.7d). In the latter case, to establish a monophyletic taxon containing both orders it is necessary to include some invertebrate taxon. Both possibilities seem equally unacceptable because of the enormous amount of convergence they imply.

Jarvik (1968b) uses 'diphyletic' in a special sense in this context, namely, to mean that the two orders of Cyclostomata have originated in different fossil groups. This contention is not incompatible with the proposition that in the classification of living animals they form either one monophyletic taxon (Figure 2.7a) or two separate monophyletic taxa (Figure 2.7b). In fact, the diagram published by Jarvik corresponds to the first alternative (Figure 2.8).

The essence of the present discussion is that 'diphyly' and 'polyphyly' are used to indicate that the characters uniting the two or more taxa to which they refer are convergent, thus making these words synonymous with 'convergence'. This point has been made by Reed (1960) in a discussion of Simpson's 'polyphyletic' extinct mammalian taxa (Simpson, 1959). And since it can only promote confusion to introduce synonyms

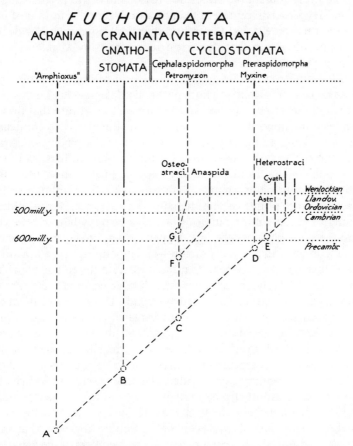

Figure 2.8. Phylogenetic classification of 'Euchordata', with special reference to Cyclostomata. (Reproduced from E. Jarvik (1968b), in *Current Problems of Lower Vertebrate Phylogeny*, by permission of the author and Almqvist & Wiksell)

for established and well-defined concepts, it may be justified to suggest that the use of these words be discontinued (cf. Ax, 1964).

In phylogenetic literature one may also meet with epithets like 'degenerate', 'specialized', 'primitive' and 'advanced'. A further list of adjectives may be found in Devillers (1973). Since these expressions clearly designate phenomena of phylogenetic significance, it is hardly possible to do without them. However, if one phylogeneticist categorically states that Cyclostomata, in contrast to Dipnoi, are not 'primitive' (Jarvik, 1968a) and most other biologists equally stubbornly claim that the former are more 'primitive' than the latter, then it is obvious that these people use the word in different senses. For this reason it may be justified to look into the intended meanings of these words and see if it is possible to introduce an unambiguous terminology.

I believe that if Jarvik's opponents contend that Cyclostomata are more primitive than Dipnoi, then they refer to the fact that through the want of characters like jaws, paired fins and other apotypic gnathostomous characters, their 'archetype' represents an earlier stage in the evolution of Vertebrata than does that of the Dipnoi. This contention implies that Cyclostomata or, rather, for reasons that will become apparent later, Hyperotreta and Hyperoartii, represent taxa of a rank superior to that of Dipnoi. It is very important in this context that the taxa in question represent a low degree of resolution; it is not at all certain that the taxon represented by *Lepidosiren paradoxa* is inferior to that of *Petromyzon marinus*.

Maybe some confusion could be avoided if, instead of arguing about which of the taxa is more 'advanced' or 'primitive', both parties could agree that Dipnoi are 'inferior' to Hyperoartii, and the latter 'superior' to the former. I am aware that these adjectives may convey a meaning opposite to the intended one, due partly to confusion with the Linnean way of ranking, and partly to the inversion of the phylogenetic hierarchy in graphic illustrations. For this reason it is pleasant to observe that, as long as we are dealing with taxa belonging to the same phylogenetic line, we may use a terminology which is simpler and quite unambiguous. Referring to T2.29, it will be seen that we may simply characterize the 'primitive', i.e. the superior, taxon as being 'older' than the 'advanced', inferior taxon, and the inferior taxon as being 'younger'. Hence, provided they belong to the same phylogenetic line, Hyperoartii are older than Dipnoi.

When Jarvik contends that Cyclostomata—or Hyperoartii—are not primitive, but that Dipnoi are, then he seems to mean that the former, in the course of evolution, have acquired a number of teleotypic characters which have ensured their survival in a diversity of forms exceeding that accomplished by Dipnoi. This situation might be described by the words 'specialized' or 'advanced', but for reasons to be discussed presently the first should preferably be used in another sense, and the second one is ambiguous— it may mean 'inferior' or 'younger', as in the sense in which it was used above. This may be avoided if, instead, we characterize Hyperoartii as 'teleoptypically advanced' and Dipnoi as 'teleotypically primitive or retarded'. Hennig's concept of 'derived' would be of little help in this case. Another possibility would be to adopt the expressions coined by Simpson (1944), characterizing the former as 'tachytelic', i.e. 'quickly evolving', and the latter as 'bradytelic', i.e. 'slowly evolving'. I doubt, however, that many phylogeneticists would gladly apply the former adjective to members of the agnathous orders.

Clearly, for the word 'degenerate' to make any sense, it must imply losses of something, presumably useful characters. That losses of this kind may occur is well documented, for instance through studies on intestinal and other parasites, as well as of sessile animals.

However, this word is often used without much substantiation; this seems to be the case when Amphioxus is thus described. Surely, without thorough knowledge of the ancestors of this animal it is impossible to make any valid statements about its state of degeneration.

Furthermore, the very notion of 'degeneration' seems to run counter to the central idea in the theory of evolution, according to which this process is distinguished by a steady increase in 'fitness', presumably exactly the opposite of degeneration. Since, as we shall see in a later chapter, losses of the kind referred to here actually do not entail reduction in 'fitness', when this concept is properly defined, it seems appropriate to discontinue the use of the word 'degeneration', replacing it by 'specialization'.

Linnean classification

Phenograms or cladograms are usually radically different from Linnean hierarchies. The most striking dissimilarity is that the number of ranks is much greater in the former than in the latter. It would appear that these recent advances must lead to the abandonment of Linnean categories, first because their number is too small to be of use, and second because it is impossible to decide in advance the number of taxonomic levels required for the classification of a given taxon, which is exactly what these categories do. However this may be, I shall in the present context discuss some classificatory problems which, it seems, may be consequences of the various conventions inherent in Linnean classification.

First is the phenomenon of polytypy or polythety, which involves that, in a Linnean taxon T_j^*† no single character is present in all (usually more than two) the T_{j-1}^* taxa of the nearest lower category which are included in the former. According to Beckner (1959), all taxa are polythetic, a situation which evidently requires that all the characters which originally defined the taxon T_j^* have been subjected to losses in one or more of the included taxa. Although this possibility cannot be excluded, it seems more likely that the problem of polythety arises through the convention of Linnean classification according to which a taxon is supposed to branch up into more than two taxa at the nearest lower categorical level.

It is possible to demonstrate a number of interesting and important differences between the consequences of Linnean and cladistic classification (Løvtrup, 1973; 1974). However, we shall deal here with two topics only, the first of which is the 'species' concept. The point of interest in this context is the purported distinction between the species and taxa at higher taxonomic levels, the elusiveness of which has baffled so many taxonomists. We shall begin by defining the species:

(D2.21) The species is the lowest category in the original Linnean system.

If it is assumed that species in Linnean classification are monophyletic and if, as proposed above, it may be possible in some instances to pursue the process of dichotomy well below the species level, we may conclude:

(T2.56) The species is a set of one to several terminal and subterminal phylogenetic taxa.

It was inferred above (T2.28) that, in a completely resolved classification, the terminal taxa are taxa under creation and, consequently, not definable. Hence we may, with

---

†A special symbol, T_j^*, is used here for a Linnean taxon, allowing for the possibility that the latter may be different from a phylogenetic taxon. In one respect these two kinds of taxa are sure to be different, for in Linnean classification the numbering of the rank of the taxa conventionally begins at the base of the hierarchy, a rule to which I have adhered in the present discussion.

reference to T2.56, conclude:

(T2.57) When a species is subdivided into inferior phylogenetic taxa, it is, barring cross-ings between the subspecific taxa, as immutably defined as the other taxa of superior rank to which the members of the species belong.

From this theorem it follows that certain taxa are distinguished by representing 'evolution in the making'. Yet these taxa are not, as is often claimed, the Linnean species, but rather the terminal taxa in the phylogenetic classification which may, or may not, coincide with the species.

Under the massive impact of neo-Darwinism, most systematists confess adherence to the 'new' systematics (cf. J. S. Huxley, 1940), according to which the properties of genetic compatibility and genetic incompatibility are important species criteria. Although it must, in principle, be deemed objectionable to introduce taxonomic standards whose application is excluded in a substantial part of the living world, we shall nevertheless discuss this question briefly because, of course, sexual reproduction is an influential evolutionary agent. It must be stressed, however, that, whereas the preceding discourse is supposedly generally valid, the remaining part of the present section pertains only to animals with bisexual reproduction.

We shall begin with the following axiom:

(A2.11) In a completely resolved classification of a monophyletic taxon, the members of which reproduce bisexually, it is possible, close to the base of the hierarchy, to draw a wavering line representing the limit of interfertility. This line, which separates all terminal, and some subterminal, taxa from the rest of the hierarchy, will never pass below the level represented by the Linnean species.

This axiom implies that, when two populations of animals are genetically incompatible, then they must be classed as different species, irrespective of whether or not it is possible to detect any other properties implying this distinction. This evidently amounts to intro-ducing genetic incompatibility as a taxonomic character, an expedient which must be adopted even in phylogenetic classification. In conformance with empirical observation, it does not assert that interfertility cannot obtain at superior levels to that of species.

According to A2.11, there must exist a considerable number of taxa, ranging perhaps from genera to subspecies, races, etc., the members of which are genetically compatible. If they do not interbreed then it must be inferred that they are wholly or partly isolated from each other. This point relates to the assertion made above that the first stage in the branching of a taxon into two twin taxa of higher rank must involve a step of isolation. This process may be active or passive (fortuitous) and it may or may not lead to immediate changes in the gene-pool of the, purportedly, small population of individuals which have become isolated from the main stock of the taxon to which they belong. However this may be, if, and when, the members of either one of the two separate populations acquire properties which correspond to the definition of 'taxonomic characters', the original taxon T_j has become separated into two taxa T_{j+1}. If T_j was classed as a species, the latter has now become fully defined, and we have two new subspecies, characterized by the absence and the presence, respectively, of these taxonomic characters.

If there is not an insuperable barrier, two different kinds of event may occur. Firstly, if the members of one of the new taxa are 'more fit in the struggle for existence' than those of the other one, then they may invade and conquer the territory occupied by the latter. If, in spite of the possibility of interbreeding, this does not take place, the less fit taxon becomes extinct. If gene exchange occurs between members of the two taxa, they will both survive, to a varying degree perhaps, in the resulting offspring. The situation envisaged here appears to represent a case of potential reversibility in evolution. Secondly, if the difference in 'fitness' is slight, no invasion is possible, but hybridization may occur where the territories border on each other. The hybrids may become genetically isolated from the parents, thus representing a new phylogenetic taxon. This case, known mainly from plants, is interesting because it involves the union of two taxa, thus constituting an exception to the cladistic theory outlined here.

Admitting the possibility of the situation presented by the first instance above, we shall make the following statement:

(A2.12) As long as genetic compatibility obtains between members of different taxa, the course of evolution is potentially reversible.

From A2.11 and A2.12:

(T2.58) The limit of interfertility represents the limit of reversibility of evolution.

This, I believe, is the particular importance of the genetical compatibility stressed so much by the adherents of the new systematics. And since species are always below the limit of reversibility, it is necessary, in T2.57, to make a reservation with respect to crossings between subspecific taxa. It should be pointed out, though, that whether reversion occurs, and at which level, is a matter of chance and that, for a number of reasons, this course of events appears to be unusual.

One other point which has caused much controversy in the discussion of phylogenetic evolution is its direction with respect to the classificatory hierarchy. As a starting point for a discussion of this question we may use the diagram constructed by Darwin (1885), which is reproduced in a slightly modified form in Figure 2.9, and some quotations representing Darwin's comments on this scheme. Thus: 'Let A[and I] represent the species of a genus [1]' (p. 90) and 'In the diagram each letter [a–n] on the uppermost line may represent a genus including several species; and the whole of the genera along this upper line form together one class, for all are descended from one ancient parent [1] and, consequently, have inherited something in common. But the three genera on the left hand [a–c] have, on this same principle, much in common, and form a subfamily [2], distinct from that containing the next two genera on the right hand [d–e], which diverged from a common parent. . . . These five genera have also much in common,

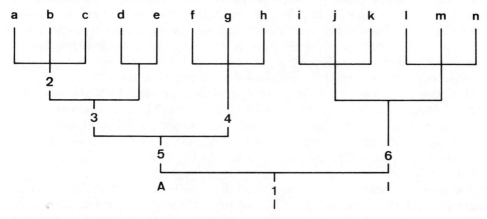

Figure 2.9. Darwin's phylogenetic dendrogram, slightly modified. Redrawn from C. Darwin (1885), *The Origin of Species by Means of Natural Selection*, Murray

though less than when grouped in sub-families; and they form a family [3] distinct from that [4] containing the three genera still further to the right hand [f–h], which diverged at an earlier period. An all these genera, descended from (A), form an order [5] distinct from [that (6) formed by] the genera descended from (I)' (p. 364).

The basic tenet in Darwin's argument and, so far as I can understand, of the present-day adherents of Darwinian theories also, is that the members of any group of organisms which fulfil certain requirements, among which is usually that of interfertility, at any period in the history of phylogenetic evolution may, and must, be classified as a species. And, as claimed by Darwin, all evolutionary events consist in nothing but the creation of new species, an outcome which ensues because 'the varieties, or incipient species, . . . ultimately become converted into new and distinct species' (p. 363).

We owe to Darwin, as demonstrated by the above quotation, the express statement of the principle that classification may be based on, and reveal, phylogenetic relationships. The acceptance of this notion leads, however, to a result different from that envisaged by Darwin even when interpreted on the basis of a Linnean classification. If, namely, the genera (a–n) form a class because they have descended from 1, then the taxonomic characters they have in common, i.e. their class characters, must have been possessed by the members of 1. Hence, what we, on the basis of our present knowledge, can state with full confidence is that these ancient organisms fulfilled all the requirements for their acceptance as members of our contemporaneous class. Likewise, we know that the members of 5 and 6, if extant, would belong to contemporaneous orders, those of 3 and 4 to families and those of 2 to a subfamily. It is regrettable that Darwin did not perceive this cladistic principle, for had he done so his authority would certainly have ensured its acceptance about one hundred years ago.

The consequence of Darwin's suppositions seems to be that the species, as the result of the continuous process of divergence, are displaced further and further away from the apex of the hierarchy. Another implication is that the only taxa which really 'exist' are the species, the superior taxa being simply sets and subsets of the species located at the base of the hierarchy. In contrast, according to the present theory a genus, a family, etc., is a taxon which, if it is monophyletic, is defined by a certain number of taxonomic characters, possessed by living animals. These properties were acquired when the taxon was created during a specific period of time in the past. The superior taxa, therefore, are not to be found at the base of the hierarchy, but higher up where, in contrast, no species are to be found. Thus, as has already been concluded above, evolution has progressed from the apex to the base of the phylogenetic hierarchy and since, through a Linnean convention, the species are located at the latter station, they represent approximately the level of evolution which has been reached at the present moment.

The proposal concerning the direction of evolution advanced here, put forward by many authors before me (e.g. Willis, 1940; Schindewolf, 1950), is not compatible with the notion implied by the above quotation concerning the origin of species from 'varieties', nor with the following citation (Darwin, 1885, p. 95): 'Thus it is, as I believe, that two or more genera are produced by descent with modification, from two or more species of the same genus'.

The ideas outlined here imply that if we go to the past, then the part of the hierarchy represented by the species did not exist, and yet interfertility clearly obtained between the animals living at that time. From A2.11 and our knowledge about the way the phylogenetic hierarchy grows with time, we may therefore conclude:

(T2.59) With the passage of time the limit of interfertility has gradually moved downwards in the phylogenetic hierarchy.

From this it follows:

(T2.60) In the course of evolution genetic compatibility has existed between members of taxa of any rank in the phylogenetic hierarchy.

From the present discussion I shall further conclude:

(T2.61) The species usually, but not always, represent the terminal, not immutably defined, taxa in the phylogenetic hierarchy.

(T2.62) The species usually, but not always, represent the potentially reversible level of evolution.

Apart from the implications of these statements, which refer essentially to the temporal or 'historical' situation of the species, there is no justification for regarding the latter as being more 'real' and less 'abstract' than any other phylogenetic taxon. Rather, as we have already concluded, each animal is the bearer of the taxonomic characters defining all the taxa to which it belongs, from kingdom Animalia through the species and down to the terminal, not yet immutably defined, phylogenetic taxon.

NUMERICAL TAXONOMY

It is not easy to penetrate the modern literature on taxonomy. If Hennig (1966) is obscure on many points, the numerical taxonomists (Sneath and Sokal, 1973) are so mathematical in their approach that a naive reader is apt to forget that he is dealing with a biological discipline.

I have the impression that much of the criticism voiced against numerical taxonomy concerns the choice of characters. It may be that the latter are often quite unorthodox, but if it is accepted that in principle all properties are potential taxonomic characters, then I fail to see the justification of this objection. At any rate, as mentioned above, Sneath and Sokal (1973) stress that 'invariant' characters should be excluded, and this may amount to following Hennig in ignoring features which may with some confidence be regarded as plesiotypic.

Taxonomic characters should preferably be independent, not only logically, as emphasized by Sneath and Sokal, but also epigenetically. In a study of the avian suborder Lari, Schnell (1970) has classified the several species on the basis of 51 skeletal measurements. Believing the growth of the skeleton to be controlled by the interaction of a very limited number of epigenetic parameters, I am convinced that these characters are far from being independent. This may be an extreme case, but our ignorance about epigenetic mechanisms implies that it is very seldom possible to decide upon the interdependence of various characters. And if, as suggested by Sneath and Sokal (1973), many characters are used, morphological ones representing different parts of the body, and non-morphological ones as well, then the possible bias on this account may be cancelled out.

Are the classifications arrived at not then phylogenetic? Sneath and Sokal distinguish strictly between numerical taxonomy and numerical cladistics. This fact is borne out, for instance, by dissimilarities between their phenograms and their cladograms. This graphical representation is, however, a matter of convention; thus, the numerical phenograms resemble strikingly the phylogenetic dendrograms in the present book.

From their book one gets the feeling that to Sneath and Sokal it is a desirable goal that different sets of characters and different computer programs should lead to the same classification. It thus appears that some unique classifications are superior to all others. Which one should this be, if not the one revealing the phylogenetic kinship?

Sometimes the result of numerical classification leads to extensive revisions of current systematics. In the work of Schnell (1970) mentioned above, it was found in the classification of 44 species of gulls that all other genera (*Catharacta, Creagrus, Gabianus, Pagophila, Rhodostethia, Rissa, Stercorarius* and *Xema*) were included in taxa containing members of the most numerous genus, *Larus*. If this shuffle does not involve some substantial advantage, for instance because it represents the phylogenetic relationships between these birds, then I, for one, fail to see the justification for the revised classification.

I am unable to evaluate the merits of the specialized phylogenetic approach to numerical cladistics developed by Camin and Sokal (1965). The method of Farris, Kluge and Eckardt (1970) seems to be based on the assumption that it is possible to distinguish apotypic from plesiotypic characters. However this may be, the basic principles involved in ordinary numerical taxonomy, the use of many characters and the principle of parsimony must, in my opinion, be employed in every attempt to establish an objective phylogenetic classification. I am therefore convinced that the hierarchical numerical phenograms are cladograms, in which the correlation coordinate is a crude measure of the periods of time during which the various taxa existed. The cladograms may be wrong in details, but this holds for all phylogenetic classification. The only way to remedy this is to expose them to continuous testing when new characters become available.

The extensive use of mathematics in numerical taxonomy is a great advantage because it permits work with large samples of characters and taxa. It is a disadvantage insofar as most systematists probably lack the training in mathematics which is necessary to understand and critically evaluate the method.

I shall finish this section with a few words about Hennig's cladistic approach. Hennig (1966) has done great service to phylogenetics by introducing a number of valuable concepts, serving primarily to show that phylogenetic classification is possible at all. One sometimes gets the impression from the literature that he has also devised a radically new methodology. This is not the case for, to the extent that he employs the principle of parsimony, his method is exactly the same as the one used by the numerical taxonomists, provided they exclude invariant characters. For practical reasons he is forced to work with much smaller numbers of characters. As a consequence, he must weight the characters and use those which seem most important. The problem of weighting has been discussed by Sneath and Sokal (1973) and they rightly conclude that equal character weighting is the only practical solution. The approach adopted by Hennig must imply that some criterion, say, complexity, is adopted for the estimation of the classificatory importance of a character. It is highly probable that this method will most often lead to the correct result, but it nevertheless comprises an element of intuition and subjectivity. Complexity may at times arise through causally connected epigenetic processes, and in such cases the first link in the sequence may be a relatively simple agent whose origin through convergence is not at all implausible. It will be argued later that the segmented axial complex in Cephalochordata and in Vertebrata is an example of this phenomenon.

SUPERPHYLETIC CLASSIFICATION

The value of a character for classificatory purposes depends, among other things, upon its degree of stability. An extremely stable property, remaining constant throughout the animal kingdom, would obviously be of no value for the classification of animals. Features which vary between the individuals of a population, for example those studied in genetics, are likewise unsuitable for systematic purposes. However, all properties lying in the range between ubiquitous and 'unique' characters, both known and hitherto unknown ones, may potentially be employed for classification. Since superior taxa have arisen before the inferior ones which they include, it must follow as a general rule, albeit not quite without exceptions, that the characters used for the definition of the former must be much more stable than those distinguishing the latter.

Linnean classification has been based exclusively on morphological characters, for reasons of expediency and convention. As a consequence it clearly follows that, apart from the characters which distinguish them as animals, the members of a phylum, the highest-ranking Linnean taxon within the animal kingdom, are in principle systematically isolated from those of all other phyla as far as morphological features are concerned. If this was not the case, the rules of classification would surely require unification of the phyla in question.

In spite of this, morphological features have been applied for the purpose of creating superphyletic systems of classification. The most well-known of these use the symmetry relations of the adult body, the nature and the mechanism of formation of the coelom, the blastopore–anus–mouth relations and the number of germ layers. As shown in a different context (Løvtrup, 1974), the proposed criteria are either so general, or so difficult to assess, that they are of limited value, especially since they give rise to different classifications. Some of them have been employed, together with certain other expedients, in attempts to solve the problem concerning the ancestry of Vertebrata, as will be discussed in the following chapter.

The presently known Linnean hierarchies are a testimony to the fact that morphological characters exhibit an enormous range of stability and yet are not stable enough to warrant the establishment of a satisfactory superphyletic classification. Hence, in order to arrive at a classification which can bridge the gap between invertebrates and vertebrates, it is necessary to use features which, though not ubiquitous, have greater stability than the morphological ones distinguishing the various phyla. According to T2.55, it appears that the only properties which comply with these demands are those that were characterized above as 'non-morphological' characters.

Some evolutionists, although in principle admitting the validity of non-morphological characters for phylogenetic classification, may still feel some reluctance towards using them for the purpose proposed here. One possible objection is that the time when the separation between vertebrates and invertebrates took place dates so far back that conservation, even of non-morphological characters, seems quite unlikely. Various arguments may be raised against this contention, but the most weighty one is probably that we expect to find similarities in features of this kind between the classes of the various invertebrate phyla and, although large discrepancies are occasionally observed, this expectation is usually fulfilled. This being the case, there is every reason to expect a successful outcome of this approach when applied to the origin of Vertebrata, for the fossil record clearly demonstrates that many of the classes in the more important invertebrate phyla, particularly those that are the subject of the present study, arose much earlier than Vertebrata.

In botanical classification many difficult taxonomic problems have been solved through the use of non-morphological characters, notably chemical ones. A few attempts have also been made to use this approach in animal systematics but, until the advent of molecular biology, with little success. These early endeavours have received little acclaim from taxonomists, possibly because of the setbacks encountered. Still, it should be emphasized that properties of this kind constitute an important reservoir of potential taxonomic characters, the mobilization of which may help to solve the problem of the origin of Vertebrata.

THREE METHODOLOGICAL PREMISES

In the present chapter various proposals concerning the phenomenon of phylogenetic classification have been advanced, controversial to some extent, maybe, but certainly bewildering merely through their number. For that reason it may be expedient here to outline three methodological premises which, together with the two stated in the introduction to the present chapter, constitute the foundation of the phylogenetic classification adopted in the following text.

The third of our premises is:

(iii) *In phylogenetic classification it is necessary to adhere to the methodological rules derivable from the theory of cladism.*

The most important thesis in this theory undoubtedly is that *every taxon, i.e. set of living animals, has a twin taxon.* This contention is of importance for both projects pursued in the present book, but particularly for the purpose of establishing the ancestry of Vertebrata. The implication of the italicized statement is that, among all monophyletic invertebrate taxa, there is one, and only one, the members of which have ancestors in common with the vertebrates which are not at the same time ancestors to the members of any other taxa among invertebrates, except, of course, those included in the taxon in question. This is the twin taxon to Vertebrata, and the pinpointing of this taxon will, of course, settle the issue about whence the vertebrates sprang.

The fourth of the main premises states:

(iv) *In phylogenetic classification it is necessary to employ as many taxonomic characters as possible, or as many as are necessary to reach a classification that is incontestable on the basis of known data.*

It should be mentioned that all of the characters need not be used for the establishment of a basic classification; some may be reserved for the testing of more complex classifications.

The fifth of the main premises asserts:

(v) *In phylogenetic classification of taxa above the phylum level it is necessary to employ non-morphological characters exclusively.*

This premise is, of course, of particular value in the search for the ancestors of Vertebrata, but it must be pointed out that non-morphological characters may, and will here, be used also in classification within the phylum.

Finally, it may be stated that a phylogenetic classification of an animal taxon constitutes a theory about the process of evolution through which the various included taxa arose. As an empirical theory the establishment of such a hierarchy is superior to any theory based, say, on fossil data, because, dealing with living animals, its empirical content, and hence its degree of falsifiability, is greater than in theories based on the slight amount of information obtainable about the properties of extinct organisms.

'The purpose of a theory of vertebrate ancestry, however, is to reveal the nonchordate predecessors of
chordates. But there appear to be no invertebrates
which can be considered the immediate predecessors
of chordates. . . . Thus the chordate clue seems to
lead us into a blind alley out of which the most
promising exit is the way back.'
(H. V. Neal and H. W. Rand, 1936, pp. 664–665)

3

ANCESTRY OF THE VERTEBRATES

In this chapter we shall deal with the long-standing problem of the origin of Chordata.
Before this problem is subjected to cladistic analysis, some pages are dedicated to a
survey of current views on this issue.

CURRENT VIEWS

Current conceptions about animal phylogenesis are embodied in phylogenetic trees.
Of these several have been published which differ in some, occasionally important,
respects, but essentially there is agreement on certain points that are crucial in the
present context. Thus, it is generally asserted that in the course of phylogenetic evolution
a bifurcation occurred, the main phyla in the one branch being Echinodermata and
Chordata, in the other Annelida, Arthropoda and Mollusca (Figure 3.1). The two groups
have been called the 'echinoderm' and the 'annelid' superphylum, respectively (Kerkut,
1960). Secondly, many authors concur in including in the phylum Chordata, in addition
to Vertebrata, two other subphyla, viz., Urochordata and Cephalochordata; some even
confer this rank and position on Hemichordata. These two or three subphyla are often
subsumed in the group Protochordata.

In the present section we shall discuss first the arguments which have been advanced
in favour of the superphyletic branching of the phylogenetic tree, secondly the justification for the affiliations between Protochordata and Vertebrata, and thirdly the theories
which have been advanced to explain the origin of Vertebrata.

The annelid and echinoderm superphyla

The reasoning behind the establishment of the two animal superphyla includes many
different contributions, and, in consequence, the result is a patchwork. That this is
seldom acknowledged is due primarily to the fact that one rarely finds a complete survey
of the basis of this classification. Fortunately, this task has been performed by Kerkut,
from whose book the following table is reproduced (1960, p. 105).

Table 3.1 comprises an impressive list of contrasting characters which, if they were
apotypic and independent, would strongly corroborate the proposed subdivision.

These stipulations, however, are vital, and I propose to investigate the characters from that point of view in order to reveal the firmness of the foundation on which this phylogenetic theory is based.

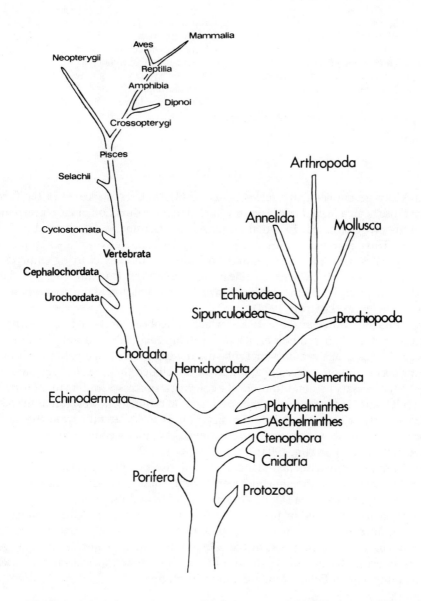

Figure 3.1. A phylogenetic tree outlining the currently accepted relationships between the animal phyla. On the left limb a branch has been grafted purporting to represent the phylogenetic affiliations within phylum Chordata. The close resemblance to a real tree has, for aesthetic and didactic reasons, decided the choice of this particular figure. (Redrawn from W. S. Hoar (1966), *General and Comparative Physiology*, by permission of Prentice-Hall)

Table 3.1
Characters distinguishing the currently accepted superphyla
(Kerkut, 1960)

Annelid superphylum	Echinoderm superphylum
Spiral cleavage	Radial cleavage
Blastopore = mouth	Blastopore = anus
Schizocoelic coelom	Enterocoelic coelom
Determinate cleavage	Indeterminate cleavage
Nervous system delaminates	Nervous system invaginates
Ectodermal skeleton	Mesodermal skeleton
Trochosphore larva	Pluteus-type larva

CLEAVAGE PATTERNS

Spiral cleavage implies that each new tier of blastomeres is located in the furrows between those of an adjacent tier (Figure 3.2), and that the direction of the corresponding 'rotation' is predetermined. In radial cleavage the blastomeres are situated on top of each other. This difference in cleavage pattern is determined by the orientation of the cleavage spindles, which are vertical and horizontal in radially cleaving embryos, but slanted in a fixed direction when the cleavage is spiral (cf. Conklin, 1903). A dextro-sinistral polarity may be taken to account for the determination of the orientation of the cleavage spindles (cf. Løvtrup, 1974).

In general, the two cleavage patterns which have been proposed as superphyletic characters are found only in holoblastically cleaving eggs. When, usually in large yolk-laden eggs, the cleavage becomes meroblastic, the typical cleavage patterns disappear. Since this occurs in both superphyla, it may be concluded that the cleavage patterns as such are not necessary preconditions for the epigenetic processes leading to the formation of the adult bodies in the respective groups. It should be noted, however, that the dextro-sinistral polarity, which determines the direction of the bilateral asymmetry in some members of the classes Gastropoda and Bivalvia, is causally connected with the spiral cleavage pattern (cf. Conklin, 1903).

Apart from the comments just made, it should also be observed that in the phylum Arthropoda, included in the annelid superphylum, spiral cleavage is unknown except, to some extent, in Crustacea (D. T. Anderson, 1973), and that both patterns occur in various phyla not embraced by the classification under discussion. What remains of the justification for considering the cleavage pattern as a taxonomic character seems to be the purely negative assertion that in the echinoderm superphylum spiral cleavage has not been observed in cases where the size of the blastomeres does not *a priori* exclude this possibility (e.g. in Echinodermata and in Amphibia).

BLASTOPORE–ANUS–MOUTH RELATIONS

The superphyletic classification founded upon the purported identity between the blastopore and the anus or the mouth was proposed by Grobben (1908) on the basis of ideas conceived by Goette. This notion has gained widespread acceptance; in fact, of

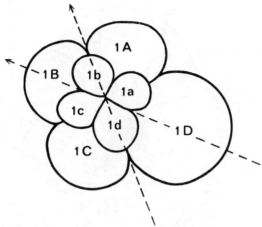

Figure 3.2. Schematic representation of the 8-cell stage of a spirally cleaving egg. The blastomeres 1a–1d are micromeres, 1A–1D macromeres. The micromeres are seen to lie in the furrows between the macromeres, a situation which theoretically may arise through rotation of the micromeres by 45°. If, say, 1d and 1D are both different from the other members of the quartet, each set of blastomeres will determine a polarity, as indicated by arrows. If the one passing through the macromeres is a dorso-ventral (or median) polarity, the other one will represent a dextro-sinistral polarity

the names coined by Grobben in this context, 'Protostomia' and 'Deuterostomia', the latter is frequently used for the echinoderm superphylum.

In the present discussion I have accepted that Vertebrata belong to Deuterostomia and thus that their blastopore becomes the anus. If, as asserted by Goodrich (1958), this proposition is highly doubtful for most, if not all, vertebrates, and even for cephalochordates, then clearly Grobben's classification must be abolished and with it support for the existence of the annelid and the echinoderm superphyla.

It will be seen that the suggested distinction depends to a large degree upon the mechanism of gastrulation. When this event occurs through invagination, as in echinoderms, at least in those species which are usually employed in embryological studies, or by a combination of invagination and epiboly, as in Amphibia, then the process of gastrulation involves the tip of the archenteron, after traversing the blastocoele, fusing with the ectoderm to form an opening.

Since the invagination always begins at the dorsal–vegetal side of the embryo, it follows that the gastrulation must end at the ventral–animal side. As the nerve cells, generally formed around the animal pole, serve to establish the head, and thus the antero–posterior axis, this opening must be the mouth. This type of gastrulation presupposes the formation of a reasonably spacious blastocoele, a phenomenon which usually depends upon the presence of a supracellular structure, a hyaline membrane or a surface coat, serving to keep the cells together. In cases where spirally cleaving eggs form a blastocoele (Figure 3.3), gastrulation may proceed through invagination, which presumably entails formation of the mouth through fusion as described above, and it

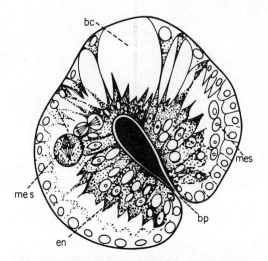

Figure 3.3. Gastrulation through invagination
in *Physa*. Optical section of a gastrula, showing
cells extending filiform pseudopodia between the
wall of the archenteron and the blastocoele wall.
bc, blastocoele; bp, blastopore; en, endoderm;
mes, mesoderm. (Redrawn from A. Wierzejski,
Z. Wiss. Zool., **83** (1905), by permission of
Akademische Verlagsgesellschaft)

may therefore be asked whether they really conform to the definition of Protostomia.

When gastrulation occurs through epiauxesis, a mechanism presupposing the absence
of a surface coat, the lips of the blastopore coincide with the edges of the expanding
layer of ectodermal cells. When the overgrowth approaches completion, generally at
the ventral side, four different possibilities exist. The 'blastopore' may close completely;
it may remain open at the anterior end, just beneath the prototroch, when this is present;
it may stay open at its posterior end, near the paratroch, when this is present; or it may
be open at both ends. All these alternatives appear to have been described in various
spirally cleaving eggs (as concerns Annelida, see the review by Dawydoff, 1959). It is
evident that in the first and in the fourth case the animals cannot be classified according
to the proposed criterion, whereas the second mode entails affiliation to Deuterostomia
and the third to Protostomia. Furthermore, in large and yolk-laden eggs the criterion
cannot be applied either, because gastrulation by invagination, as well as by epiauxesis,
becomes impossible owing to the bulk of the egg. This situation is encountered in both
superphyla, thus in Echinodermata (Fell, 1948) and Vertebrata and in Hirudinea and
Cephalopoda.

FORMATION OF THE COELOM

In coelomate animals three different kinds of coelom have been distinguished on the
basis of their mechanism of formation. These three types were described thus by
Hyman (1951, p. 22):

'1. *Schizocoel.* In the teloblastic, derived teloblastic, and lamellar modes of mesoderm
formation, the coelom arises as a split inside the mesodermic bands, plates or masses. . . .

The split expands until the mesoderm comes in contact with the body wall on one side and the gut wall on the other.

2. *Enterocoel.* In the enterocoelous method of mesoderm formation, the mesodermal sacs evaginated from the archenteron become cut off and lie in the blastocoel. Their cavity is the coelom, their walls the mesoderm. These sacs expand until they touch the body and gut wall, and the end result is therefore the same as by the schizocoelous method.

3. *Mesenchymal coelom.* The mesenchyme may rearrange itself so as to enclose a space which is then a true coelom since it is bounded on all sides by mesoderm. This method is said to occur only in the worm *Phoronis.*'

Morphologically a coelom is, in the last analysis, a cellular vesicle. As such, it may arise in two, and only two, different ways, viz., as a spontaneous formation, involving the aggregation of free cells, or as a topological deformation of a two-dimensional cell aggregate. The former principle clearly corresponds to methods 1 and 3, the latter to method 2, as outlined above. In the following pages the outcomes of these two alternative mechanisms will be designated by the names 'schizocoele' and 'enterocoele', respectively.

The superphyletic classification based upon the mechanism of coelom formation was advanced by O. and R. Hertwig (1881). This proposition obviously presupposes that the chordates form their coelom according to the enterocoelous method, and this was duly 'demonstrated' in drawings published by the founders of the classification (Figure 3.4). The consequence of a hypothesis submitted by such authorities was that, as stated in the words of Brachet (1921, p. 187, author's translation), 'the desire which many

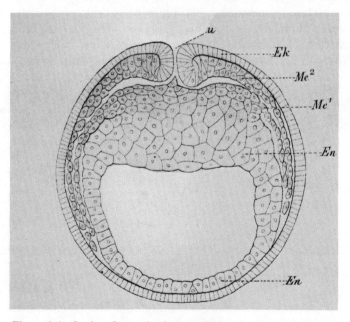

Figure 3.4. Coelom formation in the embryo of *Triton* as envisaged by the fathers of the coelom theory. Ek, ectoderm; En, endoderm; Me^1, Me^2, mesoderm; u, blastopore. (Reproduced from O. Hertwig and R. Hertwig (1881), *Die Coelomtheorie*, by permission of Gustav Fischer Verlag)

authors . . . have had to discover at any price coelomic diverticula in embryos of vertebrates, has engaged much research in an unprofitable direction and retarded a sound comprehension of the middle germ layer'. In any case, it finally turned out to be impossible to uphold the theory that in chordates, with the exception of cephalochordates, the coelom is an enterocoele.

Since, as a very general rule, early embryonic cells, particularly those in the interior, are not very adhesive, it may be inferred that enterocoele formation owes its occurrence to a supracellular structure keeping the cells together. From studies on echinoderms, where this phenomenon is easily observable, it is known that the agent in question is the hyaline membrane, also responsible for gastrulation through invagination. And, indeed, when in this phylum the eggs are large enough for the latter event to be suppressed, the coelom is formed by the schizocoelous method (Fell, 1948).

From the information presented above, it appears that enterocoele formation occurs in some echinoderms and in cephalochordates, a circumstance which may suffice to preserve a vestige of the hypothesis. It is necessary, however, to consider the case of *Branchiostoma* in more detail to make sure that even in this case the survival of the classification does not represent an attempt to preserve it 'at any price'. I have looked through several descriptions of the embryology of *Branchiostoma*, and found that in most cases the published drawings correspond closely to the theoretical prediction. One exception was found, namely, the figures published by Cerfontaine (1906), some of which are reproduced in Figure 3.5. This gives rise to a suspicion that is further strengthened by a moment's reflection on the mechanism of gastrulation in *Branchiostoma*. This process occurs through invagination, thus presupposing the existence of a supracellular structure, a membrane, the presence of which has been confirmed by independent observation. However, if this membrane was preserved during the 'enterocoelous evagination' of the mesoderm, then the latter would be expected to form two tubes at each side of the notochord. As shown earlier (Løvtrup, 1974), the process of

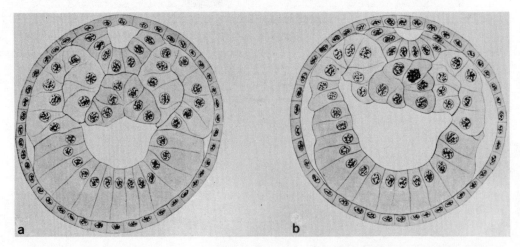

Figure 3.5. Coelom formation in *Branchiostoma*. a, transverse section from a larva with nine pairs of somites; b, transverse section near the posterior end close to the neurentic canal. A virtual myocoele is indicated by heavy lines, but it is questionable whether this expedient is justified. (Reproduced from P. Cerfontaine, *Arch. Biol.*, **22** (1906), by permission of *Archives de Biologie*)

segmentation, in chordates as well as in annelids, may be accounted for by a process of cell transformation, occurring along an antero-posterior gradient, through which solitary cells become adhesive and consequently capable of vesicle formation. The very fact that myotomal segmentation takes place in *Branchiostoma* indicates that the membrane has disappeared, and with it the prerequisite for enterocoele formation.

Hence, there is no support for the generalization that the two mechanisms of coelom formation represent superphyletic characters.

DETERMINATE AND INDETERMINATE CLEAVAGE

The distinction between determinate and indeterminate cleavage is based upon observations on the development of isolated blastomeres. In the echinoderm superphylum various famous experiments have shown that blastomeres from embryos of echinoids and amphibians may develop into normal larvae. These findings are now classical, and discussed in most textbooks.

In contrast to these results are those obtained on spirally cleaving eggs of annelids and molluscs. Here it has generally been found that no single blastomere can sustain normal development, but certain combinations comprising the D macromere may do so. These observations have been considered to reveal a fundamental difference in the organization of the two egg types.

The two most important polarities in both superphyla are the apical, or animal–vegetal, and the median, or dorso–ventral. The first of these is generally determined by variations in cortical or cytoplasmic properties along the animal–vegetal axis (cf. Løvtrup, 1974). The determination exerted by this polarity in spirally cleaving eggs has been established by the demonstration that isolated micromeres cannot sustain normal development (E. B. Wilson, 1904). In echinoids it has been shown that separation of the animal and the vegetal blastomeres results in abnormal development (Hörstadius, 1935). The differentiation patterns observed in explants of amphibian embryos clearly indicate a determination controlled by the animal–vegetal polarity (Holtfreter, 1938a, b). This assertion appears to be contradicted by the fact that 'vegetative' embryos may develop normally (Ruud, 1925), but a proper evaluation of this observation clearly depends on whether or not a complete separation between endoderm and ecto-mesoderm was really achieved. The results discussed here thus demonstrate that determination obtains with respect to the apical polarity in the echinoderm, as well as in the annelid, superphylum.

In eggs with spiral cleavage the median polarity passes through the centres of the B and D blastomeres, which means that it is impossible for one or two blastomeres, except possibly B and D, to contain the normal set-up of the median polarity (Figure 3.2). That incomplete combinations may result in almost normal development shows the occurrence of 'indetermination' (regulation) in determinate eggs.

It has often been claimed that the median polarity is determined in echinoids. I shall not discuss this question in detail here, but only note that this assertion is incompatible with observations indicating that, after isolation, the development of the first four blastomeres is approximately identical (cf. Løvtrup, 1974).

Through the findings of Chabry (1887) and Conklin (1906; 1911) that the blastomeres in the 2-cell stage of ascidian embryos do not give rise to normal embryos, the proposition was advanced that the cleavage is determinate. Later experiments, carried out with properly isolated blastomeres (Pisanò, 1949; Reverberi and Ortolani, 1962; see

Reverberi, 1961), showed that some regulation does occur, since the half embryos are essentially bilaterally symmetrical, as might be expected from eggs with a bilateral cleavage pattern,but development is clearly impaired. Isolation of groups of blastomeres at the 8-cell stage shows that the fate with respect to cell differentiation is rather rigidly determined. However, since the four animal blastomeres give rise only to epidermis, nerve cell differentiation must be established through induction. It is only the dorsal cells which may be thus induced, and the determination is therefore more fixed than in the amphibian embryo.

It would appear that the suggested determinate cleavage in urochordates supports affiliation of this taxon with invertebrates rather than with Vertebrata. Indeed, the importance of this feature rests on the fact that formerly cleavage in cephalochordates was held to be determinate; this might thus serve as a protochordate character. However, the investigations by Tung, Wu and Tung (1958) have established with certainty that cleavage is indeterminate to a considerable extent in *Branchiostoma*. The difference concerning the degree of determination in the two protochordate groups may possibly be related to the number of cells present in the embryos. Thus, gastrulation in ascidians begins at the 64-cell stage, and in a gastrula containing 128 cells there are only eight neurocytes and eight chordocytes; in *Branchiostoma* the young gastrula comprises more than 1000 cells, of which 80–86 are presumptive nerve cells and a similar number chordal cells (Drach, 1948). The small cell number in the former case may give little morpho-genetic plasticity, and little possibility for regulation through induction.

In amphibians the first division plane does not bear a definite relation to the median polarity, as determined by the grey crescent. If the division passes through the latter, the two blastomeres may develop normally after isolation, and if it is perpendicular to the median plane, the dorsal, but not the ventral, blastomere may develop into a normal embryo. This clearly shows that in this case determination obtains.

It must thus be admitted that an empirical basis exists for a distinction between deter-minate and indeterminate cleavage, although the difference is apparently quantitative rather than qualitative. In fact, it seems to depend mainly on the time required for deter-mination to become established, this process being much slower in indeterminate eggs. It may therefore be more correct to distinguish two kinds of determination, called 'static' and 'dynamic', respectively (Løvtrup, 1974).

Although static determination is more common in the annelid, and dynamic in the echinoderm, superphylum, the above discussion shows that this criterion is nonetheless of little discriminatory value. I may add that even if no experimental observations exist on this point, I would be very surprised if determinate cleavage were to be found in cephalopods.

FORMATION OF THE NERVOUS SYSTEM

According to the proposed classification the nervous system is formed by delamination in the annelid, and by invagination in the echinoderm, superphylum. It seems justifiable to observe at the outset that these two expressions are anything but well-defined. However, it seems that 'delamination', which is also used to describe certain kinds of gastrulation and germ-layer separation, represents a type of ingression of single cells from the blastula wall. From what is known about this process, it appears that formation of the nervous system in echinoderms occurs in this way. The nerve cells in molluscs also originate from this source: some remain in the body wall, while others enter the body cavity and aggregate to form ganglia. This mechanism is also known from annelids, but in oligochaetes and hirudineans the ventral nerve cords are initially situated at the edges of the ectodermal cell layer which, through epiauxesis or epiboly, or a combination of both, gradually covers the remaining parts of the embryo. When the edges meet in the midventral line, the nerve cells are overgrown by epidermal cells.

In chordates at least two different methods may be distinguished. In the first one, described in cephalochordates and actinopterygians, the cells in the middorsal region, after differentiation into nerve cells, are covered by epidermal cells which migrate inwards from the edges of the neural plate primordium. The latter is subsequently organized to form a tube. It seems to me that, so far as the basic morphogenetic mechanisms are concerned, this process is similar to, if not identical with, the one occurring in the higher classes of Annelida. The other method, typically observed in the amphibians, involves a hinge-like folding of the neural plate, the edges of which upon meeting fuse to form the neural tube. Apart from the agents responsible for the movement proper, the physical preconditions for this way of forming the axial nervous system seem to be twofold, that the cells of the neural plate are kept firmly together, and that the latter is anchored to the notochord along its median line, thereby ensuring its folding.

It will be noticed that the 'overgrowth' characterizing the former method demands that the ectodermal cells are not kept together by a surface coat. As regards the notochord, the cooperation of this entity is not required in the formation of the neural tube proper except, of course, for the process of elongation. The neural tube never becomes attached to the notochord in the cephalochordates, a circumstance which is reflected in the anatomy of the adult animal. In the actinopterygians, it appears that the formation of the chorda is relatively retarded, thus ruling out its participation in neurulation and making necessary this primitive method of neural tube formation.

From the preceding survey of the methods involved in the formation of the nervous system in the various groups it appears that just as 'delamination' does not properly describe the methods realized in the annelid superphylum, none of the most usual methods in the echinoderm superphylum corresponds particularly well to what is usually understood by 'invagination'. Furthermore, although neural tube closure through folding is found only in the latter superphylum, where it is functionally dependent upon the notochord, the methods of ingression and overgrowth are represented in both superphyla.

THE ORIGIN OF THE SKELETON

According to the classification under discussion, the skeleton is mesodermal in the echinoderm, and ectodermal in the annelid, line. We shall discuss the implications of this distinction with respect to both calcified and non-calcified skeletons.

It is true that the calcified exoskeleton found in many crustaceans and molluscs is formed by the epidermis, and hence is ectodermal, and the absence of calcified endoskeletons in the annelid superphylum is also well documented. Likewise, the skeleton in various echinoderms is found in the matrix of the dermis, and may consequently be classified as mesodermal. It must be stressed, however, that the pattern of calcification, particularly the formation of the ossicles found in some of the echinoderm classes, is so different from anything observed in the other taxa under discussion here that a comparison is hardly possible. Among the chordates, calcification is known only in the verte-brates, but the mesodermal skeletons found in this taxon, with some few exceptions, exhibit two unique features. For one thing, their main component is calcium phosphate, in the form of hydroxyapatite, whereas the preponderant mineral in invertebrates is calcium carbonate. Secondly, the calcified structures exhibit various histological features, unknown among invertebrates, which have earned them the special name of 'bone'. However, ectodermal calcification does occur in the vertebrates in the form of enamel in teeth, and in the outer layer of the scales in selachians and in various extinct fishes.

Non-calcified endoskeletons of cartilage are unknown in echinoderms and in proto-chordates, except possibly in hemichordates, but, as will be discussed later, they are found in the vertebrates and in the annelid superphylum. Once more it thus appears that a proposed character contributes little to the discrimination between the two superphyla.

LARVAL TYPES

It is submitted in the superphyletic classification that the pluteus larva is characteristic of the echinoderm, and the trochophore larva of the annelid, line. This statement is correct insofar as trochophore-like larvae are typically found in annelids and molluscs, particularly in species with spiral cleavage, and that in the various classes of Echino-dermata, except Crinoidea, the larval types derived from eggs with low yolk content, ophiopluteus, echinopluteus, auricularia and bipinnaria, exhibit similarities which warrant their being classed together (Figure 3.6). It must be pointed out, however, that another larva found in ophiuroids, holothuroids and crinoids, the vitellaria, exhibits similarities with the trochophore rather than with the pluteus (Figure 3.7). Yet, even if this point is ignored, the proposal so far only establishes a distinction between the annelid superphylum and Echinodermata, not the echinoderm superphylum. The only link which possibly may unite the two phyla of the latter group, as far as the present criterion is concerned, is the tornaria larva found in certain hemichordates. If the similarities between the latter and the pluteus are considered significant, then Hemichordata are evidently affiliated with Echinodermata, but the whole argument collapses if they are not also accepted as members of Chordata. As we shall see in the following section, the present trend among zoologists is to exclude them from this phylum.

Even the relevance of the larval form is questionable, for, as shown by Fell (1948), it is not possible, on the basis of the embryonic form, to classify Echinodermata in a way that is compatible with palaeozoological and morphological evidence. This being the case, it is certainly justifiable to question the validity of embryology for the establishment of the affinities between taxa so different as Echinodermata and Hemichordata.

As a comment to the great lability of larval forms demonstrated in Fell's paper (1948), it may be pointed out that larvae, particularly the early and primitive types discussed here, are built up by a very limited number of morphogenetic elements (cells and cell-differentiation products); thus far they unquestionably represent instances of recapitula-tion. However, the restrictions imposed by the paucity of the material substrate, the topological rules which embryogenesis must obey, and the fact that the larva must be viable and allow for further meaningful morphological changes, may entail that the number of possible larval forms is relatively limited. If this reasoning is correct, the occurrence of similar larval types in more or less distantly related lines would not be too improbable an event.

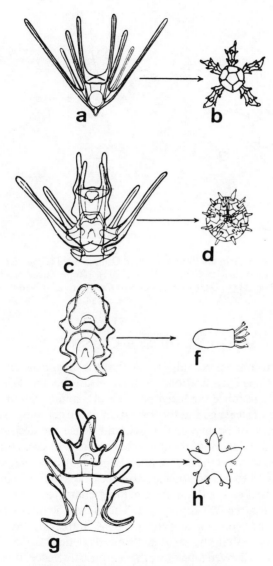

Figure 3.6. Echinoderm larvae. a, ophiopluteus, metamorphosing into an ophiuroid, b; c, echinopluteus, metamorphosing into an echinoid, d; e, auricularia, metamorphosing into a holothurian, f; g, bipinnaria, metamorphosing into an asteroid, h. (Reproduced from H. B. Fell, *Biol. Rev.*, **23** (1948), by permission of Cambridge University Press)

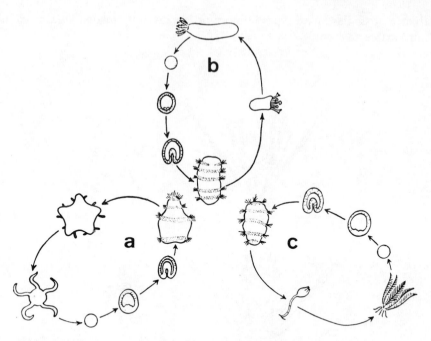

Figure 3.7. Echinoderm vitellaria larvae. a, Ophiuroidea (*Ophioderma*); b, Holothuroidea (*Cucumaria*); c, Crinoidea (*Antedon*). (Reproduced from H. B. Fell, *Biol. Rev.*, **23** (1948), by permission of Cambridge University Press)

ASSESSMENT OF THE CLASSIFICATION

It seems evident from the preceding survey that the diagnostic characters used in the traditional superphyletic classification are most unsatisfactory. For one thing, several of the characters, for instance the blastopore–mouth–anus relation, the determination and indetermination of cleavage and the formation of the nervous system, are not 'good' characters in the sense of being well-defined, and hence the decision as to whether a given character does, or does not, obtain is often a matter of great subjectivity. Furthermore, the second and third on the list are not independent since they are, at least to some degree, causally related to the presence or absence of a hyaline membrane or a surface coat. For these, as well as for the four remaining characters, it is the case that overlapping occurs between the two superphyla. Therefore, the question is not whether or not these characters are apotypic and independent. Rather, it seems that they are not taxonomic characters at all in the suggested superphyletic classification.

If this result had been arrived at in any other empirical science, this proposition would probably be accepted, leading to the conclusion that the classification does not stand up to testing, and has therefore to be refuted. This outcome would not be in accord with biological tradition, which seems to dictate that a theory, once embraced, should be preserved as long as possible. One reason for this attitude may be that the living world is so capricious and multifarious that it appears next to impossible to establish tests which can yield unambiguous and decisive answers. Even if this point is granted, it must nonetheless be stressed that reluctance to abandon theories which do not agree with empirical observations constitutes a serious impediment to progress in the biological sciences.

However, adhering to the tradition, we shall note that the superphyletic classification is highly dubious, but that it may be rescued if, with the use of more satisfactory criteria, it can be shown that Chordata are more closely related to Echinodermata than to the phyla in the annelid line.

Protochordata

The concept of 'Protochordata' may have two implications: either it is a convenient way to deal with the non-vertebrate chordates, or else it designates a phylogenetic taxon. As pointed out by Drach (1948), the latter alternative entails that the included groups are more closely related between themselves than with Vertebrata. This is not generally accepted. Usually the protochordate taxa are given the same systematic rank, subphylum, as Vertebrata. It has already been mentioned that Hemichordata are often included as a subphylum in Chordata (e.g. Rothschild, 1965), although this classification has lost support of late (e.g. Grassé, 1948; Hyman, 1959). I adopt the latter view, but in order to justify this position all three taxa will be discussed here, followed by an assessment of the validity of the classification.

Before we proceed to this task it should be mentioned that Chaetognatha have been included in the echinoderm superphylum by some systematists. Regarding this point I shall confine myself to quoting the opinion of Hyman (1959, p. 65) which, as it seems, is unanimously endorsed today: 'Discussion of the affinities of the chaetognaths therefore reaches an impasse, in that the adult anatomy shows most resemblance to aschelminths but the embryology indicates a coelomate animal with an enterocoelous mode of formation of the coelom although other resemblances to enterocoelous groups are wanting. Under these circumstances it becomes impossible to ally the chaetognaths with any other existing invertebrate group.'

Clearly, this quotation primarily reflects the impotence of the formanalytical approach. It is not at all implausible that a solution might be reached through the use of non-morphological characters, but since there is little reason to anticipate that this would demonstrate an affiliation with either Echinodermata or Chordata, this odd group may with confidence be ignored in the present context.

HEMICHORDATA

The subphylum, or phylum, comprises three classes, Enteropneusta, Pterobranchia and Planctosphaeroidea, the latter known only in the larval form. The group is defined as follows: 'The Hemichordata are solitary or colonial, more or less vermiform enterocoelous coelomates with body and coelom regionated into three successive divisions of unequal length and different structure, with an epidermal nervous system that includes a middorsal centre in the second body division and middorsal and midventral cords in the third body division, with a circulatory system that usually includes a contractile sac, without typical nephridia, with a preoral gut diverticulum, with or without gill slits, and with or without tentaculated arms borne on the second body division' (Hyman, 1959, p. 75).

Some of these characters are not specific for the group, and a consideration of the

remaining ones shows that none of them indicates any particular kinship with Echinodermata or, perhaps with the exception of the gill slits and the dorsal location of part
of the nervous system, with Chordata.

The purported affiliation with the Echinodermata is further based upon the tornaria
larva of the enteropneusts, which exhibits similarities with the larvae of some echinoderms. According to Hyman (1959, p. 197): 'This resemblance is not superficial but
extends to many details. The ciliated band takes a similar course in the tornaria and the
bipinnaria or auricularia, although the telotroch is lacking from all echinoderm larvae.
The digestive tract has the same shape and the same subdivisions into foregut, stomach,
and intestine in echinoderm and hemichordate larvae, and in both the blastopore becomes the larval anus. The greatest and most convincing resemblances, however, concern
the coelomic sacs, which in both are, in general, of enterocoelous origin.'

The characters enumerated here are all of a very general nature and, as shown above,
several of them may reasonably be related to the possession of a hyaline membrane or
similar structure which remains intact after gastrulation. The use of such features, as
well as of larval types in general, for the purpose of superphyletic classification has
already been discussed above, where the conclusion was reached that they are unsatisfactory for want of discriminatory power. This decision is supported, I believe, by the
fact that the only larva known in the class Pterobranchia, that of *Cephalodiscus*, bears
very little resemblance to types discussed above, and more to the larvae of various other
invertebrates.

The characters which originally led Bateson (1886) to suggest that Hemichordata are
related to Chordata are the two mentioned above, the gill slits and the dorsal location
of part of the nervous system, plus a structure which is represented in Hyman's definition
by the expression 'a preoral gut diverticulum', but which was interpreted by Bateson as
'a supporting structure, *i.e.* a notochord' (1886, p. 552). The story of the 'notochord' in
hemichordates seems to me to represent an instance of formanalysis at its worst, when
purely geometrical–topological relations are used for the purpose of making very far-
reaching inferences. If only Bateson had taken the trouble to define the concept of
'notochord' with respect to its histological details, and not simply as 'a supporting
structure', he would have been unable to postulate the presence of a notochord in
Hemichordata. However, this notion has now largely been abandoned, and most zoologists seem to concur with Hyman (1959, p. 105) that 'the buccal diverticulum does not
represent a notochord and probably is just what it appears to be—a preoral extension of
the digestive tract'.

As regards the second feature, the organization of the nervous system, Hyman (1959,
pp. 200–1) writes as follows: 'Resemblance between the nervous systems of hemichordates and chordates rests on a better basis. In its dorsal position, its occasionally
hollow construction, its mode of formation from the dorsal epidermis, and the occasional
presence of a neuropore, the collar cord of enteropneusts is comparable with the neural
cord of vertebrates. But this comparison fails as regards the rest of the nervous system.
In its intraepidermal position, and in the presence of circumenteric connectives and of a
main ventral nerve cord, the nervous system of enteropneusts is distinctly invertebrate.
The invertebrate features of the nervous system of hemichordates outweigh the chordate
features.'

Pharyngotremy (gill slits) is thus the only definite similarity between Hemichordata
and Chordata, and the fact that this phenomenon is found in no other animal taxon

must, of course, be given some weight. In particular, 'the branchial apparatus in both groups [Enteropneusta and Cephalochordata] has the same general construction with tongue bars, synapticules, and an arcade of trifid skeletal supports. Such identity is inconceivable except on the basis of a common ancestry. Hence a phylogenetic relationship between hemichordates and chordates is not open to question' (Hyman, 1959, p. 201).

Without rejecting this conclusion, I would like to consider the question of convergence. If pharyngotremy is really the only feature which unites Hemichordata, or rather Enteropneusta, because it is found only in this taxon, with Chordata, then it seems imperative to analyse this question in more detail. The formation of gill slits is a morphogenetic process which requires the simultaneous presence of a relatively limited number of morphogenetic elements, for example an ectodermal and an endodermal cell layer separated by mesenchyme (Løvtrup, 1974), and even the formation of the skeleton is presumably dependent upon a relatively simple mechanism. It is obvious that the pharyngotremy in the two cases presupposes certain similarities with respect to the course and the nature of the epigenetic mechanisms involved, but whether or not convergence under these conditions is a likely phenomenon depends upon the universality of the participating elements. It must be emphasized, however, that if the material of the endoskeleton is radically different in the two groups, then the case for convergence would be strengthened considerably. Since such a disparity apparently obtains (cf. p. 95), the affiliation between Hemichordata and Chordata rests on a very slender basis.

UROCHORDATA

The elevation of Tunicata to their present state as a subphylum, Urochordata, within Chordata, was a consequence of the studies on the embryology of these animals by Kowalevsky (1867; 1871), who discovered that the ascidian larva bears certain resemblances to the tadpoles of frogs, indicating that Tunicata might be related to Vertebrata. Being advanced in the wake of the publication of Darwin's book, and supplying a link between vertebrates and invertebrates which clearly was required by the theory of evolution, Kowalevsky's contention was received with enthusiasm by the early evolutionists. Although it has not remained unchallenged, it survives to the present day as the only acceptable attempt to trace the origin of the vertebrates from the invertebrate realm.

Concerning the classification established on this theory, it must be emphasized that it is based only on larval characters, a method fraught with jeopardy according to the view presented above. Furthermore, I feel bound to point out that various authors in diverse contexts have been only too willing to explain away larval peculiarities inconvenient to the preconceived classification as 'larval adaptations'. Without endorsing this concept, I shall argue that if the urochordates had not been considered an urgently needed missing link, it might well have been applied in this case.

However, among the chordate properties possessed by the urochordate larvae, the 'notochord' is surely the most important, since some of the other characters, in my interpretation at least, are simple epigenetic consequences of the formation of the 'notochord'. It is not customary in morphology to define structures according to their material composition, but only with respect to their general form and function. An attempt has been made above to define a 'notochord' on the basis of the differentiation patterns involved in its formation, and according to that the linear aggregate of cells in the tail of the urochordate larvae is not a notochord because a collagenous sheath is missing, a circumstance which incidentally may in part explain its disappearance during metamorphosis in most cases. I do not contend that it is necessary to adopt the proposed definition, but I think it is important to contemplate that this simple expedient would suffice to exclude Urochordata from the phylum Chordata.

It will be recalled that a 'notochord' is also present in adult appendicularians. This structure is 'a liquid skeleton which functionally and morphologically may be considered as a counterpart to the coelom in many coelomate invertebrate forms' (Olsson, 1965, p. 213). Consequently, it is of no concern in the present context.

The characters of the urochordate larva will be discussed in more detail in the section dealing with the assessment of the classification.

CEPHALOCHORDATA

When cephalochordates are compared with vertebrates, the similarities are very striking, but the dissimilarities are even more so, because the members of the former taxon possess a number of traits which are otherwise typical of invertebrates. The remarkable feature of this taxon is that the general body plan exhibits a number of typical vertebrate characters: notochord, dorsal axial nervous system, metameric segmentation of the myotomes and of the peripheral nervous system, gill slits and organization of the circulatory system. Superimposed upon this are a number of properties, which, according to de Beer (1928), may be characterized as either 'primitive' or 'specialized' (Table 3.2).

Table 3.2
Primitive and specialized characters in Amphioxus
(G. R. de Beer, 1928)

Primitive characters
Ciliary mode of feeding, with endostyle
Epidermis one-cell thick
Afferent nerve-fibres derived from sensory cells
Complete row of segmented myotomes from front to rear
Very slight specialization of brain
No specialized head
No paired limbs or paired sense-organs
No specialized heart
Gonads segmental, without special ducts
Nephridia; segmentally arranged
Simple and unbranched liver diverticulum

Specialized characters
Atrium
Extra large number of gill-slits, having lost correspondence with the segmentation of the body
Tongue-bars
Asymmetry of oral hood and early development
Extreme anterior extension of the notochord

This enumeration shows that the 'primitive' characters comprise both vertebrate and invertebrate features. As regards the 'specialized' characters it is seen that those relating to the number of gill slits and the skeleton of the gills suggest affiliation to Hemichordata. The extension of the notochord may be responsible for at least three of the 'primitive' traits, viz., those concerned with the myotomes, the brain and the head, and the segmentation of the gonads and nephridia may in turn be causally related to the segmentation of the myotomes.

I think it may be relevant to point out that even in cephalochordates the notochord does not correspond to the definition given above, since it has been found to be 'a highly specialized hydroskeleton, with a unique muscular mechanism capable of varying its stiffness' (Flood, Guthrie and Banks, 1969, p. 88). This circumstance does not necessarily invalidate current views concerning its homology, but it certainly increases the probability that it has originated independently of the notochord found in vertebrates.

One feature in *Branchiostoma* which points to an invertebrate affiliation is that the larva in the blastula and gastrula stages is free-swimming, owing to the ciliation (or rather flagellation; cf. Løvtrup, 1974) of the ectodermal cells.

Cephalochordata are undoubtedly of very great importance for our appreciation of certain aspects of phylogenetic evolution, but owing to their being such mosaics of 'unrelated' characters, the proper evaluation of their significance may still remain unsolved.

ASSESSMENT OF THE CLASSIFICATION

Drach (1948) has made a very penetrating analysis of the problems concerning the systematic position of Protochordata, which will be used as the basis for the following discussion. According to Drach, the group Protochordata, i.e. Urochordata and Cephalochordata, may be defined as follows (p. 545, author's translation):

'*Marine, bilateral, primitively coelomate Metazoa belonging to the phylum Chordata. As chordates*, they are characterized: (1) by the superposition in the sagittal plane of the digestive tube, the notochord and a tubular neural axis; (2) by the existence of a system of slits with respiratory function, piercing the pharyngeal wall; (3) by a muscular tail into which passes the notochord and the nervous system; (4) by the special arrangement of the embryonic nervous system, the cavity of which opens anteriorly in a neuropore and communicates posteriorly with the digestive tube through the neurenteric canal. *They differ from the other Chordata* (Vertebrata): (1) through the existence of a ciliated groove, the *endostyle*, which passes through the pharynx along a medio-ventral line; (2) through the fact that the branchial slits do not open directly to the exterior but into a large peripharyngeal cavity; but particularly: (3) through the absence of a cephalic skeleton; (4) through the absence of a brain with differentiated segments and with specialized sensory annexes; (5) through the absence of paired limbs.'

As regards the first set of characters it will be seen that those listed under (1) and (3) may be referred to the directing and stretching influence of the 'notochord', and (4) to the morphogenetic mechanism through which the neural tube is formed, again primarily effected by the notochord. It therefore appears that ultimately Protochordata are defined by the possession of two characters, namely, the 'notochord' and the gill slits. With respect to the first feature, it has already been questioned whether it really deserves the name given to it in Urochordata, even though it admittedly assumes some of the epigenetic functions elsewhere carried out by the notochord.

The quoted definition refers to larval characters, preserved in adult cephalochordates, but lost in the urochordates. The characters in common between the adults of the two groups are very few, owing to the extensive modifications occurring during metamorphosis in the urochordate larvae. Hence, 'for phylogenetic reasons, biologists have attributed more importance to the organization of the ascidian larvae than to that of the adults' (Drach, 1948, p. 545, author's translation).

Among the characters which differentiate Protochordata from Vertebrata, three of the five are negative, which are less well suited for this purpose. Of the remaining ones, Drach points out that the perforation of the pharyngeal wall occurs in different fashions, and that the peripharyngeal cavities are not homologous, in the two taxa. Finally, since an endostyle is found in the ammocoete larva of Hyperoartii, and absence of paired limbs distinguishes Cyclostomata, these characters cannot be used for the definition of

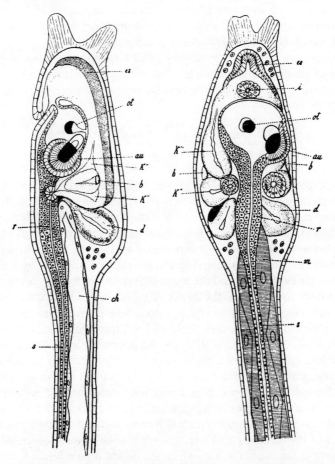

Figure 3.8. Larva of *Phallusia*. au, eye; b, blood sinus between gill slits; ch, notochord; d, intestine; es, endostyle; K, gill slits; m, tail muscles; ot, otocyst; r,s, neural tube. (Reproduced from E. Korschelt and K. Heider (1936), *Vergleichende Entwicklungsgeschichte der Tiere*, after Kowalevsky, by permission of Gustav Fischer Verlag)

Protochordata. Hence it seems impossible to establish an unambiguous definition of the group Protochordata, at least on a morphological basis.

Without questioning their kinship with Vertebrata, Drach suggests two alternatives concerning Protochordata, namely, that they constitute a monophyletic taxon of primitive chordates, or that they are two independent, i.e. monophyletic, branches within this phylum. The latter viewpoint is taken by those who class the three groups of Chordata in taxa (subphyla) of equal rank, a solution which would, of course, be unlikely if the animals were classified phylogenetically. According to this notion, all three groups, in order to constitute a monophyletic taxon of higher rank, must have had common ancestors belonging to Chordata.

With reference to the first classification, Drach enumerates a number of adult and embryonic characters which are different in the two protochordate groups. These are well known, and there is no need to list them here, except the absence of metameric segmentation in urochordates, the feature suggested as the greatest obstacle for the unification of the two taxa. In my opinion, this may not be such a decisive character if regarded from an epigenetic point of view. As already mentioned, segmentation is a morphogenetic process, based upon the gradual transformation of free cells into 'epithelial' cells in an elongated cell aggregate. This event will give rise to a series of vesicles which, to prevent their subsequent fusion, must be covered by a collagenous membrane. The elongation of the aggregates of mesodermal cells is a consequence of the notochordal stretching, and may be anticipated in both groups. Segmentation through vesicle formation requires the participation of a minimum of cells which, it seems, is not present in the urochordate larva. Instead the muscle cells, being 'epithelial' cells, form a cellular envelope around the chorda (Figure 3.8). In justification of this interpretation, which does not represent an established view, it may be mentioned that Remane (1932), having demonstrated a number of similarities between Gastrotricha and Annelida, suggested that the absence of segmentation in the former could be a consequence of their minute size, a circumstance which might entail that neither space nor cell number would allow for the morphogenetic mechanism of segmentation.

On the basis of the preceding discussion one is inclined to agree with Drach when he writes (1948, p. 551, author's translation): 'Their [Urochordata and Cephalochordata] unification in a narrower systematic unit is far from being imperative, as clearly shown by the difficulties encountered in defining Protochordata. In the absence of reliable facts permitting us to decide whether the descent of Chordata is diphyletic or triphyletic, historical (Kowalevsky) or didactic reasons must determine whether to conserve or abandon the group Protochordata.'

In the quoted statement Drach opens a very interesting, and certainly unexpected, possibility, namely, that Chordata may be 'triphyletic'. Surely, the very existence of the group as an acknowledged phylogenetic taxon presupposes that it is monophyletic, and the whole discussion about Protochordata concerns the question whether *they* are monophyletic, i.e. whether they have ancestors in common which were not also ancestors to Vertebrata. If so, they are a monophyletic group within Chordata, and ought to be ranked as a subphylum, just as Vertebrata. If not, the present subdivision in three separate subphyla would, approximately at least, reflect the phylogenetic relations.

However, if, as suggested here, what may be common to all chordates are only two attributes, pharyngotremy and a 'notochord', and all other similarities arise as epigenetic consequences of these, then it is possible to question whether these two features are so unique that they could not arise independently three times in the course of evolution. The suggestion that the 'notochord' is a convergent character is supported, I believe, by the fact that chordoid supporting structures are found in a variety of invertebrates (Ax, 1966).

If the 'chordate' characters are convergent ones then, indeed, Chordata may be 'triphyletic', if by this word is implied that the three taxa arose on separate occasions from non-chordate, i.e. invertebrate, ancestors. Should this be the case, it would be necessary to include one or more taxa of invertebrates in order to establish the mono-phyletic taxon in which all chordates are included. Submitting that this question cannot be answered on the basis of morphological features, we shall analyse it by means of non-morphological characters later in the present chapter.

Genesis of the vertebrates

If the superphyletic classification discussed above is to have any meaning, it must follow that the ancestors of the chordates are to be found in the echinoderm super-phylum, i.e. that the echinoderms and the chordates have predecessors in common which are not ancestors to members of the annelid superphylum or, indeed, to any other invertebrate phylum. This viewpoint is accepted by most zoologists today, but in the past several proposals were put forward to the effect that the vertebrates were derived from the annelid superphylum. Much of the argument about the origin of vertebrates has been concerned with rejection of the latter, rather than with presentation of observa-tions corroborating the former, contention, the reason for this largely being lack of convincing evidence (cf. Neal and Rand, 1936; Gregory, 1951). In the present section we shall briefly discuss the two sets of hypotheses. It should be mentioned that Willmer (1974) has recently propounded a view which does not conform with either of these, namely, that the nemertines may be the closest relatives of the vertebrates.

ORIGINATION IN THE ECHINODERM SUPERPHYLUM

The only attempt to trace the ancestors of vertebrates back to echinoderms seems to have been made by Garstang (1894). With a formanalyst's magic wand this author introduced a notochord and gill slits in an echinoderm larva and displaced the flagellated bands towards the dorsal side, thus creating an organism bearing some resemblance to an ascidian larva (Figure 3.9). By neoteny, through which the larval form is preserved into adulthood, this organism was purported to explain the origin of the chordates. This proposal has been met with some approval by de Beer (1951) and by Gregory (1951), but otherwise it seems to be largely neglected today. The main flaw in Garstang's suggestion is that, as discussed above, larvae in general are constructed with the partici-pation of a very limited number of differentiation patterns. Thus, in order to pass on from Garstang's hypothetical larva to the adult form one of two occurrences would seem necessary, both of which appear to be equally improbable. The first alternative would invoke the participation of the non-morphological characters of adult echino-derms, an expedient which, owing to their fundamental difference from those of the chordates, would hardly ensure the creation of members of the latter group. The other choice would be to presume survival of the larva, in a sexually mature form, until, with the benign assistance of natural selection, it had compiled a complete new set of these very stable characters, so hard to come by.

Garstang (1928) has later confined his attention to the tornaria larva of the enteropneusts, which through a similar deformation could lead to an ascidian larva and, through neoteny, to the vertebrates. The idea that urochordates might be at the base of the chordate line has also been endorsed by Berrill (1955), but most other authors argue,

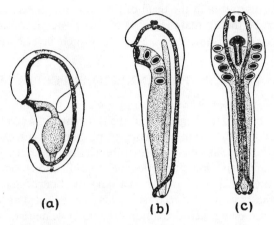

(a) **(b)** **(c)**

Figure 3.9. Conjectural derivation of a 'generalized' protochordate larva from an auricularia (echinoderm) larva. (a) auricularia larva; (b), (c) lateral and dorsal view of the hypothetical formanalytical product. (Reproduced from W. Garstang, *Quart. J. Micros. Sci.*, **72** (1928), by permission of The Oxford University Press)

for a number of reasons, that the sessile urochordates are 'degenerate' descendants of pelagic ancestors close to those of the vertebrates (e.g. Willey, 1894; Franz, 1927; Holmgren and Stensiö, 1936; White, 1946; Berg, 1947; Gregory, 1951; Eaton, 1953; Jägersten, 1955; Whitear, 1959; Bone, 1960a). Much of the argument in favour of the various views is based on anatomical features and morphological considerations, but some of it is of the teleological kind, presumably inspired by Darwinism. To exemplify this, in my opinion completely inadmissible, way of reasoning, it may be illuminating to quote Berrill (1955, p. 12), who postulates that segmentation was 'called forth by the need to maintain or improve position in the face of downflowing freshwater currents'. Surely, no needs in the world, however urgent they might be, could possibly induce segmentation in the tail of an ascidian larva if the number of cells present for this purpose is insufficient. Conversely, if by chance the number should reach an adequate level, and collagenocytes be present, segmentation might be anticipated to take place without the slightest regard to the ecological needs of the animal.

Remarkably few authors (e.g. Berrill, 1955; Bone, 1960a) accept a relation between the cephalochordates and the ancestors of the vertebrates, in spite of the striking similarity of the body plans of these two groups. However, as discussed above, *Branchiostoma* and its relatives display a number of features which to most morphologists are incompatible with this idea. Hence the conception has become prevalent that these animals have arisen from the original chordates through 'degeneration'. Presumably this concept is meant to imply that the ancestors of the cephalochordates managed to select such environmental conditions that the almighty natural selection saw fit to reintroduce in these animals invertebrate characters which their predecessors for aeons had fought so ardently to get rid of.

I believe that the valiant efforts to trace the descent of Vertebrata in the echinoderm line have not, so far, brought about any convincing results. The most serious deficiency is that, as noted in the quotation at the head of the present chapter, they contribute so little to what ought to be the ultimate purpose of theories concerning the ancestry of Vertebrata, namely, the establishment of the nature of their invertebrate forebears.

ORIGINATION IN THE ANNELID SUPERPHYLUM

The idea that the ancestors of the vertebrates were to be sought in the annelid super-phylum may be traced back to the beginning of the last century, when Étienne Geoffroy Saint-Hilaire suggested that by turning arthropods, particularly insects, upside down, a morphological organization was obtained resembling that observed in vertebrates. It must be emphasized that at the time no phylogenetic implications were involved in this proposition. During the last quarter of the past, and the first decades of the present, century several hypotheses were advanced which may be regarded as variations on this theme, with annelids also being included among the possible ancestors.

The latter alternative has been proposed by Semper (1875), Dohrn (1876), Minot (1897), Delsman (1913; 1921) and many others (cf. Neal and Rand, 1936; Gregory, 1951). The basis of this hypothesis is, of course, to be found in the fact that annelids are metameric, coelomate animals, but the various authors have advanced a number of further arguments in support of their belief. Most of these are formanalytical conversions of the type favoured by some morphologists, but quite a few relate to non-morphological features.

The arthropod origin has been propounded by Gaskell (1908) and Patten (1912), both of whom contended that chelicerates in particular exhibit a great number of traits that corroborate the relationship between them and vertebrates. Owing to the exoskeleton in the taxon of their choice, these authors could bring in palaeozoological evidence to vindicate their views.

The real difficulty for all hypotheses in this group is the fact that the nervous system is located ventrally in annelids and arthropods, dorsally in vertebrates. To supporters of the theories this circumstance seemed to cause little embarrassment for, except for Gaskell (1908), they simply supposed that the transformation began with the animals swimming upside-down. No objection can be raised against this proposition, for this way of swimming is observed in Limulus and its relatives (Fage, 1949). The next step in the argument is that when, finally, the ancestors of the vertebrates tired of these 'form-analytical' acrobatics, they changed their body plan to bring it into agreement with their inverted position. This conversion is touchingly illustrated in Patten's frequently repro-duced drawings (Figure 3.10).

The last important taxon in the superphylum is Mollusca. Various similarities between the latter and vertebrates were revealed long ago by Geoffroy Saint-Hilaire, Latreille, Laurencet and Meyranx (cf. Geoffroy Saint-Hilaire, 1830), but these animals have never attracted the attention of phylogeneticists. However, quite recently a paper was published concerning the origin of the vertebrates (Sillman, 1960) in which the suggestion is made that they may have originated from molluscs. Although the author imagines a transformation between adult forms, the argument is mainly based upon non-morphological characters.

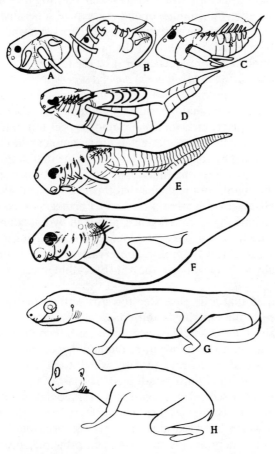

Figure 3.10. Conjectural transition between Arthropoda and Vertebrata. A, nauplius stage; B, ostracode; C, cladoceran; D, merostome; E, transitional form; F, larval fish; G, amphibian; H, mammalian. (Reproduced from W. Patten (1912), *The Evolution of the Vertebrates and Their Kin*, by permission of Churchill Livingstone)

From this cursory discussion it appears that the processes envisaged by the various authors to account for the transition between animals in the annelid superphylum and the vertebrates are hopelessly inadequate. Yet the study of previous literature indicates that a number of similarities on the non-morphological plane have been mobilized in favour of this ancestry, a circumstance which hints that, in spite of the general disapprobation it has encountered, it may still contain a grain of truth.

PHYLOGENETIC CLASSIFICATION OF VERTEBRATA AND
PROTOCHORDATA

From the preceding outline of current ideas about the origin of vertebrates it is, I believe, quite clear that everything hinges upon the accepted superphyletic classification. If this can be borne out by a cladistic analysis, then the theory as a whole is corroborated; if not, then it must be rejected. Evidently, if the prevailing position has any substance at all, then its most important consequence must be the following:

In a basic classification of some taxon in the annelid superphylum together with Echinodermata and Vertebrata, the outcome must be that the latter are the secondary twins.

It has been pointed out to me that it is such a long time since the separation between these last two taxa took place, etc., that one cannot expect too much agreement, etc. Therefore, should it turn out that more characters are common to the two other taxa, then this may easily be explained by convergence. It seems to me that those who express such opinions commit two errors. For one thing, they hold their theory so sacred that they do not want it to be falsifiable, not realizing that through this attitude they end up with having no theory at all, at least not an empirical one. Further, as a second line of defence they mobilize the phenomenon of convergence, although by not adhering to the rule of minimizing convergence they pull away the ground from all phylogenetic classification, for it convergence can explain away empirical observations which do not fit preconceived notions, it may equally well do so when they fit. Everything may be convergence, and that is that!

Hence, it is proposed that a decisive step towards settling the issue concerning the Protostomia–Deuterostomia superphyla and, consequently, the origin of the vertebrates, may be taken by establishing a basic classification according to the plan outlined above, following the directions set down in our three methodological premises. From the first two of these it follows that in the present case, where the three taxa are so distantly related, the quantitative method must be adopted. The last of the premises implies that the classification must be based solely upon non-morphological characters.

The annelid superphylum comprises three major phyla, viz., Annelida, Mollusca and Arthropoda, each of which has been alleged to possess progenitors in common with Vertebrata, as shown in the preceding discussion. In principle, any of these might be used to test the superphyletic classification, but a choice must be made, and since, according to the cladistic rules, it is completely free, I have chosen the phylum Mollusca.

The reason for this choice is the simple fact that the study of comparative physiology shows a remarkable number of features to be common to this phylum and Vertebrata. Yet, as has been seen above, among the advocates of the origin of vertebrates from the annelid superphylum, Mollusca have never been a favourite candidate. This does not necessarily mean that this taxon cannot be the one looked for in the search for the forebears of the vertebrates, should the present notions turn out to be falsified. Nevertheless, I propose to pursue the present quest one step further, by attempting to establish a classification involving the arthropod taxa Chelicerata and Crustacea also.

In the last section in this chapter an attempt will be made to assess the validity and some consequences of the classification which issues from my endeavours. Some of the conclusions arrived at in the present section have been presented in separate publications (Løvtrup, 1975a, b).

Mollusca

The cladistic method ostensibly operates with monophyletic taxa, and this implies in principle that the three taxa dealt with in the present basic classification should meet this requirement. I imagine nobody will question that Vertebrata are a monophyletic taxon, but can we be sure that this holds also for Echinodermata and Mollusca? Personally I shall not be surprised if one, or both, of these taxa turns out to be non-monophyletic. However, the answer to this question, if obtainable on the basis of existing knowledge, would require special analyses, thus leading far beyond the scope of the present work.

This point is not of particular significance in the case of Echinodermata because, as we shall see, so few characters are common to this taxon and any of the other ones. The situation is different with respect to Mollusca, because so many are found both in this taxon and in Vertebrata. In fact, we shall find that most of the characters occur in either of the purportedly monophyletic taxa Gastropoda and Cephalopoda, or in both. Since any set of three arbitrary taxa may be used in basic classification, I might have chosen one of these taxa instead of Mollusca. The conclusion arrived at would be exactly the same even if the number of characters available for settling the issue would be somewhat lower.

It is possible to form an opinion about the degree of non-apotypy to be expected in the present basic classification. For one thing, the plesiotypic characters common to the two least advanced taxa, which otherwise constitute a serious impediment to a basic classification, probably need not concern us. Apparently there are very few characters in common between Echinodermata and Mollusca. As regards convergence, we have seen above that this phenomenon may be anticipated to occur in all three pairs of taxa, and therefore it should not significantly affect the outcome of the analysis.

The problem of the classification of Protochordata will also be approached in an attempt to demonstrate whether or not the three subphyla of Chordata possess more characters in common than the maximum number any of these shares with one of the invertebrate taxa investigated.

The conclusions regarding these two problems will be presented in separate subsections.

TAXONOMIC CHARACTERS*

The non-morphological characters whose distribution will be subject to investigation may crudely be classed as being biochemical, physiological and histological (differentiation patterns), respectively. They may also be grouped according to the functions they assume; this has been done in Table 3.3, in which it will be noted that in some instances the heading only corresponds approximately to the contents of the group.

The various characters have been chosen mainly on the basis of surveys found in comprehensive treatises. The only criterion which has been adopted in the selection of characters from these sources is that their presence, or absence, must be established in all of the three main taxa under investigation, namely, Echinodermata, Mollusca and

*The aim of the following survey, up to the cumulative score, is principally to establish the presence or absence of characters in the three phyla. It does not pretend to supply any information of interest from the point of view of comparative biochemistry, physiology or histology, so for the reader well versed in these subjects it may suffice to study Tables 3.4–3.27 to get an impression of the characters used in the classification.

Table 3.3
Survey of characters examined in the search for the ancestors of Vertebrata

Connective tissues	Food Supply
Glycosaminoglycans	Feeding and digestion
Sialic acids	Liver and pancreas
Collagen	Excretion
Cartilage	Osmoregulation and kidney function
Chitin	Nitrogen excretion
Skeletons	Nervous regulation
Energy supply	Visual pigments
Haemoglobins	Photoreceptors
Phosphagens	Chromatophores
Respiration	Sensory organs
Heart, blood and circulation	Nervous system
	Humoral regulation
	Hormones

Reproduction
  Egg jelly coats

Vertebrata, in the latter case particularly in one or more of the lower taxa, Cyclostomata, Chondrichthyes and Osteichthyes, and preferably also in Urochordata and Cephalo-chordata. Among these, all characters present in two, but not in all three, of the main taxa have been recorded without discrimination.

In some cases a measure of subjectivity cannot be avoided in deciding whether or not a certain character is present, or whether a similarity really implies the presence of the *same* character, but it may be presumed that, when many features are investigated, any possible bias will be eliminated.

After the survey of the various characters the cumulative score is summed up.

Glycosaminoglycans

This class of substances is best known from studies on vertebrates in which the following substances have been discovered: hyaluronate, chondroitin, chondroitin 4-sulphate (C4-S), chondroitin 6-sulphate (C6-S), dermatan sulphate, corneal and skeletal keratan sulphate, heparan sulphate and heparin. The two keratan sulphates are distinguished by their degree of sulphation, which is higher in the skeletal than in the corneal type (Mathews, 1967). No data exist for protochordates, so in the present context we shall confine our attention to the question of whether one or more of these substances have been demonstrated in either echinoderms or molluscs.

It should be mentioned that studies have been made upon other substances related to the glycosaminoglycans, for example glycoproteins and mucins, but the results are not sufficiently comprehensive and clear-cut to be of use for comparative purposes.

Echinodermata

Various representatives of the class Holothuroidea have been analysed for the presence of glycosaminoglycans. Thus, Motohiro (1960, quoted by Mathews, 1967) found fucan sulphate and a chondroitin-like substance in *Cucumaria*. Maki and Hiyama (1956, quoted by Hunt, 1970) obtained similar results on *Stichopus*, except that they report the presence of glucosamine rather than galactosamine. Katzman and Jeanloz (1969) have isolated fucan sulphate and oversulphated chondroitin sulphate from the body-wall of *Thyone*.

These findings have been confirmed in my laboratory on *Stichopus* and *Mesothuria*; in Asteroidea and Ophiuroidea it appears that only sulphated glycoproteins are present (unpublished observations).

Yamagata and Okazaki (1974) have reported the presence of a dermatan sulphate isomer in the larva of *Pseudocentrotus*. We have made analyses on adults, but the material did not permit exhaustive analyses. However, since the presence of hexuronic acid and hexosamine was recorded, together with sulphate and neutral sugars, our results do not exclude the presence of either chondroitin or dermatan sulphate (unpublished observations). Kinoshita (1969) has demonstrated the presence of heparan sulphate in the larvae of two Japanese echinoid species.

Mollusca

The biochemistry of the glycosaminoglycans in Mollusca has been quite extensively studied, presumably because of the presence of 'cartilage' in various members of the taxon. The 'cartilage' of the odontophore of *Busycon* contains a matrix of glucan sulphate (Lash and Whitehouse, 1960a). Similar compounds have been found in the mucus secreted by *Charonia* (Soda, 1948, quoted by Hunt, 1970), *Buccinum* (Hunt and Jevons, 1966) and *Pecten* (Doyle, 1967). Substances of this type are of little interest in the present context because they are unknown in vertebrates.

Of more importance are the studies of the glycosaminoglycans present in the skin, the eye and the cranial cartilage of cephalopods. Chondroitin has been isolated from the skin of *Ommastrephes* by Anno, Kawai and Seno (1964). Srinivasan *et al.* (1969) have shown that the skin of *Loligo* contains chondroitin and oversulphated chondroitin sulphate in the proportion 1:4. The latter substance is electrophoretically similar to that found in shark cartilage.

Anseth and Laurent (1961) have found the most important sulphated glycosaminoglycans in the corneal stroma of various vertebrates and of *Loligo* to be keratan sulphate and chondroitin sulphate. Hunt (1970) argues that chondroitin may be present in the squid.

In the cranial cartilage of *Loligo* Lash and Whitehouse (1960b) observed the presence of chondroitin sulphate and keratan sulphate, and similar results were obtained by Anno, Seno and Kawaguchi (1962) on cartilage from *Ommastrephes*, but the keratan sulphate content was very low. The latter authors further suggested the presence of hyaluronate in this tissue. The chondroitin sulphate is oversulphated, to a large extent apparently because both the 4- and 6-positions of the galactosamine are esterified, thus making the substance, partly at least, a C4,6-S (Kawai, Seno and Anno, 1966). In cartilage from *Loligo*, Mathews, Duh and Person (1962) observed only oversulphated chondroitin sulphate. This material appears to be very similar to mammalian C4-S, except for the higher sulphate content and minor differences in the infrared spectrum. Material isolated from the giant squid, *Architeuthis*, and from *Octopus*, respectively, was found to contain less sulphate, but to be of higher molecular weight than that in *Loligo* (Mathews, 1967). These results on cephalopods have been confirmed (Rahemtulla and Løvtrup, 1975). In the same paper the presence in *Anodonta* of hyaluronate, chondroitin, chondroitin sulphate and heparan sulphate was reported.

Vertebrata

As regards the composition of their glycosaminoglycans, the members of Hyperotreta appear to assume a unique position. Thus, the skin of *Paramyxine* and *Eptatretus* contains a mixture of glycosaminoglycans, among which may be mentioned hyaluronate and oversulphated dermatan sulphate (Seno, Akiyama and Anno, 1972). Through hydrolysis of the latter substance, these authors were able to isolate a trisulphated disaccharide, in which the iduronate is sulphated in the 2- or 3-position and the galactosamine in the 4- and 6-position.

In the notochord of *Eptatretus* the main glycosaminoglycan is a substance which, on the basis of its composition, may be called dermatan 4,6-sulphate (Anno *et al.*, 1971), thus very similar to that found in the skin. Besides these substances, the skin of *Myxine* was found to contain ordinary dermatan sulphate and an unknown glycosaminoglycan (Rahemtulla *et al.*, 1976).

The skin of *Petromyzon* contains hyaluronate plus equal amounts of dermatan sulphate and C4-S (or C6-S), while the notochord has C4-S (Mathews, 1967). In *Lampetra* only hyaluronate and dermatan sulphate could be demonstrated in the skin (Rahemtulla *et al.*, 1976).

The corneal stroma of vertebrates generally contains keratan sulphate, chondroitin and C4-S (Balazs, 1965). No information is available concerning the eyes of cyclostomes.

In vertebrate cartilage keratan sulphate usually occurs together with one of the chondroitin sulphates. According to Mathews (1967) the cartilage of *Myxine* and *Eptatretus* contains at least two acid mucopolysaccharides which, apart from being oversulphated, resemble C4-S, while in *Petromyzon* the cartilage contains C6-S.

Score

From the preceding survey it appears that oversulphated chondroitin sulphate is a common feature in the three major phyla, at least as concerns some classes. On the other hand, hyaluronate, chondroitin and keratan sulphate are present only in Mollusca and Vertebrata.

Yamagata and Okazaki (1974) claim that their observations on *Pseudocentrotus* constitute the first demonstration of dermatan sulphate in any invertebrate. It should be mentioned that Ashhurst and Costin (1971a, b) submit that this substance occurs in some insects (*Locusta* and *Galleria*), but this result could not be confirmed by analyses on the blowfly *Calliphora* (Höglund, 1976). Although this point needs further scrutiny, I shall in the present context accept the presence of dermatan sulphate as a character uniting Echinodermata and Vertebrata.

The sulphated glycosaminoglycans in the lower classes of Mollusca and in Holothuroidea, glucan sulphate and fucan sulphate, respectively, exhibit some resemblance, the significance of which is difficult to judge (Table 3.4).

Table 3.4
Distribution of glycosaminoglycans

Mollusca–Vertebrata
Hyaluronate
Chondroitin
Keratan sulphate
Echinodermata–Vertebrata
Dermatan sulphate

Sialic acids

The sialic acids, various derivatives of neuraminic acids, are particularly known from their occurrence in glycoproteins and glycolipids in vertebrates. It would not be necessary to discuss these substances in the present context if their distribution in invertebrates had not been used by Warren (1963) as a support for the annelid–echinoderm superphyletic classification. The observations upon which this argument is based are shown in Table 3.5. It will be seen that these substances do not occur ubiquitously in any of the phyla, including Chordata, since they are absent in uro-chordates. Warren (1963) is inclined to doubt the significance of the findings of sialic acids in the 'digestive gland' of *Homarus* and *Loligo*, because they were largely, if not entirely, in the free state, suggesting an exogenous source. On the other hand, in gastropods, where Warren did not observe any sialic acid, Inoué (1965) has found this compound in a mucopolysaccharide, horatin sulphate, from *Charonia*. In *Chiton* it is possible to demonstrate sialic acid as a component in the glyco-proteins (unpublished observations). Finally, Lunetta (1971) has found sialic acid in the mucin surrounding of the eggs of *Physa*. Under these circumstances it must be concluded that sialic acid is a plesiotypic character which cannot be used for the purpose of demonstrating the phylogenetic relationship between the various taxa.

Collagen

The collagens are a class of proteins which may be easily identified by various criteria. Among the most important ones may be mentioned: (1) the amino acid composition, distinguished by the

Table 3.5
Distribution of sialic acids

Phylum and class (subphylum)	Genus	Location, when present
Annelida	Seventeen species	Not detected
Arthropoda		
Insecta	Three species	Not detected
Crustacea	Eight species	Not detected
	Homarus	Digestive gland
Chelicerata	Two species	Not detected
Mollusca		
Polyplacophora	*Chaetopleura*	Not detected
Polyplacophora	*Chiton*[1]	Whole animal
Gastropoda	Six species	Not detected
Gastropoda	*Charonia*[2]	Digestive gland
Gastropoda	*Physa*[3]	Egg mucin
Bivalvia	Nine species	Not detected
Cephalopoda	*Loligo*	Digestive gland
Echinodermata		
Crinoidea	*Nemaster*	Viscera
Holothuroidea	*Thyone*	Viscera
	Synapta	Viscera
Echinoidea	*Strongylocentrotus*	Viscera
	Arbacia	Viscera
	Echinarachnius	Viscera
Asteroidea	*Asterias*	Viscera
	Henricia	Viscera
Ophiuroidea	*Ophioderma*	Whole animal
	Unidentified	Whole animal
Hemichordata	*Dolichoglossus*	Whole animal
Chordata		
Urochordata	Eight species	Not detected
Cephalochordata	*Branchiostoma*	Whole animal
Vertebrata		
Cyclostomata	Ubiquitous	Many tissues
Selachii	Ubiquitous	Many tissues
Pisces	Ubiquitous	Many tissues
Amphibia	Ubiquitous	Many tissues
Reptilia	Ubiquitous	Many tissues

1, Unpublished observations; 2, Inoué (1965); 3, Lunetta (1971). The remaining data are from Warren (1963).

fact that about one-third of all residues are glycine and one-fifth the amino acids proline and hydroxyproline, (2) the typical wide-angle X-ray diffraction pattern and (3) the axial periodicity in the range 60–70 nm revealed by the electron microscope (Ramachandran, 1963). The latter feature is absent in some collagens, for example in *Lumbricus* and *Ascaris* cuticular collagens. Collagen occurs in most, if not all, phyla of the animal kingdom, and in many organisms it constitutes the overwhelmingly major fraction of the total protein content (J. Gross, 1963).

From a comparative point of view, the amino acid composition would seem an important parameter. In general, it appears that the range of variation for individual amino acids is much higher in invertebrate than in vertebrate collagens (Table 3.6). This state of affairs may very well be a result of mutations accumulated with time, since the origin of many invertebrate phyla dates much further back than that of Vertebrata. Unfortunately, it is hardly possible to base any con-

Table 3.6
Amino acid composition of collagen (residues per 1000)

| | Range[1] | | | Thyone briareus[3] | |
	Invertebrates	Vertebrata	Thyone sp.[2]	Unfraction-ated gelatin	Fraction 2
Glycine	286–324	301–359	306	312	328
Glutamic acid	77–110	62–94	110	112	115
Aspartic acid	56–97	43–55	**62[4]**	**47**	**45**
Arginine	21–57	36–53	54	56	54
Proline	13–280	97–130	109	108	109
Alanine	56–114	81–126	**71**	**135**	**130**
Valine	13–34	15–27	30	30	30
Phenylalanine	5–12	11–19	**9**	**14**	**16**
Leucine	18–37	18–36	22	21	19
Hydroxyproline	24–165	63–108	**60**	**36**	**35**
Threonine	18–52	16–26	39	36	31
Lysine	9–37	13–36	8	7	7
Isoleucine	13–24	9–20	13	12	12
Serine	22–105	25–70	**54**	**45**	**40**
Histidine	0–8	3–11	3	4	2
Methionine	0–12	4–14	2	2	0
Hydroxylysine	0–30	4–23	11	12	12
Tyrosine	0–11	1–9	8	8	6
Cysteine	0–6	—	3	2	0

1, J. Gross (1963); 2, Piez and Gross (1959); 3, Katzman *et al.* (1969); 4, Numbers printed in bold type represent amino acids for which significant differences obtain in the two (?) species of *Thyone*.

clusions of phylogenetic significance on the available data. This is indicated in Table 3.6, where results obtained on *Thyone* on two separate occasions are markedly different with respect to the estimated content of five amino acids. Species differences or unsatisfactory analytical methods may explain these observations, but that incomplete separation of collagen is often a source of error is seen by comparing the results on unfractionated and fractionated gelatin from *Thyone briareus* in Table 3.6. The differences are not very great in most cases, but they are important, for instance, with respect to the sulphur-containing amino acids, particularly cysteine, which is usually considered to be absent from collagen (Katzman, Bhattacharyya and Jeanloz, 1969).

D. F. Travis (1970) has electron microscopically demonstrated the presence of fibrils with the typical axial repeating period in *Strongylocentrotus*, *Asterias* and *Ophiopholis*. In the sutural connective tissue of *Arbacia* collagen has been demonstrated histochemically (Moss and Meehan, 1967). Against these findings may be quoted the work of Klein and Currey (1970), in which it is asserted, on the basis of hydroxyproline analyses, that there is no collagenous matrix in the calcified tissues of *Strongylocentrotus*. Although their contention apparently cannot be substantiated, their experimental results are to some extent corroborated by the values reported for glycine and hydroxyproline in the matrix proteins of Echinoidea by Degens, Spencer and Parker (1967). But their assertion seems to be contradicted by the demonstration of a hydroxyproline-containing material in the skeletal matrix of *Paracentrotus* embryos (Pucci-Minafra, Casano and La Rosa, 1972).

Some of the generalizations which can be made about the amino acid composition in various collagens (Pikkarainen and Kulonen, 1969) are that the content of acidic amino acids is distinctly higher in invertebrates than in vertebrates (Figure 3.11) and that the content of hydroxyamino acids is conspicuously lower in endothermic than in all other animals (Figure 3.12). These authors also tried to estimate the summed differences in amino acid composition between collagens from various species. As shown in Figure 3.13, the difference between *Loligo* and *Mytilus* is larger than, or as large as, those between the former and *Petromyzon* and *Myxine*, respectively.

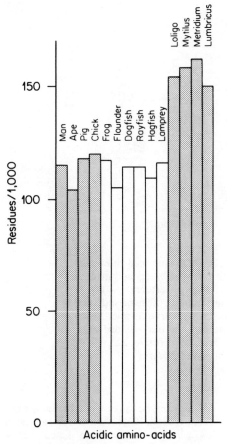

Figure 3.11. Content of aspartic acid and glutamic acid in collagen from various vertebrates and invertebrates. (Redrawn from J. Pikkarainen and E. Kulonen, *Nature*, **223** (1969), by permission of the authors and *Nature*)

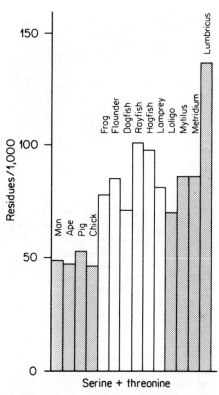

Figure 3.12. Content of serine and threonine in collagen from various vertebrates and invertebrates. (Redrawn from J. Pikkarainen and E. Kulonen, *Nature*, **223** (1969), by permission of the authors and *Nature*)

It has generally been found that the carbohydrate content of collagen is much higher in the invertebrates than in the vertebrates (Bailey, 1968). In the latter case glucose and galactose are the preponderant, if not the only, monosaccharides, whereas in invertebrate collagens wide arrays of carbohydrate residues have been reported. These results seem to depend upon incomplete separation; as demonstrated by Katzman *et al.* (1969), the gelatin from *Thyone* can be separated into three fractions, of which two have an amino acid composition typical of collagen and contain only glucose and galactose, while all the other carbohydrates (fucose, mannose, hexosamine and hexuronate) are present only in the third, non-collagenous fraction.

Unfortunately, it thus does not seem possible, from any of the published data concerning the properties of collagen, to make any inferences of importance for the present problem.

Epidermin

The discussion of keratin and keratinization may seem to be beyond the scope of the present book insofar as keratin is generally thought to occur only in terrestrial vertebrates. However, in

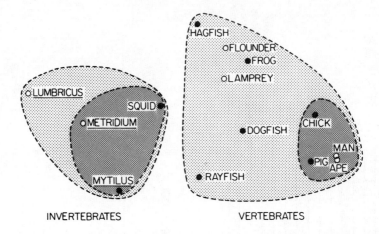

Figure 3.13. Two-dimensional presentation of the summed differences in the composition of collagen from various vertebrates and invertebrates. The outer dashed curves circumscribe vertebrates and invertebrates, respectively. Inside 'Invertebrates', the mesodermal collagens are surrounded by a second dashed curve. Inside 'Vertebrates', the homoiothermic animals occupy the dark-shaded area. (Redrawn from J. Pikkarainen and E. Kulonen, *Nature*, **223** (1969), by permission of the authors and *Nature*)

recent years it has been proposed that the fibrous proteins may not be the essential component of the process of keratinization (Mercer, 1961; Matoltsy, 1962; Rudall, 1968). Rather, it is suggested that this phenomenon is similar to that of sclerotization, which, as known from the cuticles of insects, involves the stabilization of the state of aggregation of rather inert chitin fibrils through the tanning, by the action of quinones, of proteins in the surrounding matrix (Pryor, 1962). Similarly, the important agent in keratinization may be a protein which, through the formation of disulphide bridges, gives rise to a hardened matrix in which fibrous proteins may be present or absent, as the case may be.

The terminology concerning the fibrous proteins is anything but unambiguous. Previously the name 'keratin' was used, but in view of the propositions just quoted this is obviously a misnomer. Rudall states that 'we should continue to use the term prekeratin for the fibrous protein of the monofibrils because it forms such a large constituent of the final product we are familiar with as keratin' (1968, p. 564). However, that name implies that this type of fibrous protein occurs only in potentially keratinizing cells, i.e. in cells that are able to synthesize the proteins responsible for keratinization. Since this is not necessarily true, it would seem more appropriate to adopt the word 'epidermin'. This name has been proposed by Rudall (1952) for the class of fibrous epidermal (ectodermal) proteins in mammals. It is easily seen that this definition hardly allows for a distinction between 'prekeratin' and 'epidermin'. In the present context I propose to use 'epidermin' as a name for the class of all fibrous ectodermal proteins except those that can be shown to belong to the collagens, as is the case in annelids and nematodes.

Echinoderms are covered with an epidermal cuticle in which no structural proteins can be observed (Hyman, 1955). In the tunic of the urochordates the main structural elements are cellulose and collagen (Rudall, 1955), although the presence of 'pseudokeratin' has been reported in *Cynthia* (Tsuchiya and Suzuki, 1962). The 'cuticle' of the cephalochordates seems to consist of a layer of 'mucoprotein', containing no formed elements (Olsson, 1961). Therefore the only groups which need be discussed are Mollusca and Vertebrata.

Mollusca

The epidermal proteins in the exoskeleton, comprising the periostracum, the prismatic and the nacreous layers, have been subjected to extensive investigations (Hare and Abelson, 1965; Degens *et al.*, 1967; Wilbur and Simkiss, 1968; Bricteux-Grégoire, Florkin and Grégoire, 1968; cf. also Hunt, 1970).

From these investigations it is apparent that in contrast to the collagens, the tertiary structure of which clearly imposes a number of restrictions on the acceptable mutations, the epidermins cannot be characterized on the basis of the amino acid composition. This point is particularly borne out by the comprehensive work of Degens *et al.* (1967), which shows, for instance, that the glycine content ranges from 91 to 526 per 1000 in the class Gastropoda. Considerable differences are observable even within the same species, so a great number of chemical 'subspecies' clearly exists. In spite of this variability it was possible, leaning on the accepted classification, to establish a phylogenetic tree for Mollusca on the basis of these findings (Ghiselin *et al.*, 1967).

As a conclusion of their work, Degens *et al.* (1967, pp. 576–577) state that 'the solubility tests indicate a gelatinous nature for the bulk of the proteinaceous matter which is not unlike that of collagen. However, based on X-ray work (lack of wide-angle diffraction patterns), electron micro-graphical studies (absence of cross-striated fibrils) and on chemical grounds (amino acid composition) the organic matrix more closely resembles the so-called k–m–e–f group of proteins (keratin–myosin–epidermin–fibrin)'.

Hunt (1970) has paid particular attention to the problem of the stabilization of the non-calcified material in the periostracum and the operculum. He has convinced himself that no keratinization takes place, and even the occurrence of sclerotization is difficult to demonstrate. According to Brown (1952), the byssus threads, the shell hinge and the periostracum in *Mytilus* consist of quinone-tanned proteins. The supporting structures of the gills are not sclerotized, however, and they do not contain chitin.

As regards banding patterns demonstrable by electron microscopy, it should be mentioned that in fibrous ribbons isolated from the egg capsules of *Buccinum* a periodicity of 96 nm has been demonstrated, closely similar to that of 'prekeratin', 105 nm (Rudall, 1968).

Vertebrata

Hardened proteins occur in the teeth in both cyclostome orders, and histochemical methods have revealed the presence of sulphide groups (Barrnett, 1953; Dawson, 1963). Whether or not it is justified to speak of 'keratin' in this case is clearly a matter of definition, but the presence of epidermal fibrous proteins may be inferred, although nothing is known about their chemical composition. Tonofibrils can be observed in the skin (Blackstad, 1963).

In the next lowest vertebrate group, Chondrichthyes, epidermal calcification occurs in the form of enameloid in teeth and scales. It has been postulated that the matrix of enameloid contains collagen, but amino acid analyses now appear to have disproved this proposition (Levine *et al.*, 1966). In their relatively high content of aspartic acid, serine, glutamic acid and glycine they resemble, according to these authors, the proteins of human and bovine enamel. They are also similar to the proteins in the organic matrix of the shells in cephalopods; as shown in Table 3.7 only six out of 17 amino acids fall outside the range observed in this class. Among these six amino acids are hydroxyproline and hydroxylysine, which are absent in Cephalopoda but occur sporadically in the shell proteins of other molluscs. It may be mentioned that, from the data of Degens *et al.* (1967), it appears that of 34 species of Gastropoda only 18 species have a similar or better score; for Bivalvia the corresponding numbers are 13 and 8, respectively. The non-collagenous nature of ichthylepidin, a protein present in teleost scales, has been confirmed on the basis of amino acid analyses and X-ray diffraction patterns (Seshaya, Ambujabai and Kalyani, 1963).

Studies of the fibrous layers of the skate egg capsule, called ovokeratin, indicate the presence of collagen, but amino acid and X-ray analyses show that a large amount of a different type of protein is present (Gross, Matoltsy and Cohen, 1955; Gross, Dumsha and Glazer, 1958).

The problem of the nature of protein in enamel and enameloid matrices has been subject to much discussion. It is now reasonably certain that results indicating the presence of collagen are

Table 3.7
Amino acid composition of calcified proteins in Cephalopoda and
Selachii (residues per 1000)

	Cephalopoda (4 species)[1]	Selachii (*Sphyrna*)[2]
Glycine	153 (89–262)	180
Glutamic acid	65 (42–128)	119
Aspartic acid	98 (75–128)	96
Arginine	23 (10–56)	43
Proline	55 (15–205)	59
Alanine	125 (74–216)	**72**[3]
Valine	36 (20–64)	45
Phenylalanine	30 (20–45)	26
Leucine	44 (29–67)	**73**
Hydroxyproline	—	**4**
Threonine	34 (18–64)	47
Lysine	20 (5–83)	48
Isoleucine	21 (17–26)	**40**
Serine	95 (62–146)	84
Histidine	5 (0–35)	14
Methionine	—	**17**
Hydroxylysine	23 (15–75)	**8**
Tyrosine	23 (15–75)	25
Cysteine	7 (0–65)	2

1, Degens *et al.* (1967), mean and range listed; 2, Levine *et al.* (1966); 3, Numbers printed in bold type represent amino acids which in *Sphyrna* lie outside the range observed in Cephalopoda.

due to unsatisfactory separation between enamel and dentine. From a comparative and phylogenetic point of view it is very important to know whether the enamel proteins can be classified as 'keratins' or 'eukeratins'. This view has been taken by some authors on the basis of amino acid analyses and X-ray diffraction analyses (Frank, Sognnaes and Kern, 1960), but clearly the agreement is not quite satisfactory, among other reasons because the cystine content is rather low (Piez, 1962; Schiffmann, Martin and Miller, 1970). Yet, in view of the above discussion on the nature of keratinization it may be questioned whether the structural proteins in keratin, the epidermins, must necessarily have a high content of cystine.

Score

On the basis of their amino acid composition, Rudall (1968) has suggested that certain silk fibroins, elastin and resilin, have been derived from collagen in the course of phylogenetic evolution. Collagen is usually elaborated by mesodermal cells, but in nematodes and annelids, and possibly in urochordates, it is produced by ectodermal cells. In view of this fact it is possible that the epidermins also are related to collagen, although more distantly than the proteins mentioned above. Certain of the observations discussed in this section may support this contention.

The structural proteins present in ectodermal calcified matrices, as encountered in the shells of molluscs and in the scales of chondrichthyan and actinopterygian fishes, are clearly distinct from the collagens, and there may be some justification for classing them together as epidermins (Table 3.8).

Table 3.8
Distribution of epidermal structural proteins

Mollusca–Vertebrata
Epidermin

Cartilage

Person and Philpott (1969, p. 2) define cartilage as 'an endoskeletal animal tissue of gristle-like consistency . . ., relatively rigid and resistant to a variety of forces such as those of compression, tension and shearing. . . . Histologically, cartilage is a form of connective tissue composed of polymorphic cells suspended in colloidal gel matrices . . . which are metachromatic to basic dyes. . . . Chemically, cartilage is especially characterized by high contents of collagen, acidic polysaccharides and water'. In a footnote, Person and Philpott add: 'The possibility that some cartilage may be of ectodermal origin should also be considered. We have seen ectodermal tissues in some tunicates (*Corella, Ciona*) and also in the cephalochordate *Branchiostoma* which, in haematoxylin and eosin stained sections, have the appearance of cartilage'. Since detailed histo-chemical and chemical studies have not as yet been made, the question of whether or not such tissues can be called cartilage must be deferred for the present. It is of interest, particularly for the discussion in the section dealing with skeletons, to note that here 'ectodermal' is contrasted with 'endoskeletal', the implication clearly being that the latter is of mesodermal origin. It is not obvious, however, which criterion has been used to establish the ectodermal nature of the tissues observed in protochordates.

The formation of cartilage must depend somehow upon the interaction between collagen and some matrix. Since collagen apparently always occurs in a matrix, without necessarily forming cartilage, it seems to follow that it is the chemical nature of the matrix substances rather than that of collagen which is of decisive importance for the formation of cartilage.

Recent work (Toole, Jackson and Gross, 1972) has shed some light on this problem. Thus it appears that during the initial phase of chondrogenesis a matrix of hyaluronate is elaborated, presumably by the presumptive chondrocytes. In this matrix the cells can migrate freely, but the next step in cartilage formation, the aggregation of the chondrocytes, is suppressed by hyaluronate. This therefore means that the cells must begin producing both hyaluronidase, serving to abolish the primary matrix, and then a new glycosaminoglycan, presumably chondroitin sulphate, which can form the secondary matrix needed for cell aggregation and chondrogenesis. It thus appears that hyaluronate and chondroitin sulphate are necessary, if not sufficient, conditions for cartilage formation.

As pointed out by Person and Philpott (1969), cartilage has traditionally been considered as a typical vertebrate feature, so much so that similar tissues encountered in invertebrates were often believed not to be true cartilage. It must be stressed that this view was not wholly general, for according to Person and Philpott quite a few authors openly referred to the presence of cartilage in invertebrates. However, Echinodermata are not among the phyla in which this tissue can be demonstrated, and since its occurrence in protochordates has been reported only in the footnote by Person and Philpott (1969) quoted above, the present discussion may be confined to Mollusca and Vertebrata.

Mollusca

The early observations on the occurrence of cartilage in invertebrates were all confined to Cnidaria and the three phyla of the annelid superphylum.

In Mollusca the tissues which conform to the above definition of 'cartilage' are found in the odontophores of gastropods and the cranium and other structures in cephalopods. As already noted, the 'cartilage' of the odontophore in *Busycon* does not contain chondroitin sulphate, but rather glucan sulphate (Lash and Whitehouse, 1960a).

Histologically, the cranial cartilage of cephalopods is so similar to some types of vertebrate cartilage as to be 'virtually indistinguishable from them' (Person and Philpott, 1969, p. 8). A characteristic feature of cephalopod cartilage is the presence of chitin, as demonstrated by Halliburton (1885). It should also be mentioned that cartilaginous scales have been found in the integument of cephalopods (Person, 1969).

In trying to decide whether or not true cartilage occurs in invertebrates, Person and Philpott (1969) leave the question open as regards Cnidaria and Gastropoda. However, their answer is affirmative with respect to Annelida (*Eudistylia*), Arthropoda (*Limulus*) and Cephalopoda.

Vertebrata

The occurrence of cartilaginous tissue in the endoskeleton of vertebrates, from cyclostomates to mammals and birds, is so well known that there is no need to discuss it on this occasion.

Score

The preceding survey leads to the unavoidable conclusion that the only cases where cartilage has been convincingly demonstrated are in members of the annelid superphylum and in the subphylum Vertebrata. The observations of Person and Philpott (1969) quoted above suggest the possibility that cartilage may be present in Protochordata (Table 3.9).

Table 3.9
Distribution of cartilage

Mollusca–Vertebrata
Presence of cartilage
Mollusca–Urochordata–Vertebrata
Presence of cartilage
Mollusca–Cephalochordata–Vertebrata
Presence of cartilage

I shall take the opportunity to quote some interesting statements relating to the observed distribution of cartilage. In their review, Person and Philpott (p. 11) write: 'Until now the principal focus of inquiry in relation to the question of the origin of cartilage has been the endoskeleton of the ancestral vertebrates, the ostracoderms. . . . However, recognition once again of the existence of cartilaginous tissues in the invertebrates means that cartilage, as a tissue type, may have originated in these animals rather than in the vertebrates.' Furthermore, in their study of cartilage in *Eudistylia*, Person and Mathews (1967, p. 250) observed that 'if the cartilage and "osteoid-like" components of the skeletal complex of *Eudistylia* were mineralized, then one would have a structure strongly resembling the outer armor of the vertebrate ancestors, the ostracoderms'.

Chitin

The absence of chitin in Echinodermata and Vertebrata is a characteristic to which, so far, no exceptions have been found. Equally well established is the presence of this substance in the major phyla of the annelid superphylum, Annelida, Arthropoda and Mollusca (Rudall, 1955; Jeuniaux, 1963). The attempt made by Jeuniaux (1963) to corroborate the accepted phylogenetic tree on the basis of the distribution of chitin appears to fit reasonably well. This result depends to a large extent, however, on the presumed absence of chitin from Protochordata. Even though this is a negative feature, it has been hailed with enthusiasm because it supports the frail scaffolding of the echinoderm superphylum. We shall in the present section discuss the occurrence of chitin in Mollusca and Protochordata.

Mollusca

In Mollusca chitin is usually present in the organic matrix of the shell and in other ectodermal structures such as the radula in gastropods and cephalopods, and also in the cuticle lining the oesophagus and the stomach in many species. Chitin seems to be absent from mesodermal tissues in molluscs, except cephalopods (Halliburton, 1885), but is present in arthropods (Rudall, 1955).

Protochordata

Chitin has been looked for in vain in the tunic of Urochordata (Rudall, 1955). However, Peters (1966), who investigated five species of Ascidiaceae, found in all cases that the peritrophic membrane contains chitin. Several tests were employed, but in particular it was ascertained that the microfibrils which could be demonstrated electron microscopically were not made up of cellulose. These observations have been accepted by Jeuniaux (1971).

It has traditionally been asserted that the gill bars of *Branchiostoma* are 'composed of a chitin-like substance' (de Beer, 1928, p. 12), but the occurrence of chitin in the subphylum was emphatically disclaimed by Jeuniaux (1963). Fortunately for the present discussion this question has been taken up anew by Sannasi and Hermann (1970). Using all available tests, physico-chemical and colour reactions, enzymatic and chemical hydrolysis and identification of the ensuing products, these authors demonstrated that the gill bars contain chitin. This result may be of particular interest because it shows this compound to be located in an endoskeleton. On the other hand, the branchial skeleton in Enteropneusta (Hemichordata) shows a typical collagen diffraction pattern (Rudall, 1955). It should be mentioned that Rudall and Kenchington (1973) contest the results of Peters (1966) and Sannasi and Hermann (1970), on the basis of X-ray diffraction studies. If they are right, the purported affinity between Hemichordata and Cephalochordata may be supported, but this clearly does not sustain the validity of the phylum Chordata.

Score

The absence of chitin is a character in common between Echinodermata and Vertebrata.

As regards the Protochordata, the positive results reported above may be interpreted as an affinity between Urochordata and Cephalochordata and the annelid superphylum, while the negative results discount this proposition. Since this single character cannot change the final outcome, the former alternative has been accepted in Table 3.10.

Table 3.10
Distribution of chitin

Echinodermata–Vertebrata
Absence of chitin
Mollusca–Urochordata
Presence of chitin
Mollusca–Cephalochordata
Presence of chitin

Skeletons

It is an unfortunate, but rather usual, habit among biologists to contribute to terminological confusion by using established names in different, and often contradictory, senses. This is to some extent the situation within the subject of the present section, and we must therefore begin the discussion with some remarks on the question of terminology.

According to Neal and Rand (1936, p. 673), the concept 'endoskeleton' is defined as 'the internal skeleton as distinguished from the dermal skeleton' and 'exoskeleton' as 'that part of the skeleton which is derived from the skin'. The latter definition clearly implies 'exoskeleton' = 'dermal skeleton'. A very similar stand is taken by Hyman (1942). Weichert (1965, p. 380), discussing the skeleton in vertebrates, states that 'the word *endoskeleton* is used to denote internal skeletal structures. . . . The term *dermal skeleton* is used in referring to [dermal scales, etc.] and their derivatives. Sometimes the dermal skeleton is spoken of as the exoskeleton, but the latter is more properly used in connection with the skeleton of invertebrates'.

It is not difficult to characterize the shells in molluscs and the carapaces in arthropods on the basis of these definitions; they are clearly exoskeletons. However, the situation is less simple with regard to the skeleton of echinoderms, about which Hyman (1955, p. 4) writes: 'The dermis contains and produces the skeleton, which is thus an endoskeleton'. Yet according to the definition of Neal and Rand, as endorsed by Hyman, it should be classed as an exoskeleton or a dermal skeleton, and, following Weichert, only the former name is applicable. Weichert's proposal to restrict the use of 'exoskeleton' to invertebrates is unfortunate, because the 'dermal' skeleton of vertebrates, as exemplified by teeth and scales as well as by the calcified coverings in various extinct forms, is usually overlaid by a layer produced by epidermal or ectodermal cells. It seems quite unjustifiable to include this element in the dermal skeleton, and the most appropriate name for it seems to be 'exoskeleton'. The only way to resolve this terminological confusion is probably to distinguish between three kinds of skeleton, two of which are mesodermal, namely, the 'dermal skeleton', located in the dermis, and the 'endoskeleton', located beneath the dermis, while one is ectodermal and is called 'exoskeleton'.

Referring to the quotations above, it appears that there is unanimity with respect to the definition of endoskeletons. Calcified endoskeletons are known only in vertebrates, and hence are of no comparative value in the present context. Uncalcified endoskeletons are encountered in both invertebrates and vertebrates in the form of cartilage, as discussed and evaluated in a preceding section; in the former, chitinous endoskeletons also occur.

Dermal skeletons may be calcified, as in echinoderms and in dermal or membrane bones and in the mesodermal parts of scales and teeth found in some vertebrates, or uncalcified, as found for instance in cyclostomes and chondrichthyans and in the dermal cartilaginous scales in cephalopods (Person, 1969).

The distinguishing feature of exoskeletons is that they are derived from the ectoderm. They may occur as independent structures, but often they are fused with dermal or internal skeletons of mesodermal origin to form 'integumental skeletons'. The former alternative is represented in the calcified form by the previously mentioned invertebrate shells and carapaces, and in the uncalcified form as various 'horny' formations, radulae and horny teeth, as well as the scutes and shells found in reptiles. The carapace of crustaceans and the 'exoskeletons' in ostracoderms are presumably calcified integumental skeletons, but so are teeth covered with enamel, which is clearly an ectodermal product.

Uncalcified mixed skeletons appear to be rare, but according to the view expressed by Person and Mathews (1967) quoted in a preceding section, the integumental skeletal complex of *Eudistylia* may belong to this type.

Echinodermata

Echinoderms possess a dermal skeleton in which the inorganic phase is calcium carbonate. As already discussed, the studies of D. F. Travis (1970) have demonstrated that the matrix fibres are collagen fibrils. The mineralized parts of the endoskeleton are either ossicles or plates. The plates in *Strongylocentrotus* consist of trabeculae arranged in a three-dimensional network, in the holes of which sclerocytes are located. The crystals in each trabecula are composed of ordered, sheet-like layers. As observed by Glimcher (1960) in vertebrate collagen, crystallization occurs first inside the fibrils (D. F. Travis, 1970). According to this author, the structure of the plates 'resembles that of cancellous or spongy bone' (p. 265), although there is no vascularization. A similar pattern of calcification is found in ophiuroids, and a comparison with that obtaining in holothuroids and asteroids suggests that the reticular form of the plates in the former classes essentially represents a three-dimensional extension of the ossicles occurring in the latter. This contention is borne out when the width of the dermal layer occupied by each of the two types of calcification is considered (Hyman, 1955). Basement lamellae do not appear to be present in the dermis of echinoderms.

Mollusca

The skeleton in molluscs is a purely ectodermal exoskeleton. The matrix usually contains chitin, but this substance may be totally absent (Degens *et al.*, 1967; Wilbur and Simkiss, 1968). A structural protein is present which has been classified above as an epidermin. The inorganic phase is always calcium carbonate. In the prismatic structure of part of the shell, and in the very low content of organic matrix, the molluscan shell resembles enameloid, the ectodermal component of vertebrate integumental skeletons (D. F. Travis, 1970).

Uncalcified ectodermal skeletons occur in the form of radulae in gastropods and cephalopods, and jaws in the latter class. The peculiar anatomy of the radula, as compared with teeth, whether calcified or not, may reflect the fact that it arises through the interaction between two epidermal layers; teeth are the outcome of mesodermal–ectodermal cooperation (cf. Moss, 1968).

Molluscs have no proper dermis, the space between the epidermis and the viscera being largely occupied by connective tissue and muscles. This is not the case in cephalopods, where there is a compact fibrillar and non-vascularized dermis, in which the chromatophores are localized (Lang, 1900).

Protochordata

Urochordata are covered with a tunic, located outside the epidermis, but no dermis is present. The unusual fact that the fibrillar element in the tunic is cellulose has already been mentioned. The vascularization of the tunic is a noteworthy feature (Brien, 1948).

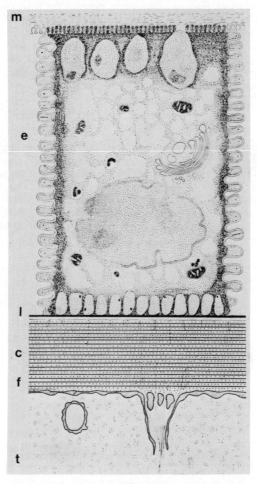

Figure 3.14. Diagrammatic representation of the skin of *Branchiostoma* (12,000 ×). c, dermis, consisting of piles of collagen fibres; e, epidermis; f, 'fibrocyte' layer; l, basement lamella; m, mucous surface layer; t, loose connective tissue. (Reproduced from R. Olsson, *Z. Zellforsch.*, **54** (1961), by permission of the author and Springer-Verlag)

The skin of *Branchiostoma* has been subject to a comprehensive study by Olsson (1961). The 'cuticle' was found to contain mucous substances together with protein. Inside the epidermal cells, close to the outer membrane, were found some filaments, with a mean diameter 9 nm, which were suggested by Olsson as being keratinous (epidermin ?). Below the epidermis there is a basement membrane and a basement lamella consisting of collagen fibrils arranged in alternating plies, beneath which is found loose connective tissue (Figure 3.14). Calcification is unknown in proto-chordates.

Vertebrata

Except for the otoliths, which consist of calcium phosphate (Carlström, 1963), no calcification occurs in cyclostomes. In *Myxine* the dermis is mainly occupied by collagen fibrils. Three layers may be distinguished, of which the one immediately beneath the basement membrane is characterized by the fact that the fibrils course at random. In the two deeper layers the fibrils are arranged in plies, parallel within one ply and perpendicular to those in the adjacent ones. The outermost of these layers consists of thinner plies than the deeper ones. The 100 μm thick dermis is not vascularized, but cells, chromatophores and 'fibroblasts' are present. Below the dermis is a sub-cutaneous tissue containing fat cells (Blackstad, 1963). Except for the finer details, this description also holds for the ammocoete larva of *Petromyzon* (Johnels, 1950).

Score

As is normal in invertebrates, the inorganic phase in the skeletons of echinoderms and molluscs consists of calcium carbonate.

A dermal skeleton is present in echinoderms and vertebrates, but the structural details are so different in the two taxa that it is not possible to make any inferences about affinities on this basis.

However, as mentioned above, the ectodermal exoskeletons in molluscs and vertebrates exhibit a number of similarities, which may be of significance even if the difference with respect to the inorganic phase, among other things, suggests that we are not dealing with a directly inherited character. The possession of 'horny' ectodermal skeletons is also a point of similarity between these two groups.

Finally, a dermis consisting of dense layers of collagen is present in cephalopods, cephalo-chordates and cyclostomes (Table 3.11).

Table 3.11
Distribution of characters related to skeletons

Echinodermata–Mollusca
$CaCO_3$
Mollusca–Vertebrata
Calcified ectodermal skeleton
Horny ectodermal skeleton
Structure of dermis
Mollusca–Cephalochordata–Vertebrata
Structure of dermis

Haemoglobins

The word 'haemoglobin' is sometimes used as a common name for the oxygen-binding iron–porphyrin proteins found in blood, muscles and other tissues. This meaning is implied in the heading to this section, but in what follows we shall distinguish, when possible, between 'haemoglobin' and 'myoglobin'. The distribution of haemoglobin and myoglobin in the annelid and echinoderm superphyla is shown in Table 3.12. As mentioned in the caption, occurrence is very erratic in the various taxa; this is particularly true for haemoglobin. The latter substance is known from other invertebrate phyla, distributed in a fashion which certainly makes little sense on the basis of present classification. Even if future systematic modifications may change the situation somewhat, it seems unavoidable to conclude that haemoglobin has arisen, or been lost, independently several times.

Table 3.12

Distribution of haemoglobin and myoglobin in the annelid and echino-derm superphyla

Haemoglobin present	In blood cells	In plasma
	Echinodermata	Annelida
	Holothuroidea	Polychaeta
		Oligochaeta
	Annelida	Hirudinea
	Polychaeta	
		Arthropoda
	Mollusca	Crustacea
	Bivalvia	Insecta
	Vertebrata	Mollusca
	All classes	Gastropoda
Haemoglobin absent	Urochordata	
	Cephalochordata	
Myoglobin present	Annelida	
	Oligochaeta	
	Mollusca	
	Polyplacophora	
	Gastropoda	
	Scaphopoda	
	Bivalvia	
	Vertebrata	
	All classes	
Myoglobin absent	Arthropoda	
	Urochordata	
	Mollusca	
	Cephalopoda	

Positive findings have been recorded on the class level. This does not imply that the presence of the sub-stance in question is a class characteristic; in most instances it is rather confined to a smaller taxonomic unit. The table, which is largely based on data compiled by Prosser and Brown (1961) and by Read (1966), may be incomplete since, in spite of the large amount of work done in this field, very few attempts have been made to produce comprehensive reviews.

Echinodermata

As appears from Table 3.12, haemoglobin is found only in Holothuroidea, but not in all members of the class. According to Florkin (1960a), it is present only in those species which have no test. The molecular weight has been estimated to 23000, a value suggesting that it occurs in the mono-meric form in the blood corpuscles. The peptide pattern of the haemoglobin in *Neothyone* shows that it is quite different from that of the vertebrates (Figure 3.15). Myoglobin has not been demonstrated.

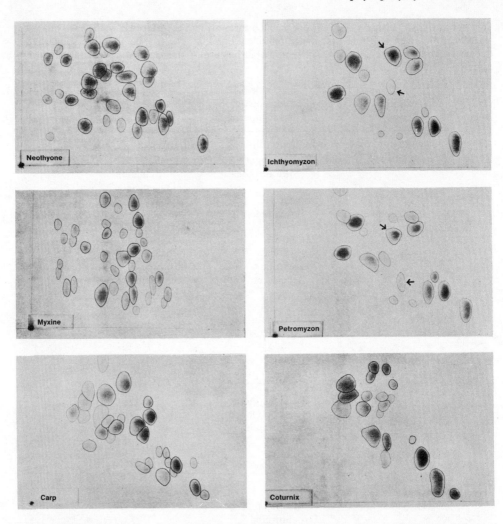

Figure 3.15. Peptide patterns for purified haemoglobin from *Neothyone* (*Thyonella*), *Myxine*, *Ichthyomyzon*, *Petromyzon*, *Coturnix* (Japanese quail) and *Cyprinus*. (Reproduced from C. Manwell (1963), in *The Biology of Myxine*, by permission of Universitetsforlaget, Oslo)

Mollusca

The usual respiratory pigment in the blood of molluscs is haemocyanin, but haemoglobin does occur in a few species of Gastropoda and in several of Bivalvia, in the latter case in blood corpuscles. Myoglobin is generally present in four of the molluscan classes, but remarkably enough it is absent in Cephalopoda (Read, 1966; and Table 3.12).

Roche and Fontaine (1940) first demonstrated that the contents of certain amino acids in haemoglobin (and chlorocruorin) in annelids are distinctly different from that observed in vertebrates, with the exception of cyclostomes. This comparison can now be extended with data on myoglobin from molluscs (Table 3.13) showing that the arginine content is typically high in annelids and low in molluscs and vertebrates, while the opposite holds for lysine. Although the values are less divergent for histidine, it nevertheless appears that this amino acid is lower in the invertebrates, cyclostomes and *Thunnus* than in the remaining vertebrates. The concentration of cystine varies too much to allow for any generalizations. It is also seen in Table 3.13 that among the molluscs myoglobin occurs in the monomeric or the dimeric form. In vertebrates myoglobin is always monomeric, while haemoglobin, except in the cyclostomes, is tetrameric.

Table 3.13
Amino acid contents in various haemoglobins and myoglobins

	Genus or species	Arginine	Histidine	Lysine	Cystine	MW
Haemoglobin	*Homo*[1]	3·3–4·0	8·1–8·4	8·0–9·6	—	68 000
Haemoglobin	*Homo*[2]	2·8	8·3	9·3	—	68 000
Myoglobin	*Homo*[2]	2·5	12·9	22·3	—	17 000
Haemoglobin	*Equus*[1]	3·6	8·7	8·5	0·5	68 000
Haemoglobin	*Equus*[3]	3·6	8·1	8·3	0·7	68 000
Myoglobin	*Equus*[1]	2·2	8·5	15·5	—	17 000
Haemoglobin	*Thunnus*[4]	5·6	4·1	8·6	—	68 000
Haemoglobin	*Petromyzon*[3]	3·5	3·4	7·5	4·4	17 000
Haemoglobin	*Lampetra*[2]	2·9	1·0	10·4	—	17 000
Haemoglobin	*Lampetra*[7]	3·7	1·6	10·1	—	19 000
Haemoglobin	*Myxine*[8]	1·6–5·6	2·7–5·9	11·9–14·9	—	20 000–28 000
Myoglobin	*Busycon canaliculatum*[5]	3·4	5·0	15·5	1·6	31 000
Myoglobin	*B. contrarium*[6]	3·0	4·4	13·6	1·3	34 000
Myoglobin	*Aplysia*[5]	4·1	1·2	6·8	—	22 600
Myoglobin	*Siphonaria*[6]	3·6	2·4	11·0	0·6	17 500
Haemoglobin	*Phacoides*[5]	2·1	1·9	10·5	0·7	14 700
Haemoglobin	*Arenicola*[3]	10·0	4·0	1·9	4·1	Around 30 000 000
Haemoglobin	*Glycera*[3]	9·6	5·4	4·9	3·4	
Haemoglobin	*Lumbricus*[3]	10·1	4·7	1·7	1·5	
Chlorocruorin	*Sabella*[3]	9·6	2·4	3·6	1·6	

1, Prosser and Brown (1961); 2, Allison *et al.* (1960); 3, Roche and Fontaine (1940); 4, De Marco and Antonini (1958); 5, Read (1966); 6, Read (1968); 7, Braunitzer and Fujiki (1969); 8, Paléus and Liljequist (1972). The values from Allison *et al.* (1960), Roche and Fontaine (1940) and those recalculated from Read (1966; 1968), Braunitzer and Fujiki (1969) and Paléus and Liljequist (1972) represent percentage weight. Those from Prosser and Brown (1961) are stated to represent percentage of total nitrogen. Note that one chlorocruorin is included.

Vertebrata

Since neither myoglobin nor haemoglobin appears to occur in Protochordata, all that remains to be discussed is the situation in Vertebrata. In this subphylum myoglobin and haemoglobin are ubiquitous apart from some phylogenetically unimportant instances. Haemoglobin is always present in blood corpuscles. The amino acid composition of vertebrate haemoglobin and myoglobin (Table 3.13) has already been evaluated.

As shown in Figure 3.15, the peptide patterns of haemoglobin in *Ichthyomyzon* and *Petromyzon* are quite similar to those of the higher vertebrates, while that of *Myxine* is as different from these as from that of *Neothyone*.

The unique position of Hyperotreta *vis-à-vis* Hyperoartii is borne out by amino acid analyses on haemoglobin (Braunitzer and Fujiki, 1969; Paléus and Liljequist, 1972), but even so the values from *Myxine* are seen to lie in the predicted range (Table 3.13).

Score

The amino acid composition of haemoglobin in holothuroids is unknown, unfortunately, but the results illustrated in Figure 3.15 suggest that it is very different from that of the vertebrates. The presence of myoglobin in four classes of Mollusca and the similarities in amino acid composition between this protein and the myoglobin and haemoglobin in vertebrates may be regarded as two features uniting these two groups.

The absence in protochordates of myoglobin, haemoglobin and any other respiratory pigments suggests affiliation with the invertebrate phyla rather than with Vertebrata (Table 3.14).

Table 3.14
Distribution of haemoglobin

Mollusca–Vertebrata
  Myoglobin
  Amino acid composition
Echinodermata–Urochordata–Mollusca
  Absence of haemoglobin
Echinodermata–Cephalochordata–Mollusca
  Absence of haemoglobin

Phosphagens

The systematic distribution of the phosphagens, arginine phosphate (AP) and creatine phosphate (CP), has for many years been the showpiece of comparative biochemistry. The early observations suggested that AP was a typical invertebrate, CP an equally typical vertebrate, phosphagen. The investigations of Needham *et al.* (1931) and Baldwin and Needham (1937) concerning the occurrence of CP in the echinoderm superphylum showed the presence of this substance in Echinoidea and Ophiuroidea, Hemichordata and Cephalochordata, and its absence in the remaining classes of Echinodermata and in Urochordata. These findings were taken to support Bateson–Grobben's theory postulating the origin of vertebrates from the echinoderm superphylum.

In the last couple of decades this problem has been the subject of renewed studies from which, partly owing to improved techniques, a much more varied picture has emerged. One thing which has been emphasized is that the mere occurrence of a substance, although of interest *per se*, must nevertheless be supplemented with information about the way in which it is acquired by the organism, through synthesis or otherwise. This question has been scrutinized by Stephens, Van Pilsum and Taylor (1965). As shown in Table 3.15, these authors investigated the presence of the enzyme *L*-arginine: glycine amidino transferase, catalysing the reaction

$$\text{arginine} + \text{glycine} \rightleftarrows \text{ornithine} + \text{guanidinoacetate,}$$

of the product guanidinoacetate and of the methylation product of the latter, creatine, plus creatine phosphate. The amidino transferase is 'a key enzyme in the apparently unique synthesis pathway of creatine which is found in mammals' (Stephens *et al.*, 1965, p. 574). The investigated animals are seen to fall into three subgroups, one in which creatine and CP are present but no enzyme, one in which the latter occurs together with guanidinoacetate, and one where all analyses were negative. It would appear that one more group might be anticipated, viz., one in which both enzyme and creatine were present, but, fortuitously perhaps, no representative of this alternative, otherwise typical of Vertebrata, was among those investigated.

The presence of creatine in animals unable to synthesize it seems rather surprising, but Stephens *et al.* (1965) showed convincingly that in *Glycera*, at least, this substance may be accumulated from the surroundings even when it is present in very low concentrations. It is of interest that *Saccoglossus* (Hemichordata) belongs to those animals which have creatine without being able to synthesize it.

In the following subsections we shall investigate the possible phylogenetic implications of the distribution of phosphagens in Echinodermata, Mollusca and Protochordata in relation to the universal presence of CP in Vertebrata.

Table 3.15

Distribution of creatine and creatine precursors in some invertebrates (Stephens *et al.*, 1965)

Phylum	Genus	Amidino transferase activity[1]	Guanidino-acetate[2]	Creatine + CP[3]
Annelida	*Glycera*	—	—	500
	Diopatra	—	—	100–200
Hemichordata	*Saccoglossus*	—	—	30
Urochordata	*Styela*	—	—	15– 35
Cnidaria	*Metridium*	100	+	—
Annelida	*Nereis*	250	+	—
	Lepidonotus	900	+	—
Mollusca	*Venus*	300	+	—
Echinodermata	*Thyone*	—	—	—
	Leptosynapta	—	—	—
Annelida	*Amphitrite*	—	—	—
Sipuncula	*Golfingia*			
Mollusca	*Spisula*			
Urochordata	*Ciona*	—	—	—
	Aplidium	—	—	—

1, μg guanidinoacetate acid formed per g tissue in six hours; 2, chromatographic demonstration; 3, mg creatine per 100 g wet weight.

Echinodermata

As already mentioned, Needham *et al.* (1931) and Baldwin and Needham (1937) found CP in Echinoidea and Ophiuroidea, in the former case together with AP, while AP alone was represented in the remaining three classes of Echinodermata. This has been largely confirmed by later investigations (Table 3.15), but with some remarkable supplementary observations. Thus, according to Roche, Thoai and Robin (1957), AP is also present in two investigated species of Ophiuroidea (*Amphipholis* and *Ophiothrix*), establishing the interesting generalization that AP is present in all classes. Furthermore, these authors found that in Asteroidea and Holothuroidea CP was present in the mature male genital glands, and they suggested, on the basis of their observations in many other invertebrates, that CP is present in flagellated, but absent in non-flagellated (amoeboid), spermatozoa. It is difficult to evaluate the significance of this finding, which may indicate that creatine is either synthesized or specifically accumulated by the spermatozoa. Rockstein (1971) has published results showing that both AP and CP may be present in all the four major classes of Echinodermata.

Mollusca

Like the annelids, the molluscs are distinguished by their ability to synthesize a variety of guanidino compounds (Florkin, 1966b). However, whereas the former have been able to use several of these as phosphagens (Roche *et al.*, 1960), only AP has ever been demonstrated in the latter phylum. Only the three major classes, Bivalvia, Gastropoda and Cephalopoda, have been investigated (Roche *et al.*, 1957; 1960; Ennor and Morrison, 1958; Robertson, 1965; Stephens *et al.*, 1965), and the number of species in each is not very large, so it cannot be excluded that further search may lead to revision of this generalization, but for the time being we are forced to uphold it. It must be stressed, however, that in *Venus* (Bivalvia) the 'key enzyme' in creatine production has been demonstrated by Stephens *et al.* (1965), as shown in Table 3.15. Since the enzymatic methylation of guanidinoacetate to creatine is potentially a simple matter, the reason for the lack of CP must be found elsewhere. Apparently the explanation lies in the specificity of the phosphorylating enzyme, which in Mollusca is a rather specific ATP: *L*-arginine phospho-transferase, reacting only with arginine (Campbell and Bishop, 1970). This finding is in striking contrast to observations made on the corresponding enzyme in annelids, which may phosphorylate a number of different guanidino derivatives (Watts and Watts, 1968).

It may finally be noted that, regrettably enough, Roche *et al.* (1957) did not investigate the nature of the phosphagens present in the sperm of any molluscs.

Protochordata

The work of Stephens *et al.* (1965) confirms the previously observed absence of CP in urochordates, with the amendment that creatine is used as a phosphagen in *Styela*, although it cannot itself synthesize the compound. Unfortunately, we do not know whether this situation exists in cephalochordates, or whether creatine is actually elaborated by the animals.

Score

The original suggestion concerning the phylogenetic significance of the distribution of phosphagens has been much criticized (e.g. Yudkin, 1954; Roche *et al.*, 1957; Kerkut, 1960; Rockstein, 1971). The recent studies by Stephens *et al.* (1965), stressing the distinction between uptake and synthesis of the creatine used for CP formation, only strengthen the grounds for this criticism. It therefore seems impossible to use the distribution of phosphagens as a superphyletic character.

It may be mentioned, however, that the observations on urochordates do not support their affiliation with the vertebrates, whereas the active synthesis of guanidino compounds in annelids and molluscs may in my opinion suggest a relationship between Vertebrata and the annelid superphylum.

Respiration

The subject of respiration covers several aspects, of which the most important may be the possible nature of specialized structures involved in oxygen uptake, the mechanical transport of the oxygenated blood or body fluids to the various parts of the body, the form in which the oxygen is transported (physical solution or chemical binding) and, finally, the regulation of oxygen uptake. Of these only the first is dealt with here, while some of the other problems are discussed in later sections.

Echinodermata

A large part of the oxygen uptake in echinoderms occurs through the body wall, but various kinds of structures, respiratory tree, tentacles, tube feet, peristomeal gills, etc., although not involved exclusively in oxygen uptake, may serve to increase the rate of exchange with the environment. The peristomeal gills present in most echinoids are so small that their contribution to oxygen uptake must be very modest. The respiratory tree in holothuroids is, however, of great importance, being responsible for about 50 per cent of the oxygen uptake (Farmanfarmaian, 1966). The active contractions which ensure renewal of the sea water in the respiratory tree are coordinated and controlled by the nervous system.

The haemal system does not appear to be involved in the transport of oxygen. This function is exerted rather by the coelomic fluid, propelled about by cilia and by muscular movements in various organs. Haemoglobin-containing cells are present in the coelomic fluid of some holothuroids; their function is not clearly established, but some evidence suggests that they may actually be involved in the storage and transport of oxygen. Most echinoderms seem to be oxygen conformers, i.e. their oxygen consumption varies with the ambient oxygen tension, but this is probably not the case with holothuroids.

Mollusca

Cutaneous respiration is common in molluscs; in certain species it is the only means of oxygen uptake. As a curiosity we may mention the development of lungs in Pulmonata. However, the most usual respiratory organ in molluscs is the ctenidia, or gills, highly branched and vascularized structures. The renewal of the water bathing the gills is ensured by currents created by cilia covering parts of the gill surface. In cephalopods this exchange is ensured by muscular respiratory movements under nervous control; the concentration of carbon dioxide has been found to influence their frequency (Ghiretti, 1966).

Protochordata

Oxygen uptake in urochordates occurs through the ciliated pharyngeal gills and probably through the vascularized tunic also. It may be assumed that branchial and cutaneous respiration also occur in cephalochordates.

Vertebrata

In Vertebrata respiratory currents produced solely by ciliary movements occur only in the ammocoete larva. In all other cases muscular movements, in the gills, the mouth, etc., ensure that a current of water passes over the gill surfaces. Nervous control of the respiratory rate has been observed in many cases.

Score

Nervous control of muscular respiratory movements occurs in Holothuroidea, Cephalopoda and Vertebrata. Although the mechanisms involved are much more sophisticated in the two last groups, this feature is common to all three taxa. The presence of vascularized, ciliated gills is a trait which unites Mollusca, Protochordata and Vertebrata (Table 3.16).

Table 3.16
Distribution of characters related to respiration

Mollusca–Vertebrata
Ciliated gills
Mollusca–Urochordata–Vertebrata
Ciliated gills
Mollusca–Cephalochordata–Vertebrata
Ciliated gills

Heart, blood and circulation

It would be of some interest to include, in the present context, a section on the comparative aspects of the biochemistry and physiology of muscles, but the observed variation is so great that it seems impossible to arrive at any important conclusions. However, it should be mentioned that the muscles of cephalopods and crustaceans in many ways resemble those of the vertebrates, for instance with respect to innervation, composition, etc. Fortunately, some information exists concerning the cardiac muscle, so the latter may be subject to comparative evaluation, together with various features related to the mechanism of blood circulation and the formed elements of the blood.

Echinodermata

A haemal system is present in all classes of Echinodermata, least developed in asteroids and most advanced in holothuroids. The haemal system usually consists of networks of lacunae in the gut wall, branched off from two longitudinal sinuses running along the margins of the intestine which may communicate with the axial complex. The channels of the haemal system are called 'lacunae' rather than 'vessels' because they lack an inner lining (Hyman, 1955).

There is not conclusive evidence of unidirectional vascular circulation in echinoderms, and the adult animals do not have a heart comparable to those found in the annelid superphylum, in Protochordata and in Vertebrata (J. M. Anderson, 1966; Farmanfarmaian, 1966). Rhythmic contractions, presumably myogenic, may be observed in the haemal system of *Stichopus*, an activity that is inhibited by acetyl choline (ACh) and accelerated by adrenaline (Takahashi, 1966).

The fluid in the haemal system is often referred to as 'blood', but since there may be communication with the coelomic cavities this designation is questionable.

Mollusca

A more or less perfect circulation of blood, propelled by one or more hearts, distinguishes most members of the annelid superphylum. High levels of development have been reached in some crustaceans and molluscs, but a closed circulation, involving the presence of a capillary network, is found in invertebrates primarily in annelids and in cephalopods.

The systemic heart in cephalopods consists of two auricles which pump the blood into the ventricle, whence it is conveyed to the cephalic artery and the posterior aorta. After passing through the tissues of the body, it is pumped through the gills by the branchial hearts and the oxygenated blood returned to the systemic heart (Hill and Welsh, 1966). Clearly, the efficiency of oxygenation is better in cephalopods than in most vertebrates and it is clear that this partly explains the speed of movement displayed by these animals and also the fact that animals of enormous size have evolved within this class.

Cross-striated fibres are found in the branchial hearts and in the ventricle, but not in the auricles. 'The myocardial fibers of molluscs are arranged in a complex network of branching and anastomosing trabeculae' (Hill and Welsh, 1966, p. 131), which in the cephalopods has a very compact structure and is traversed by capillaries (Alexandrowicz, 1960b). Aortic and atrio-ventricular valves are a common feature in many molluscs, as is a well-developed pericardium which, when closed, may be of importance for the diastolic filling of the heart (Krijgsman and Divaris, 1955).

The active parts of the circulatory system in cephalopods are, according to Johansen and Martin (1962), the systemic ventricle, the branchial hearts and certain rhythmically propulsive vessels. A certain passive pumping action is accomplished by the respiratory movements and by a 'windkessel', consisting of the larger arteries and the afferent branchial vessels. The latter effect is explained by the presence of elastic fibres in the walls of these vessels (Jullien, Cardot and Ripplinger, 1957). Owing to the presence of capillaries the circulation is essentially closed, apart from the existence of a few venous sinuses (von Buddenbrock, 1967).

The atria and the branchial hearts are innervated from the cardiac ganglia. The ventricle is innervated by the cardiac nerve, which contains some fibres from the cardiac ganglia, some from the visceral ganglia. Nerve cells have never been demonstrated in molluscan hearts, except in *Sepia* (Alexandrowicz, 1960b), and many lines of evidence suggest that the heartbeat proper is myogenic, but the presence of a pacemaker near the atrio-ventricular junction is indicated in the hearts of certain molluscs (Hill and Welsh, 1966). The nervous supply to the heart should then primarily be cardioregulatory, either inhibitory, or acceleratory, or both. The latter situation is usual, but not general, in cephalopods (Hill and Welsh, 1966).

It has been established that ACh inhibits the heartbeat in most molluscs, and the available evidence suggests that it is the normal regulatory agent. Among excitatory substances, 5-hydroxytryptamine (5-HT) is the most effective, affecting either amplitude, or frequency, or both. From what is known about neurotransmitters in molluscs, it might be reasonable to infer that this substance is the natural activator. However, the situation is more complicated than this, for besides 5-HT, another cardioexcitor, substance X, has been demonstrated in certain gastropods (Hill and Welsh, 1966). Furthermore, dopamine is present in some ganglia in various molluscs (Dahl *et al.*, 1962), and observations suggest that it may have a cardioinhibitory function (Hill and Welsh, 1966).

As far as blood cells are concerned, several types have been described in molluscs, intimating the presence of leucocytes, lymphoid cells, macrophages and eosinophilic granular amoebocytes, names that suggest a similarity to the blood cells found in vertebrates. As discussed earlier, erythrocytes are present in certain bivalves.

Protochordata

The heart in urochordates is a simple tube-like structure, surrounded by a pericardium. The wall usually consists of one layer of muscle cells, the basal parts of which contain myofibrils which may be cross-striated. The cells are arranged circularly or spirally, and in some cases longitudinal fibres may be present (Krijgsman, 1956). According to Brien (1948), the cells in some cases may form syncytia. An endocardium is not really present, nor are valves.

Contractions take the form of peristaltic waves passing along the tube. The pressure produced is very low, probably about 2 cm H_2O, but the resistance to be overcome is not great, the blood being pumped from the anterior end of the heart into the mantle and the branchial sac which contain a system of anastomosing sinuses or lacunae, and then to the visceral mass and back to the heart (Krijgsman, 1956). It is sometimes asserted that a capillary net is present in the haemocoelic mesenchyme or visceral mass, but according to Brien (1948) these vessels are not true capillaries, since no endothelium is present, their delimination being the work of 'mesenchyme' cells.

The direction of beat is periodically reversed, a fact which is attributed to the presence of two pacemakers, one at each end of the heart, but all parts of the heart may beat autonomously. It has not been possible to decide whether the beat is myogenic or neurogenic on the basis of either histological or pharmacological observations. The urochordate heart is remarkably inert to drugs of various kinds, including ACh and adrenaline, and to changes in the ionic composition, and the effects which have been obtained do not warrant any definite conclusions. A conductive system is probably present in the heart, but no extracardiac nervous regulation seems to be present. On the basis of these various observations Krijgsman (1956, p. 309) 'is inclined to consider the heart as having myogenic, non-innervated pacemakers. This reminds us of the embryonic hearts of a number of arthropods and of the exceptional heart mechanism in adult *Artemia* and *Eubranchipus*'.

Branchiostoma has no heart, but circulation in the essentially closed circulatory system is ensured by contractions in some of the larger vessels. The contractions are irregular and slow. The existing minimum coordination necessary for circulation occurs through propagation, involving a contraction which provokes a passive dilatation of neighbouring parts, which in turn elicits their contraction, a mechanism also known from Mollusca. The direction of flow is occasionally reversed, an occurrence which is possible owing to the absence of valves. No pacemakers have been demonstrated (Drach, 1948; von Buddenbrock, 1967).

The blood cells in protochordates do not seem to exhibit any remarkable features.

Vertebrata

The blood circulation in cyclostomes displays many dissimilarities as compared with the higher vertebrate classes, but also between the two orders.

In *Myxine* (Johansen, 1963) the propulsion of the blood is effected not only by the systemic heart, but also by a number of accessory hearts. The former is enclosed in a pericardium which opens to the coelom and consists of three chambers, sinus venosus, atrium and ventricle. Valves are present between these compartments, at the exit from the ventricle and at various other locations in the circulatory system. The ventricular myocardium is of a dense, trabeculate structure, a feature which is common to most of the lower vertebrates.

The heart pumps deoxygenated blood to the gills, the striated muscles of which assist in pumping the blood towards the dorsal aorta. The gills thus function as branchial hearts. There is also a caudal heart, of a unique construction, and a portal heart. The vascular system of *Myxine* contains true capillaries at many locations, but characteristically arteries and veins are connected through sinuses.

The heart in the hagfish is aneural, the beat thus being myogenic. A diffuse, pressure-sensitive pacemaker seems to be present. A cardioaccelerator effect arising from increased pressure or tension may be observed. The heart is not vascularized, but the spongy nature of the myocardium may allow for nutritional supply directly from the blood passing through the heart (Jensen, 1961).

This situation is similar to that which is found in the lower molluscs, although there the endo-cardium is missing (Hill and Welsh, 1966). The heart of *Myxine* is remarkably insensitive to various drugs, including adrenaline, noradrenaline and ACh. Part of the explanation of this phenomenon seems to be that relatively large amounts of catecholamines are present in the heart, as indicated by sensitivity following treatment with reserpine, a substance that causes a release of stored catecholamines (Bloom, Östlund and Fänge, 1963).

The presence of a wide repertoire of specialized cells, including of course erythrocytes, has been described by Holmgren (1950).

The heart in Hyperoartii, consisting of the same three compartments as in Hyperotreta, is completely enclosed in a cartilaginous pericardium, and no accessory hearts are present. The heart is innervated by nervus vagus and contains ganglion cells; a network of sympathetic nerves is found in the sinus venosus. Stimulation of vagus results in an acceleration of the heartbeat. The heart of *Petromyzon* reacts to catecholamines with a slight positive inotropic and chronotropic effect. ACh also increases the rate of beat, but with a negative inotropic response (Bloom *et al.*, 1963).

It should finally be mentioned that in chondrichthyans and in osteichthyans no sympathetic innervation obtains; various kinds of stimulation, for instance directly of the vagus, lead to standstill or slowing of the heartbeat. Perfused ACh has only a negative chronotropic effect: apparently the inotropic fibres are not cholinergic (Prosser and Brown, 1961).

Score

From the data reviewed above it appears that the influence of ACh may be common to all three taxa, and so may the presence of sinuses and the fact that the hearts are myogenic. If we assume that adrenaline is the natural cardioaccelerator substance in Echinodermata, then this seems to be the only character which the latter have in common with Vertebrata and not with Mullusca.

Otherwise, it must be concluded that the so-called 'haemal system' in Echinodermata has very little in common with a circulatory system such as is found in Mollusca, particularly in Cephalo-poda, and in Vertebrata. Many similarities do, indeed, obtain between these two of the three main taxa. Thus, we may first observe the possession of a chambered heart, in which the myocardium

Table 3.17
Distribution of characters related to heart,
blood and circulation

Mollusca–Vertebrata
  Chambered heart
  Myocardial structure
  Pericardium
  Accessory hearts
  Blood vessels
  Valves in hearts and vessels
  Elastic fibres in arteries
  Capillaries
  Nervous regulation of heart
  Specialized blood cells
Echinodernata–Vertebrata
  Adrenaline
Mollusca–Urochordata–Vertebrata
  Pericardium
  Vessels
Mollusca–Cephalochordata–Vertebrata
  Accessory 'hearts'
  Vessels
  Capillaries

---

has a trabeculated structure, surrounded by a pericardium. The prrsence of accessory hearts is a typical feature in Cephalopoda and Hyperotreta, and valves are common in Mollusca and Vertebrata. We may further note the presence of true blood vessels containing elastic fibres, thus enabling them to participate in the propulsion of blood through their function as 'windkessels'. Blood capillaries are present in cephalopods and in all vertebrates. Nervous regulation of the myogenic heart occurs in Mollusca and Vertebrata, except Hyperotreta, and has reached a state of perfection in Cephalopoda the equivalent of which is not encountered below Amphibia among Vertebrata. Finally, the presence of several types of blood cells with affinities to those found in Vertebrata should be mentioned. Only a few of these characters are found in Protochordata, as is seen from Table 3.17.

Feeding and digestion

Ciliary filter mechanisms for the collection of food are found in both protochordates and in the ammocoete larva of Hyperoartii, and even in some of the extinct early vertebrates (Stensiö, 1958). This fact, and especially the structural similarities between the endostyle in the living taxa, have generally been used as arguments in favour of kinship between the protochordates and the vertebrates. The circumstance that filter-feeding also occurs quite generally in echinoderms might then, of course, be taken to support the present superphyletic classification. However, filter-feeding is also common in molluscs, particularly in gastropods and in bivalves, and incidentally in many other invertebrate taxa. In fact, Orton (1914) pointed out the great structural similarities between the filter mechanisms in molluscs and cephalochordates, and even went so far as to talk about an 'endostyle' in the former case. It therefore seems that this argument is anything but convincing.

A number of other feeding mechanisms are known in echinoderms, none of which bears any resemblance to those found in vertebrates, but possibly to some found in molluscs. A very important structure for feeding in many molluscs is the radula, formed by a sclerotized ectodermal differentiation product.

Digestion may be intracellular or extracellular. In the former case, representing the most primitive stage, food particles are taken up by amoebocytes in the gut wall through phagocytosis and degraded by intracellular enzymes. In the latter case enzymes, secreted into the digestive tract, hydrolyse the food into small molecules that are subsequently resorbed by the gut cells. Both kinds of digestion occur in echinoderms and in molluscs (Yonge, 1937), except in cephalopods where digestion is exclusively extracellular (Bidder, 1950). The same holds for urochordates and vertebrates (Yonge, 1937) and for cephalochordates (Barrington, 1938).

From this survey it is seen that the characters discussed here do not contribute significantly to the establishment of phylogenetic relationships.

Liver and pancreas

The tissues of the liver and the pancreas represent two very typical endodermal differentiation patterns in the vertebrates. In the present section we shall trace the occurrence of similar differentiation patterns in the other animal groups.

Echinodermata

The only tissue in echinoderms which may remotely be compared with that of liver occurs in the pyloric or hepatic caeca in asteroids. In these are found secretory and storage cells; the latter may contain lipids and small amounts of glycogen. Absorption also occurs in the caeca (J. M. Anderson, 1966).

Mollusca

All molluscs, as well as members of the other groups in the annelid superphylum, possess digestive diverticula (hepatopancreas), which are assumed to produce various digestive enzymes and to be involved in absorption, phagocytosis, food storage and excretion (Owen, 1966).

In cephalopods two different tissues can be distinguished, in some cases united in one organ and in some cases separate, to which the names 'liver' and 'pancreas' are applied. It appears that the pancreas elaborates a number of digestive enzymes, a function similar to the exocrine function of the vertebrate organ. The liver also secretes digestive enzymes, but at the same time it serves as an excretory organ (Bidder, 1966), as well as for storage of lipids and glycogen (Arvy, 1960). Food absorption may occur in the livers of *Octopus* and *Sepia*, but not in *Loligo* (Bidder, 1966).

Protochordata

The most detailed investigation of the structure and function of the digestive system of *Branchiostoma* has been published by Barrington (1938). Regarding the midgut diverticulum, which has been characterized by various authors as a homologue of the liver in vertebrates, Barrington made a careful scrutiny in order to settle this point. Among the characteristic liver functions, storage of glycogen and, to some degree, of fat, urea production and formation and excretion of bile pigments, this author could find only some indication of fat storage. The histology of the diverticulum exhibits many similarities with the midgut proper, and various lines of evidence indicate that the lining cells produce and secrete digestive enzymes. Altogether, Barrington (1938, p. 307) arrived at the conclusion that 'there is as yet no sound support for the interpretation of the diverticulum as homologous with the Craniate liver'.

Comparing the diverticulum with similar formations present in some, but not all, ammocoete larvae, and showing that cells of this type elaborate trypsin, Barrington (p. 308) asserted that 'it would be possible to compare the diverticulum of Amphioxus with the pancreas of the Craniata at least as plausibly as with the liver'.

Glycogen accumulation has been observed in certain cells in the intestine of *Ciona* (Yonge, 1925), but there is no indication that the 'liver' present in stolidobranchiates (Barrington, 1965) has any of the functions normally attributed to the vertebrate liver; rather, it appears to be a secretory organ (Berrill, 1929), thus perhaps similar to the midgut diverticulum in *Branchiostoma*.

Vertebrata

The liver, with associated gall bladder, is already well developed in *Myxine*, suggesting that the main functions of the vertebrate liver have been assumed in this animal. The endocrine and exocrine functions of the pancreas are separated, the tissues responsible for the latter being present in three different locations (Adam, 1963a).

Score

The presence in asteroids of a tissue with some of the functions of the vertebrate liver is acknowledged, but both the anatomy and the physiology of the hepatopancreas in molluscs is clearly much closer to that of the latter. In particular, the subdivision into two distinct tissues, as found especially in cephalopods, must be accepted as a point of affinity with the vertebrates, even though many dissimilarities exist. Nothing like a liver is found in protochordates. It may possibly be granted that pancreatic tissue is present, but in its primitive state it must be considered to be as close to that of molluscs as to that of vertebrates (Table 3.18).

Table 3.18
Distribution of liver and pancreas tissue

Mollusca–Vertebrata
Liver tissue
Pancreas tissue
Mollusca–Urochordata–Vertebrata
Pancreas tissue
Mollusca–Cephalochordata–Vertebrata
Pancreas tissue

Osmoregulation and kidney function

The present section is concerned with the non-gaseous exchange between animals and their surroundings. This function must for obvious reasons occur either through the integument, limited parts of the body surface or a more or less specialized kidney. From the terrestrial vertebrates it is known that some of the main functions of the latter organ are the regulation of water content, or tonicity, and the ionic composition of the blood, and the excretion of nitrogenous wastes. The history of the kidney may be traced back to various kinds of nephridia in the lower invertebrates. Since many of the latter are supposed to excrete their nitrogen as ammonia, which requires no special precautions, and are isosmotic with their environment, it appears that the original 'kidney' function may have been to regulate the ionic composition (Morris, 1960).

Echinodermata

The members of this phylum are generally considered as exclusively marine and stenohaline. Representatives of the four main classes are found in brackish water, but these are exceptional cases which probably constitute relics surviving from periods when the ambient medium was of higher tonicity. Active penetration into brackish water never occurs (Binyon, 1966). Experiments have shown that a certain tolerance obtains towards diluted media, a fact which may warrant classification of these animals as euryhaline. However this may be, all evidence suggests that echinoderms are osmoconformers, adjusting their osmotic pressure in accordance with that of the surroundings. This equilibration supposedly takes place across the integument, since they lack a morphologically differentiated excretory organ.

Some ionic regulation of the perivisceral fluid may occur in asteroids, involving a slight increase in potassium and calcium and a reduction in magnesium. In echinoids only calcium is regulated, often substantially above the sea-water value, while little regulation is observed in holothuroids. More extensive ionic regulation occurs in the ambulacral fluid, involving potassium in particular but also calcium and magnesium, and to a very slight extent sulphate, suggesting an activity residing in the cell walls of the tube feet. An exchange reaction between ammonia and potassium cannot be excluded (Binyon, 1966).

It has been found that in many animals, including representatives of echinoderms, the intracellular osmotic pressure is made up partly by amino acids and taurine, partly by salts. Under hypotonic conditions, the intracellular concentration in euryhaline animals, whether osmoconformers or osmoregulators, is adjusted to that of the extracellular fluids. Since part of this equilibration involves the organic nitrogenous substances, an excessive reduction of the intracellular salt concentration is prevented (Florkin, 1962).

Mollusca

Observations on various species of Bivalvia, Gastropoda and Cephalopoda show that the blood of marine molluscs is isosmotic with sea water within 1–2 per cent. Some marine species, particularly among bivalves, are euryhaline and may be encountered at quite low salinities. Certain molluscs have successfully colonized brackish and fresh water. Those living in the former habitat are usually osmoconformers. In fresh-water forms osmoregulation occurs, but the osmotic pressure of the blood is very low (Robertson, 1966b).

Ionic regulation takes place in the three classes mentioned above, particularly in cephalopods, which regulate all ions. The potassium concentration is always higher than in sea water, often twice as high, calcium is slightly higher and sulphate either unregulated or lower, except in one investigated species of *Mytilus*. Sodium and chloride are sometimes lowered, particularly in cephalopods, but this may be a secondary effect of the regulation of other ions (Robertson, 1966b). Protonephridia are found in the larvae of many molluscs, but they always disappear during subsequent development.

In adult molluscs an essential part of ionic regulation is carried out by the renal organs, which function through a combination of filtration, secretion and reabsorption. Renal fluid is produced as an ultrafiltrate in the pericardium. Reabsorption of water occurs in terrestrial gastropods, but in marine molluscs the urine is hypotonic owing to reabsorption of salts (Martin and Harrison, 1966). In cephalopods glucose, potassium, calcium, magnesium and chloride are reabsorbed, while sodium and sulphate are secreted (Robertson, 1966b). Active secretion of p-aminohippuric acid and phenol red has been demonstrated in cephalopods and gastropods (Martin and Harrison, 1966). In fresh-water forms salts are lost from the body in association with ionic regulation, even though this loss is cut through reabsorption in the renal organ. Replacement is ensured through absorption, particularly of sodium and chloride, by epithelial cells, possibly located in the gills. A similar state of affairs may be observed in many other members of the annelid superphylum that have penetrated into fresh water (Krogh, 1939; Morris, 1960).

In spite of osmotic equilibrium between the cells and the internal medium the ash content is lower in the tissues, because about half of the intracellular osmotic pressure in various molluscs is accounted for by organic nitrogenous substances. In the lower molluscs amino acids are prevalent, but in cephalopods trimethylamine oxide (TMAO) and betaine occur in the muscles in high concentration; in the nerve cells isethionic acid, a hydroxy analogue of taurine, is of particular importance.

Protochordata

In urochordates the body fluids are in osmotic equilibrium with the external medium. Ionic regulation occurs to some extent, thus the potassium may be slightly higher and the sulphate concentration considerably lower. No tubular kidney is present (Robertson, 1954).

The presence of nephridia in cephalochordates is one of the features which has been very difficult to account for by those who believe in an affinity between Protochordata and Vertebrata. It has been shown that dyes may be excreted by the nephridia, but otherwise no information is available concerning their function. No osmoregulation seems to occur (Robertson, 1963), but some ionic regulation is likely.

Vertebrata

Among vertebrates, the only animals having an ionic concentration identical with that of sea water are the members of the order Hyperotreta (McFarland and Munz, 1958). Hagfishes do not tolerate great changes in the tonicity of the external medium, especially not dilutions. Ionic regulation occurs, the values for calcium, magnesium and sulphate being significantly lower in the blood than in sea water (Robertson, 1963).

In *Myxine* both a pronephros and a mesonephros are present. The function of the former is not quite certain; it appears to be haemopoietic, but may also be an endocrine gland (Holmgren, 1950; Fänge, 1963). Of particular interest in the present context is its association with the pericardium, a feature known from molluscs. The segmented mesonephros contains glomeruli and is atubular; it is generally believed that the various regulative functions are executed by the ureter (Fänge, 1963). The literature is slightly inconsistent about these activities, but it appears that the urine is isotonic with the serum (Munz and McFarland, 1964; Rall and Burger, 1967). The former authors found that potassium, magnesium, sulphate and phosphate are higher in the urine, indicating that secretion occurs, and the lower concentration of glucose was taken to demonstrate reabsorption of this substance. Rall and Burger contend that secretion of calcium, magnesium and phenol red occurs through the bile. The very low value of magnesium in the blood is a typical vertebrate feature (Robertson, 1954).

Bellamy and Jones (1961) have shown that the tissue cells of *Myxine* contain organic nitrogenous substances, notably amino acids, TMAO and some unknown substances, accounting for about half of the osmotic pressure.

In the other aquatic vertebrates, except chondrichthyans and *Latimeria*, the blood is either hyposmotic in marine animals or strongly hyperosmotic in fresh-water animals.

Score

Some ionic regulation occurs in all the groups discussed, but Echinodermata, Urochordata and most members of the lower classes of Mollusca, and probably also Cephalochordata, are distinguished by their relatively low power of ionic regulation, which is mainly confined to potassium. Cephalopods and some gastropods have a greater power of ionic regulation, although inferior to that found in the hagfishes.

Common to vertebrates and some molluscs is the fact that a special organ, a 'kidney', is involved in the regulation, as are the processes of filtration, absorption and secretion. Besides the kidney, the gills also take part, but this can be considered a specialization of a common property of epithelial cells, and these may generally be engaged in ionic regulation.

The similarity between the composition of the muscle cells of *Myxine* and those of molluscs and decapod crustaceans, both with respect to ionic composition and the content of amino acids and TMAO, has been pointed out by Robertson (1963; 1966a), but this mechanism of intracellular osmoregulation seems to be matched by that observed in echinoderms.

Finally, we should note the presence of protonephridia in cephalochordates, a phenomenon which definitely suggests affiliation with the annelid superphylum (Table 3.19).

Table 3.19
Distribution of characters related to osmoregulation and kidney function

---

·Mollusca–Vertebrata
    Regulation of several ions
    'Kidney'
    Filtration
    Absorption
    Secretion
Mollusca–Cephalochordata
    Protonephridia

Nitrogen excretion

In the animal kingdom several methods are used to eliminate the excess nitrogen contained in the food. When these are considered against a phylogenetic background, it becomes clear that evolutionary innovations represent in part gains in metabolic efficiency (excretion of ammonia rather than of amino acids) and in part modifications which have been interpreted as detoxification mechanisms (excretion of urea and urate).

This is clearly very promising for the purpose of establishing phylogenetic relationships, and in the present section we shall investigate the extent to which this anticipation is met by actual observations.

Echinodermata

Among all marine invertebrates ammonia is the most important nitrogen waste product. Echinoderms appear to be no exception to this rule, ammonia generally being reported as the major nitrogen fraction excreted or present in the perivisceral fluid, but substantial amounts of amino acids are also found, as well as some urate (Delaunay, 1931; Boolootian, 1961). According to some investigators significant amounts of urea are also present (Cohnheim, 1901; Sanzo, 1907; Myers, 1920; Delaunay, 1931), but other authors could not corroborate this finding (Van der Heyde, 1923; Przylecki, 1926).

J. B. Lewis (1967) has recently studied nitrogen excretion in *Diadema*, finding that 61 per cent of the nitrogen in the surrounding sea water is ammonia and 26 per cent amino nitrogen. Urea, urate or other purines and creatine-creatinine could be demonstrated neither in the ambient water nor in the perivisceral fluid. The observation of these substances in other species of Echinoidea suggests that, when present, they occur in minute concentrations.

It is difficult to evaluate these contrasting reports, but the fact that uricase and allantoinase have been found in some echinoderms (Florkin, 1947) affords an explanation for the observed absence of purines, and the suggestion that these enzymes are not ubiquitous suffices to explain the presence of urate.

The situation is more complicated as regards the presence of urea. This substance may be formed either from dietary arginine, through uricolysis or through the ornithine cycle. When occurring in small amounts it can be accounted for on the assumption that arginase is present, but in some cases more than 10 per cent of the nitrogen has been reported to be urea (Delaunay, 1931; 1934) and, unless the food contains very large amounts of arginine, this value implies either uricolysis or presence of the ornithine cycle or some other mechanism for urea production. No investigations of this point have ever been made, but, since the most recent reports unanimously show that very little urea is present, it seems justified to conclude that echinoderms are not ureotelic.

Mollusca

Potts (1967) has written a comprehensive survey of excretion in molluscs, in which he points out that analyses have generally been limited to those substances known to be excreted in verte-brates. This may bias the present comparison to some extent, but is somewhat counteracted by the fact that this attitude has been prevalent in all studies of nitrogen excretion.

In conformance with their habitat, most aquatic molluscs excrete ammonia. In cephalopods the urine may contain up to about 75 per cent of the total non-protein nitrogen (NPN) as ammonia (Emmanuel and Martin, 1956). Potts (1965) found that ammonia is formed from the glutamine in the kidney, a process otherwise known in mammals (Campbell and Bishop, 1970). The latter authors comment that 'the octopus appears to be the first instance where there is an indication of the mammalian mechanism in an invertebrate and, should this prove true, it would be of con-siderable evolutionary significance' (1970, p. 182). This argument is hardly tenable, since much may have happened on the way from Cyclostomata to Mammalia, but as the same pathway is apparently present in the lower vertebrates, being demonstrable in fresh-water fish (Makarewicz, 1963), its occurrence may indeed be of importance for the tracing of phylogenetic relationships.

Excretion of ammonia has also been demonstrated in gastropods, both in marine and fresh-water species (Duerr, 1968; Bayne and Friedl, 1968; Haggag and Fouad, 1968). Ammonia pro-duction has even been reported in terrestrial snails (Haggag and Fouad, 1968; Speeg and Campbell, 1968), but this seems to be an unusual phenomenon. The latter authors were able to show that at least part of the ammonia arises through breakdown of urea by the action of urease, and that it is lost in the gaseous form. Ammonia production has also been demonstrated in bivalves (Lum and Hammen, 1964; Hammen, Miller and Geer, 1966). In *Modiolus* the ammonia nitrogen is twice as high as the amino-nitrogen fraction, and no urea is found. In *Crassostrea* about 65 per cent of total NPN is ammonia, 13 per cent urea, 5 per cent amino nitrogen and 17 per cent un-identified.

Urea excretion has been reported in both cephalopods and gastropods (Delaunay, 1931; Baldwin and Needham, 1934; Potts, 1967; Campbell and Bishop, 1970) and, as just mentioned, in bivalves. The findings in cephalopods were questioned by Potts (1967). Since urea excretion in *Helix* was somewhat at variance with their generalizations, Baldwin and Needham (1934) tried to account for this result by reference to the presence of arginase. The amounts of urea formed in certain cases are so large, however, that this explanation cannot be generally valid. In fact, some or all of the ornithine cycle enzymes may be demonstrated in several gastropods (Horne and Boonkoom, 1970). Yet, although urea is often excreted, it is very seldom the major nitrogen excre-tion product, so the animals are not ureotelic. For this reason Campbell and Bishop (1970) suggest that the primary function of the ornithine cycle in molluscs is to supply arginine, and that its utilization for nitrogen excretion is a later event. However this may be, the fact remains that the pathway used for urea synthesis in vertebrates is present in certain molluscs.

Excretion of purines has been reported in cephalopods, either in the form of urate in *Octopus vulgaris* and *Sepia* (Delaunay, 1931) or as hypoxanthine and guanine in *Octopus honkongensis* (Emmanuel, 1956; 1957). In view of the data discussed above, these substances seem to be the end-products of purine metabolism.

The presence of urate in terrestrial gastropods has been known for a long time and was suggested as representing an adaptation to the water shortage of their environment (Needham, 1938). The validity of this generalization has been contested by Duerr (1967), who found very high urate contents in several fresh-water and marine snails. Without entering into this controversy, it may just be mentioned that in certain cases, for example *Helix*, the nitrogen excretion may be accounted for almost completely by urate, although some creatine and creatinine was also observed (Jezewska, Gorzkowski and Heller, 1963). In other gastropods xanthine and guanine were found (Jezewska, 1968). Urate has been demonstrated in the kidneys of bivalves, but no definite data are available with respect to its excretion. Actually, in many species one or more of the uricolytic enzymes, including urease, have been found, so it is likely that the purines are often converted into ammonia before excretion (Florkin, 1966b).

It should be mentioned that excretion of amino nitrogen has been recorded in members of the three major classes of Mollusca. In some cases the values, expressed as per cent of NPN, are quite small, but they may amount to one-third of the total. Potts (1967), who has dealt with this question, points out that the highest values are found in marine species, in which amino acids account for about half the osmotic pressure of the tissues. It is thus possible that the amino nitrogen is not excreted, but simply lost from the superficial cells.

TMAO occurs sporadically in the tissues of bivalves and gastropods, but consistently in cephalopods (Norris and Benoit, 1945; Bricteux-Grégoire *et al.*, 1964; Robertson, 1965; cf. also Florkin, 1966b). This substance has never been recorded as an excretory product in molluscs, but according to Forster and Goldstein (1969, p. 322) 'a large fraction of that usually designated "undetermined" nonprotein nitrogen, at least in marine species [of fish], is TMAO', and this may also be the case in molluscs.

Finally, it is important to note that in the eggs of two fresh-water prosobranchs, *Marisa* and *Pomacea*, Sloan (1954) has demonstrated the formation of ammonia, urea and urate, the latter being the most important excretion product.

Protochordata

The only work in which the nitrogen excretion of urochordates has been studied quantitatively appears to be that of Goodbody (1957). Using three species of Ascidiacea (*Ascidiella*, *Ciona* and *Molgula*), this author found that, of the total NPN, ammonia may account for 90–95 per cent. This result shows that they are ammonotelic and also, since clearly little amino nitrogen is excreted, that the handling of their food materials is effcient. The absolute values for nitrogen excretion were found to be similar to those observed in various invertebrate phyla.

In many urochordates concretions are accumulated in the tissues, sometimes in vesicles around the digestive tract and sometimes in a renal sac. These concretions have generally been assumed to contain urate and other purines. This question has also been studied by Goodbody (1965), who found that in some cases (e.g. *Ascidiella*) the main constituent is calcium carbonate, but in others it is urate (e.g. *Ascidia* and *Phallusia*). Determinations of various uricolytic enzymes in 20 different species showed xanthine oxidase to be present in seven cases, and in one (*Polycarpa*) even uricase although, curiously enough, accumulation of urate could be demonstrated in this species. It thus appears that the end-product of purine metabolism in urochordates is either urate or else guanine and other purines.

Nothing is known about nitrogen excretion in cephalochordates, but there is reason to believe that they also are ammonotelic.

Vertebrata

Very little is known about nitrogen excretion in cyclostomes. However, the concentration of urea in the blood of *Myxine* is quite low (H. W. Smith, 1932; Robertson, 1954) and in *Lampetra* it is lower still (Robertson, 1954), so it is reasonable to presume that these animals are largely ammonotelic.

Owing to the use of urea in their osmoregulation, chondrichthyans are distinctly ureotelic, but small amounts of ammonia are excreted (Prosser and Brown, 1961). Osteichthyans are mostly ammonotelic, but often the excreted ammonia is only a small part of total NPN, and since in many cases substantial quantities of nitrogen remain unidentified, it is clear that the question of nitrogen excretion is far from being resolved (Prosser and Brown, 1961).

No attempts have been made to demonstrate the ornithine cycle enzymes in cyclostomes. The amounts of urea excreted by certain fish are so large that the presence of this pathway has been suggested (Grafflin and Gould, 1936). However, attempts to demonstrate the various enzymes in teleost livers have, apart from arginase, been unsuccessful (Cohen and Brown, 1960), suggesting that the main pathway of urea synthesis in these animals is uricolysis.

In some cases allantoicase is absent so that allantoic acid is the end-product of purine metabolism, but in some chondrichthyans and osteichthyans the enzymes responsible for the conversion of urate to urea have been found. However, in contrast to certain molluscs, urease is never present and hence urea is not transformed into ammonia (Florkin and Duchateau, 1943). Owing to the occurrence of uricolysis in chondrichthyans and osteichthyans, urate or other purines are excreted only in very small amounts (Prosser and Brown, 1961), and this situation may also be assumed to exist in cyclostomes.

TMAO is present in the muscle of *Myxine* (Robertson, 1966a) and in the blood and tissues of marine chondrichthyans and of some, but not all, osteichthyans (Norris and Benoit, 1945). As in cephalopods, it is supposed to be involved in intracellular osmoregulation (Forster and Goldstein, 1969). This substance has been reported among the excretion products in some cases (Prosser and Brown, 1961). The excretion appears to be low in chondrichthyans, where TMAO, like urea, is actively absorbed by the renal tubules in *Squalus* (Cohen, Krupp and Chidsey, 1958), in contrast to the situation prevailing in *Lophius* (Teleostei), where this substance is secreted (Forster, Berglund and Rennick, 1958).

Creatine and creatinine are also among the nitrogenous excretion products reported in teleosts.

Score

It appears that ammonia excretion is common to all the taxa, and this feature is therefore of no comparative value. The excretion, or loss, of amino acids is undoubtedly a primitive feature, being present only in echinoderms, bivalves and some gastropods. A complete uricolytic system may also be found in all three taxa, and possibly even in protochordates. The ornithine cycle is present in gastropods and in some, if not all, of the lower vertebrates. The use of TMAO in intracellular osmoregulation seems to unite Mollusca and Vertebrata.

The relatively confused picture which thus emerges is unquestionably partly a matter of loss mutations, involving various enzymes. Nevertheless, the relations set out in Table 3.20 seem to be reasonably well established.

Table 3.20
Distribution of characters related
to nitrogen excretion

Echinodermata–Mollusca
Amino acid excretion
Mollusca–Vertebrata
Ornithine cycle
Trimethylamine oxide

Visual pigments

Most animals are photosensitive owing to the presence of pigments, usually in specialized cells which function as photoreceptors. This differentiation pattern is absent in urochordates, indeed no records of photosensitivity seem to exist. In cephalochordates 'eye-spots' are present in the spinal cord and in the cerebral vesicle. In spite of this, the animals are not particularly photosensitive, even though intense light may provoke swimming movements (Barrington, 1965). The discussion may therefore be confined to Echinodermata, Mollusca and Vertebrata.

Echinodermata

Apart from some asteroids, the echinoderms have no specialized photoreceptors and it must be inferred that the easily observed photosensitivity resides in integumental (dermal) receptors. In the 'optic cushions' of *Marthasterias*, β-carotene and esterified astaxanthin are the most prominent pigments (Millott and Vevers, 1955). However, the role of these structures in photoreception is anything but settled (Yoshida, 1966).

Studies with monochromatic light in two species of *Diadema* have shown a peak of the action spectrum around 460 to 470 nm (Yoshida and Millott, 1960). This value corresponds to the absorption maximum of echinochrome A, which is present in abundance in the skin and the nervous system, but there is no evidence for the notion that this substance is acting as a visual pigment.

Mollusca

The most efficient visual systems known in the animal kingdom are those involving rhodopsins. These substances, found in the retinal sensory cells, are complexes between either one of the two carotenoid derivatives, retinene$_1$ and retinene$_2$, aldehydes corresponding to vitamin A$_1$ and A$_2$, respectively, in association with certain specific proteins called 'opsins'. Among all the possible isomers of the retinenes only one, the 11-*cis* or neo-b, can form visual pigments (Wald, 1960).

Visual systems of this type, based on A$_1$, are found in certain arthropods and molluscs. In *Loligo* it has been found that the light-induced reaction may be represented as follows (Hubbard and St. George, 1958):

$$\text{Rhodopsin} \overset{\text{Light}}{\rightleftarrows} \text{Metarhodopsin.}$$

In the eye of Limulus, Hubbard and Wald (1960) have found the photoreaction to be:

$$\text{Rhodopsin} \overset{\text{Light}}{\rightleftarrows} \text{Retinene}_1 + \text{opsin.}$$

Since the pigment can be derived only from vitamin A$_1$, the animals must possess the enzymes required for converting this substance to the neo-b isomer and reducing it to retinene. No alcohol dehydrogenase has been found in any invertebrate eye, so the mechanism of this conversion remains obscure. It is obvious, however, that under these circumstances the reversible oxidation and reduction of the visual pigment cannot be involved in the phenomenon of photic perception as is the case in the vertebrates (Wald, 1960).

Vertebrata

In the vertebrates visual systems containing either vitamin A$_1$ or A$_2$ or both are found in all taxa, except possibly the order Hyperotreta. The distribution between the two forms is such that a correlation between habitat and visual pigment is indicated. The phylogenetic tree established on this basis appears to be rather unorthodox (Wald, 1960), a fact that should not be used to challenge its validity; I believe that there are more weighty arguments mobilized for this purpose in the chapter on the divergence of Vertebrata.

The mechanism of the light-induced reaction is more complex than in invertebrates:

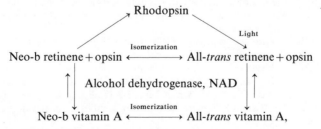

showing that alcohol dehydrogenase plays an important role.

Score

On the basis of the data which have been cursorily reviewed above, Wald made the following statement: 'The visual systems of few arthropods and molluscs have yet been analyzed chemically, but all of them so far examined are based upon vitamin A_1 and retinene$_1$ alone. It should be noted at once that this basic similarity in composition to the systems of vertebrate vision does not involve any genetic connection. The eyes of arthropods, molluscs and vertebrates are entirely separate developments, each the product of its own evolution. No anatomical or embryological homologies connect these organs with one another; nor does anyone now suppose that either arthropods or molluscs were ancestors of the vertebrates. That all three phyla, with their very different eyes, should have come to the same substance, vitamin A—indeed, neo-b vitamin A—as the foundation of their visual systems poses in its most acute form the question: what constitutes the particular *fitness* of this molecule for vision?' (1960, pp. 338–9).

This proposition is interesting for several reasons. Thus, the presence of the same visual system in the two invertebrate phyla is not taken to be related to a common origin, although they both belong to the annelid superphylum. Secondly, the firm assertion that the three visual systems are not genetically related is not in any way based upon empirical observations, but upon a belief in the correctness of the established classification. Finally, in stressing the fitness of the system, Wald adheres to the neo-Darwinian credo that the 'best of all possible' mutations always prevail and that all that remains for natural selection is to pick them out.

From an unbiased point of view it would certainly not be possible to maintain that the observed similarities necessarily discredit the present system, but it must nevertheless be emphasized that the convergent evolution of the complicated metabolic mechanisms required for the workings of a visual system of this type must be very improbable, and that their occurrence thus suggests a clear affinity between the annelid superphylum and Vertebrata.

Although carotenes may be involved in the light perception of echinoderms, the observations made so far hardly indicate any kind of relationship to the other taxa (Table 3.21).

Table 3.21
Distribution of visual pigments

Mollusca–Vertebrata
Rhodopsins

Photoreceptors

The cells for photoreception in various animals often display a very typical ultrastructure and, since a certain correlation obtains between their form and the systematic classification of the animals in question, it is very tempting to use the photoreceptors for the establishment of phylogenetic affinities. In fact, the observable variation in form is quite considerable, but common to all is an extensive enlargement of the cellular surface, an expedient which is thought to increase the light sensitivity of the receptor. Usually this is achieved through the formation of microvilli, but whorls of double membranes (in Ctenophora) and stacks of membranous discs (primarily in Vertebrata) are also found (Figure 3.16).

In an article on the evolution of photoreceptors, Eakin (1968) has tried to bring some order to this variety of form by distinguishing two types, called 'ciliary' and 'rhabdomere' photoreceptors, respectively. The former kind typically has a flagellum, originating in a centriole, from the shaft of which the microvilli and discs mostly originate. In the rhabdomere type no flagellum is present, although there may be centrioles. As shown in Figure 3.16, Eakin proposes that the two forms are distributed so that the ciliary, i.e. flagellated, type is present in the echinoderm, the rhabdomere type in the annelid, superphylum.

Echinodermata

Photoreceptors have been found only in asteroids, and observations made by several authors agree in affirming the presence of microvilli, but also (Figure 3.16) that they do not originate from the flagellated shaft (Eakin, 1968).

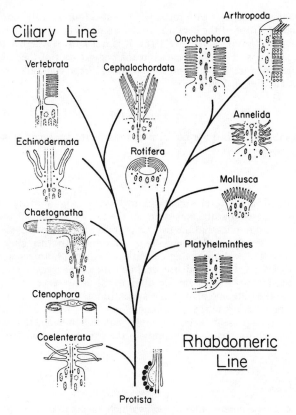

Figure 3.16. Distribution of photoreceptor organelles in the animal kingdom. 'Ciliary line' represents the echinoderm superphylum, 'Rhabdomeric line' the annelid superphylum. (Reproduced from R. M. Eakin (1968), in *Evolutionary Biology*, Vol. 2, by permission of ASP Biological and Medical Press (North-Holland division))

Mollusca

In the receptors of cephalopods, flagella and centrioles are conspicuously absent. In gastropods the rhabdomere type is most common, but rudimentary flagella may be observed in the embryonic optic vesicle (Eakin, 1968). In *Onchidium* both flagellated and rhabdomere photoreceptors occur (Yanase and Sakamoto, quoted by Eakin, 1968), and the former possibly in *Viviparus* also (A. W. Clark, 1953). Among bivalves, two species of *Pecten* have been investigated. Flagellated photoreceptors are present in both, but in one, *P. maximus*, the rhabdomere type also occurs (Miller, 1958; Barber, Evans and Land, 1967). Both kinds of photoreceptors have been observed in annelids, and there are also membrane discs in *Branchiomma* (Krasne and Lawrence, 1966), but in arthropods only the rhabdomere type is found (Eakin, 1968).

Protochordata

In the ocellus in larvae of *Ciona* the retinal cells consist of two parts, an inner section containing 50–100 'filaments' and an outer segment composed of stacked double membrane discs, the latter resembling those of vertebrate photoreceptors (Dilly, 1964). Centrioles are present, but they do not form a flagellum, and their place is taken by the filaments mentioned. Furthermore, in the cerebral vesicle of the larvae cells with microvilli are also found (Dilly, quoted by Eakin, 1968).

In *Branchiostoma* the roof of the cerebral vesicle contains ependymal cells with flagellar processes with microvillous branches (Figure 3.16). It is not known whether these cells are photoreceptors. In the eye cups of Hesse, located in the ventral wall of the spinal cord, the photoreceptors are of the rhabdomere type, and similar cells (Joseph cells) are found in the dorsal wall of the cerebral vesicle (Eakin, 1968).

Vertebrata

The vertebrate subphylum is quite uniform as regards the presence of photoreceptors of the flagellated type.

Score

It appears that there are several exceptions to Eakin's generalization. In one paper (1965, p. 376) this author suggests 'that the dichotomy of the two lines [annelid and echinoderm] is not absolute'. This statement implies that the two superphyla are not monophyletic, but if this is true, then they cannot be used for phylogenetic generalizations. Another alternative considered by Eakin (1965, p. 367) is 'the evolution of an adventitious ciliary photoreceptor in the rhabdomeric line'. Yet, should this be so, we are evidently dealing with a convergent character which cannot be used for the establishment of phylogenetic affinities. The presence of flagellated receptors in vertebrates might be the outcome of such an 'adventitious' event, in which case they could well have originated from ancestors in the rhabdomere line.

On the basis of the data presented above I think it can be concluded that the structure of photoreceptors cannot be used for testing the validity of the current superphyletic classification.

Chromatophores

Chromatophores are pigment-containing cells which, owing to their location in the integument, contribute to the colouration of animals. Through their ability to disperse and condense the pigment they allow changes in colour.

The chemistry of the pigments might potentially be of comparative value, but so far as can be judged from the literature this approach is not very profitable. Thus, melanin, carotenoids and porphyrins are ubiquitous, and other compounds, such as echinochrome, are found only in a single phylum (Fox, 1953). Hence the present survey will be confined to the structure and function of the chromatophores present in the three main taxa. This differentiation pattern is not represented in Protochordata.

Echinodermata

Very little is known about the structure of chromatophores in the echinoderms, but according to Millott 'the chromatophores of *Diadema* are not cells of the usual sort. Indeed, some of them are pigment-bearing coagula housed in an extensive system of channels' (quoted by Yoshida, 1966, p. 441).

The photoreaction has no relation to the background illumination; rather the chromatophores disperse in light and contract in darkness. It can be shown through a method of microphotostimulation that the individual chromatophores react in this way to photic stimuli (Yoshida, 1966), but the involvement of local nervous reflexes or humoral agents cannot be excluded (Millott, 1954).

Mollusca

In the phylum Mollusca, chromatophores and their physiology have been studied mainly in cephalopods. One reason for this may be their particular structure and size. They are not actually cells, but small organs, consisting of a pigment-containing bag to which are attached a number of radial smooth muscle cells which contract in order to expand the bag and thus disperse the pigment.

The control of the chromatophores is governed mainly by visual impulses which impinge upon the brain where they are relayed to various centres, excitatory, controlling and inhibitory, as the case may be. Each of the radial muscle cells is innervated by a nerve fibre (G. H. Parker, 1948). The nervous control allows for very rapid colour changes (Nicol, 1964). ACh causes expansion of the chromatophores, 5-HT contraction. Some hormonal control, involving the latter substance and tyramine, is indicated by the fact that *Octopus* blanches when the posterior salivary gland is removed, and also by the observation that when the circulatory systems of two individuals are made confluent they both assume the same colour (Sereni, 1930).

Very little is known about chromatophores in other molluscs, but it appears that in gastropods two or more types occur, some whose structure is similar to that observed in cephalopods and some which resemble those found in crustaceans and in vertebrates (Nicol, 1964).

Vertebrata

As already mentioned, the structure of vertebrate chromatophores does not exhibit any specialized features. It is of interest that, just as in echinoderms, the chromatophores in young amphibian larvae react through direct light stimulation, dispersing in light and condensing in darkness (Barrington, 1963).

In the ammocoete larva and in adult brook lampreys chromatophores exhibit a diurnal rhythm, the animals being dark by day and pale by night. This cycle, although induced by light, is controlled by pituitary hormones, and not through a direct action upon the chromatophores. The light receptor in the ammocoete larva is the pineal eye (Young, 1936).

Hormonal regulation also seems to occur in chondrichthyans. In teleosts the regulation is partly controlled by the sympathetic nervous system and partly through a melanocyte-stimulating hormone (MSH). The function of the latter seems to be to keep the chromatophores in a background adaptation, since nervous impulses can effect only transitory responses (Healey, 1948). Of some interest is the isolation of melatonin (N-acetyl-5-methoxy-tryptamine) from the ox pineal gland (Lerner, Case and Heinzelman, 1959). This 5-HT derivative has proved to be a very powerful blanching agent on the frog skin.

Score

One recorded similarity between Echinodermata and Vertebrata is the direct response of chromatophores to photic stimulation. However, since no similar investigations have been made in Mollusca, or even Crustacea, this feature cannot be used for comparison. Central control, exerted through nerves or hormones or both, is common to the annelid superphylum and Vertebrata. In its simplest form, as encountered in lampreys, this mechanism does not involve adaptation to the background, but only a response to illumination. More complex systems are found in crustaceans, where a complicated hormonal control occurs; in cephalopods, distinguished by the primarily nervous regulation, although 5-HT may participate; and in vertebrates, where adaptation takes place mainly through hormonal control in lampreys and chrondrichthyans, while the nervous system is predominant in teleosts. The substance involved in the regulation of the chromatophores in Mollusca and Vertebrata is 5-HT or its derivative melatonin. I do not think that very much importance can be placed upon the last feature, but since this does not change the final result, it has been included in Table 3.22.

Table 3.22
Distribution of characters
related to chromatophores

Mollusca–Vertebrata
Central control
5-HT and derivatives

Sensory organs

An important function of the nervous system consists in dealing with stimuli received from the environment. The free endings of single nerve cells may receive inputs of this kind, but more varied perception of, and response to, environmental changes require the presence of receptors reacting upon stimulation by specific physical and chemical environmental parameters. It is quite obvious that the progress of evolution has been associated with the acquisition and improvement of sensory receptors, and hence a survey of these may give important information concerning phylogenetic relationships.

Echinodermata

Sensory cells in the echinoderms are bipolar, spindle-shaped cells situated among the epithelial cells. They are assumed to be plurisensible, being stimulated by touch, chemical substances and light. The effect of light is taken to involve a photochemical change of a skin pigment (J. E. Smith, 1965). As mentioned above, it is still not clear whether the optic cups in the asteroids are of importance for photosensitivity (Yoshida, 1966).

Mollusca

The sensory receptors are well developed in the molluscs, reaching the highest state of perfection in cephalopods. In this class primary sensory neurons are found in and under the superficial epithelium; some of these have branching free nerve endings, which may be sensitive to touch. In the muscles multipolar heteropolar cells have been observed, the axons of which pass to the stellate ganglion. These cells are supposed to be stretch receptors (Alexandrowicz, 1960a).

An olfactory organ, responding to chemicals dissolved in water, is found in a pocket behind the eyes. Isolated arms also react to chemical stimuli, so it may be assumed that there is a general cutaneous chemical sensitivity (Bullock and Horridge, 1965).

The anatomy and histology of the statocysts exhibit a high level of organization in the cephalopods. On the basis of their structure, as well as of experiment, it has been inferred that the macula is sensitive to changes in position and the crista to changes in acceleration (Boycott, 1960; Young, 1960).

The remarkable similarities in design and function between the eyes of cephalopods and vertebrates have often been commented upon; the remarkable dissimilarity in the orientation of the retina cells has been accounted for by the well-known difference in the epigenesis of the eye in the two cases.

In the gastropods it may be expected that there are receptors responsive to all, or nearly all, of the physical and chemical parameters discussed here, but at a lower level of perfection. In the other classes the sensory reception is less developed, qualitatively as well as quantitatively.

Protochordata

In the tunic of urochordates bipolar sensory cells may be observed, sometimes arranged in clusters. These cells may be mechanoreceptive, and possibly also chemoreceptive. Statocysts and simple eyes are also found (Bullock and Horridge, 1965).

Various anatomical structures have been supposed to be photosensitive in *Branchiostoma*, but no definitive observations support responsivity to changes in illumination except that, as already mentioned, intense light will induce swimming movements (Barrington, 1965). No statocysts are present. Groups of hair cells in the velum and on the buccal cirri are supposed to function as chemical receptors (Weichert, 1965), and it may be surmised that mechanoreceptors are present in the epidermis.

Vertebrata

Over the entire body of *Myxine*, but especially in the tentacles, nerve fibres ramify among the epidermal cells, sometimes associated with specialized sensory cells. Their function has not been established, but the fact that they occur in large numbers in the tentacles suggests that they are touch receptors (Blackstad, 1963). Even light receptors are supposedly present in the skin. Lateral line organs, so common in aquatic vertebrates, are absent or poorly developed in the hagfish, but somewhat better developed in the lampreys.

The olfactory organ is well developed in *Myxine*, suggesting that chemoreception is very efficient, and various observations support this point.

The ear of *Myxine* contains a single canal in the form of a ring, whereas the lampreys have two semicircular canals, thus representing an approach towards the condition in higher vertebrates. In cyclostomes the functional contribution of the ear has not reached the level met with in cephalopods. It is possible, however, that it may pass information about direction, as well as about acceleration and position (Ross, 1963). Anatomically, the eyes of the hagfish are poorly developed, but they are nonetheless sensitive to light (Kobayashi, 1964). A study of the eyes in three hagfish genera has demonstrated that evolution within the order has been associated with a reduction of visual capability (Fernholm and Holmberg, 1975). The lamprey eye is more developed, although still 'degenerate' (de Beer, 1928).

Score

In the various theories which envisage the origin of the vertebrates from the arthropods many attempts have been made to homologize sensory organs between the two groups (cf. Hanström, 1928). This author also discusses the importance of olfactory, visual and static centres in the phylogenetic evolution of the brain. In particular it is pointed out that among invertebrates olfaction is most important for distance reception in the lower forms, for example in annelids, most molluscs and many crustaceans, whereas in the cephalopods, the higher crustaceans and the insects vision is the most important sense. This switch from olfaction to vision must obviously have occurred independently several times, and a similar change is observable in the evolution of the vertebrates.

As far as the three main taxa are concerned, it is evident that in Mollusca and Vertebrata all three senses, the static, the olfactory and the optic, are of great importance for their input of information from the environment. The conditions in Echinodermata are extremely primitive in comparison with those in the other groups, and may be homologous to the epidermal receptors found in the latter. Furthermore, stretch receptors are found in cephalopods and, incidentally, in crustaceans.

The situation in Protochordata is not quite clear but, apart from the statocysts and eyes found in certain urochordates, specialized sense organs do not seem to be present. As concerns sensory receptors, these animals can at most be at the level represented by the lower molluscs. The final score is shown in Table 3.23.

Table 3.23
Distribution of sense organs

Mollusca–Vertebrata
Optic sense organs
Olfactory sense organs
Static sense organs
Proprioceptors
Mollusca–Urochordata
Optic sense organs
Static sense organs

Nervous system

It does not seem possible to find in the literature any data concerning the function of the nervous system which can be used in the present context. In these circumstances the following discussion is confined to anatomical, and particularly to histological, features.

Echinodermata

For various reasons the application of traditional histological methods has not been especially successful in the study of the nervous system of echinoderms (Hanström, 1928; J. E. Smith, 1965; 1966). According to the latter author the nervous system embodies both primitive and advanced features. The primitive characters, which bear resemblance to the nervous system in Cnidaria, are 'scattered sensory cells of one (bipolar) type; and the occurrence of small-celled internuncially conducting neurons, basiepithelial in position, which give rise, in part, to a fibre system with a net arrangement' (J. E. Smith, 1965, p. 1522). This organization is found both in skin and gut linings. The advanced traits *par excellence* are the central nervous system, consisting of radial cords and a circumoral nerve ring, and the multifibred motor strands.

The nervous system in echinoderms is aganglionic, and contains no nerves, the main conducting pathway being basiepithelial, unsheathed medullary strands. The nerve cells are very small and their fibres very fine. The sensory neurons are bipolar and located in the epithelium of the body wall. In Asteroidea, and possibly in the other classes as well, the internuncial neurons in the superficial plexus are multipolar, isopolar cells. In the deeper layer of the plexus are found bipolar neurons, arranged in parallel and in series, and this is essentially true for the radial cords also, although the number of neurons is much larger here. Contact between the nerve cells is established through fibre-to-fibre synapses. No glia cells are present, and the neurons are consequently unsheathed.

Little is known about the motor nerves in Echinodermata. Various kinds of evidence intimate that peripheral excitation of the musculature occurs directly from the skin plexus, by bipolar neurons. The hyponeural motor system which controls the movements of the tube feet consists of bipolar and unipolar nerve cells.

Mollusca

The variation in the nervous system within, and particularly between, the classes in Mollusca is so great that it is impossible to present a comprehensive survey. For this reason attention will be focused on the cephalopods (Bullock and Horridge, 1965).

In the order Dibranchiata the central nervous system is a veritable brain, vascularized and enclosed in a cartilaginous capsule. As far as the histology is concerned, 'even a cursory inspection of microscopic sections of a dibranchiate brain reveals a diversity of texture and differentiation of parts rivalling in complexity the brains of fishes and amphibians and exceeding those of higher arthropods' (Bullock and Horridge, 1965, p. 1447). The general design incorporates a superficial rind or cortex packed with the somas of unipolar neurons which form synaptic contacts *via* axonal collaterals in the internal neuropile. Besides the latter formation the interior of the brain also contains tracts and islands of cells.

The brain neurons can be divided into several groups according to their various properties; in the present context it is of particular interest to note that they vary considerably with respect to size. As in other invertebrates the most common nerve cell type is unipolar, but bipolar and multipolar neurons are found in the cell islands. The multipolar heteropolar form is represented by the first-order giant cells in *Loligo*, which send out several dendrites and one giant axon. Glia cells are present, both in the neuropile and in the form of sheaths around the axons, but myelination does not occur. The giant fibres are often indented by the irregular folds of glia cells (Tauc, 1966). Astrocyte-like cells are found in members of Cephalopoda alone among invertebrates.

The peripheral nervous system consists of nerves and ganglia, both of which are covered by a sheath comprising an inner layer of cells, the perilemma, and an outer neurolemma (perineurium), consisting of connective tissue, part of which is collagen. A rather large number of ganglia of varying size is found in cephalopods. We shall not deal with their function here, except to mention that the ganglia supplying the viscera were formerly considered to constitute a sympathetic nervous system (Hanström, 1928), a view which has now been abandoned (Bullock and Horridge, 1965). The distinction of an autonomous system similar to that found in vertebrates is not possible, probably partly for morphological reasons, but the nervous system in the arms and suckers is known to possess a large degree of autonomy. The digestive tract is innervated by a plexus similar to that found in vertebrates. Peripheral nerve nets apparently do not occur in Cephalopoda; this primitive feature is found in the other classes, except possibly Gastropoda (Bullock and Horridge, 1965).

Bipolar sensory cells are found in the skin, but in the muscles multipolar, heteropolar receptors have been found, the axons of which pass to the stellate ganglion (Alexandrowicz, 1960a).

Protochordata

The central nervous system in urochordates consists of a cerebral ganglion or a dorsal cord with several ganglia. Histologically these structures display the typical invertebrate design, with a central core of fibrous matter and a cortex with nerve cell bodies, a circumstance which may be related to the fact that the nerve cells are chiefly unipolar. A certain variation in cell size occurs, and giant cells are present (Brien, 1948). The number of nerve cells in the ganglia is very small. Glia cells are present in the core of the ganglion. The nerves are often mixed and contain connective tissue (Bullock and Horridge, 1965). Some muscle layers in the body wall are centrally innervated, others by a peripheral nerve net (Florey, 1951).

The central nervous system of *Branchiostoma* is tubular as in the vertebrates. From the spinal cord emerge the dorsal nerves, containing many fibres and covered by a sheath of connective tissue. The dorsal root nerves carry somatic sensory nerves from cutaneous receptors and visceral sensory and motor fibres. The sensory elements are derived from at least three types of sensory neuron, the most abundant being bipolar cells. In the ventral roots somatic motor fibres emerge in loose bundles. Dorsal ganglia are absent. A nerve plexus is present in the integument. Giant fibres are present in the spinal cord, arising from multipolar heteropolar giant neurons and forming junctions with the somatic motor and sensory systems (Bone, 1960b; Barrington, 1965).

Besides ependymal cells two other kinds of glia cells are present, one of which may be related to Schwann's cells (Bone, 1960b). The spinal cord is not vascularized (Nansen, 1887).

Vertebrata

Myxine has a central nervous system subdivided into brain and spinal cord, and with separate dorsal and ventral roots. The nerve fibres are not myelinated, hence there is no clear distinction between white and grey matter. Different types of glia cells occur in the grey matter, but the astrocyte-like cells do not seem to be in contact with capillaries (Bone, 1963). Dorsal ganglia are present. The dorsal nerves contain axons originating in skin and musculature. In *Myxine* it is questionable whether the dorsal roots contain autonomic efferent fibres, but in *Petromyzon* they do (Johnels, 1956). Through the ventral roots pass the axons of three distinct types of neurons, which vary with respect to size, position, etc. They are typically multipolar heteropolar, and the largest type is supposed to innervate the fast muscles.

Giant cells are observed in the spinal cord. Their course is similar to that of the corresponding neurons in *Branchiostoma*. Synaptic connections cannot be demonstrated on the soma of the giant cells nor of any other type of neuron in the central nervous system in *Myxine*. Bone (1963) proposes that synapses may occur between dendrites or between these and axons, a situation which seems to be halfway between the typical invertebrate and the typical vertebrate synapses. The peripheral nerve fibres are unmyelinated, but they, as well as the somas of the ganglion cells, are ensheathed by Schwann's cells. The nerves are covered by a collagenous perineurium. Peripheral neurons are not confined to ganglia, but may be present in the nerves or in the integumental plexus (Retzius, 1890), and the cells may accordingly be either unipolar or bipolar.

Score

We have observed that although something which passes under the designation of a 'central nervous system' is present in echinoderms, it is very simple and very different in design from that found in the other groups. If we disregard the form of the CNS, which, when tubular, may be referred to the presence and influence of a notochord, there is no doubt that in vascularization and in other details (cortex, islands of cells, etc.) there are great similarities between that of cephalopods and vertebrates, and from observations of behaviour it appears that this similarity extends to function as well. The CNS of the cephalochordates resembles that of the vertebrates, but no particular affinity is revealed by the cerebral ganglion in the urochordates. Ganglia are absent in echinoderms, and if we consider the cephalic ganglion or the dorsal cord as a CNS, then ganglia are also absent in urochordates and in cephalochordates.

Nerves are absent in echinoderms, present in all the other taxa. Since *Myxine* is the only vertebrate considered, myelination does not occur in any case, but sheathed nerves and ganglion cells are present in cephalopods and Hyperotreta. We may take this as indicating the presence of oligodendroglia, but astrocyte-like glia cells are also present in both taxa. The two kinds of glia in cephalochordates will be assumed to be equivalent to these. No kind of glia is present in echinoderms. The multipolar heteropolar neuron, the most common cell type in vertebrates, is unknown in invertebrates outside the annelid superphylum, where it is present in cephalopods, although in small numbers. This cell type may be more usual in cephalochordates, but is rare in urochordates. Neurons of varying size are found in all taxa except Echinodermata, where only very small nerve cells occur. Furthermore, in cephalopods, protochordates and the lower vertebrates coordinating giant cells are present, confined to the CNS.

Nerve nets are present in echinoderms, protochordates and in the lower molluscs, but absent in the other groups (Table 3.24).

Table 3.24
Distribution of properties related to the
nervous system

Echinodermata–Mollusca
  Integumental plexus
Mollusca–Vertebrata
  Central nervous system
  Ganglia
  Nerves
  Oligodendroglia
  'Astrocytes'
  Multipolar heteropolar neurons
  Neurons of varying size
  Giant coordinating neurons
Echinodermata–Urochordata–Mollusca
  Integumental plexus
Mollusca–Urochordata–Vertebrata
  Nerves
  Oligodendroglia
  Multipolar heteropolar neurons
  Neurons of varying size
  Giant neurons
Echinodermata–Cephalopoda–Mollusca
  Integumental plexus
Mollusca–Cephalochordata–Vertebrata
  Nerves
  Oligodendroglia
  'Astrocytes'
  Multipolar heteropolar neurons
  Neurons of varying size
  Giant neurons
Cephalochordata–Vertebrata
  Central nervous system

Hormones

The study of vertebrate endocrinology has been pursued ardently during most of this century, and an enormous body of knowledge has been accumulated. Among the most important discoveries made in this field may be mentioned the intricate regulatory mechanisms which control the synthesis and secretion of hormones and the very exciting fact that the specificity of action in many cases depends upon interaction between a hormone and receptors on the surface of the target cells, whereas the intracellular reactions are unspecific to the extent that they are elicited by cyclic AMP.

The field of invertebrate and protochordate endocrinology is relatively less penetrated, and experimental, as contrasted to histochemical, work is for various reasons lagging much behind. However, of late the amount of research devoted to invertebrates has shown a remarkable increase, allowing the question of hormones to be discussed in this comparative survey.

It must be emphasized that, when the chemical composition of a hormone is known from studies in one animal group, it may be possible to demonstrate its presence in other taxa without thereby necessarily establishing that the function is similar in two cases. It is therefore important to distinguish between chemistry and function in the present context.

Three different types of endocrine glands may be distinguished, namely, neurosecretory cells, transformed nerve ganglia and epithelial endocrine glands. Neurosecretory cells may be associated with neurohaemal organs (Highnam and Hill, 1969). Neurosecretions clearly represent the most primitive type of hormones, since neurosecretory cells are demonstrable in all phyla possessing a nervous system, at least judging from histochemical observations (Gabe, 1966). In contrast, the other types of glands, as well as neurohaemal organs, are found only in the annelid superphylum and in Vertebrata (Highnam and Hill, 1969).

Echinodermata

Neurosecretory cells have been described in various echinoderms by Unger (1962), Fontaine (1962) and Imlay and Chaet (1965). Of great interest is the fact that extracts from radial nerves of various species contain substances which affect the maturation and shedding of the gametes (Unger, 1962; Chaet, 1964a, b; Kanatani, 1964). It has been established that the active agents are polypeptides and that they act by releasing a second substance from the ovary (Schuetz, 1969). This has been identified as 1-methyladenine (Kanatani *et al.*, 1969). Nerve extracts have also been found to affect locomotion and to produce colour changes. Furthermore, the presence of a substance influencing osmoregulation and diuresis has been claimed in two species of *Asterias* (Chaet and McConnaughy, 1959; Unger, 1962). Dodd and Dodd (1966) have observed that a compound with properties similar to oxytocin may be extracted from members of this genus. Referring to the following discussion of hormones in protochordates, it appears unlikely that the substance in question bears any relation to the vertebrate hormone.

Wilson and Falkmer (1965) isolated from the pyloric caeca of *Pisaster* a protein with several properties similar to those of vertebrate insulin. Surveying a variety of marine invertebrates, Hagerman, Wellington and Villee (1957) did not find oestradiol or oestrogen in any of the investigated echinoderms. The presence in *Strongylocentrotus* of very small amounts of oestradiol-17β and progesterone was reported by Botticelli, Hisaw and Wotiz (1961), and Hathaway (1965) found that the sperm of sea urchins can convert oestradiol-17β to oestrone.

Mollusca

As already mentioned, the annelid superphylum is distinguished by the possession of neurohaemal organs, transformed ganglia and epithelial glands. The last type is found only in Arthropoda and Mollusca, and in the latter phylum it is definitely established only in Cephalopoda (Highnam and Hill, 1969) and in Gastropoda (Boer, Douma and Koksma, 1968). On the whole it must be admitted that the endocrinology of arthropods is much better investigated than is that of molluscs, and for that reason references will occasionally be made to the former phylum in the present context.

All living cephalopods, except *Nautilus*, have optic glands, epithelial endocrine organs lying on the optic stalks just internal to the optic lobes. The optic glands are innervated from the subpedunculate lobe, but there is no evidence that the nerves in question are involved in a neurosecretory function (Wells, 1960).

Normal maturation of the ovary in *Octopus* is associated with an enlargement of the optic glands. Precocious maturation can be obtained experimentally in various ways, all of which suggest that there is nervous inhibition of the activity of the glands. The function of the secretion from the optic gland in the female is vitellogenetic, causing yolk deposition in the oocytes (Wells, 1960). Apparently no hormone production occurs in the ovary, but the testis secretes a hormone with negative feedback action on the optic gland, suggesting that the latter is a 'pituitary analogue' (Wells and Wells, 1969). In this context it may be mentioned that, although not generally present in invertebrates, Sertoli cells have been found in a mollusc (Yasuzumi, Tanaka and Tezuka, 1960); this cell type has been shown to synthesize steroids in vertebrates (Lofts, 1968). The question of hormonal influence on secondary sexual characters in cephalopods has not been settled experimentally, some results indicating an influence, others not (Durchon, 1967). The latter author also reviews a number of attempts to produce effects through injection of vertebrate sexual hormones, but they are dismissed as being of little informative value.

Various lines of evidence indicate that vitellogenesis and other reproductive events are controlled by hormones in gastropods (Laviolette, 1954; Meenakshi and Scheer, 1969), and similar results have been obtained in bivalves. The presence of steroid sexual hormones has been demonstrated in *Spisula*, but not in *Busycon* and *Loligo* (Hagerman *et al.*, 1957). They also occur in *Pecten*, where the concentration is considerably higher than that observed in *Strongylocentrotus* (Botticelli *et al.*, 1961), and the conversion of oestradiol-17β to oestrone takes place in the oyster (Hathaway, 1965).

Hormonal control of osmoregulation in gastropods has been observed by Lever, Jansen and De Vlieger (1961) and by Vicente (1963), and similar observations on bivalves were reported by Lubet and Pujol (1963) and Nagabushanam (1963).

The very large branchial glands in cephalopods are said to have functions similar to those of the adrenocortical tissue in vertebrates, but evidence on this point is lacking. The posterior salivary glands are endocrine, their secretions containing substances like 5-HT, tyramine and octopamine. This suggests a function similar to that exerted by the adrenal medulla in vertebrates (Bacq and Ghiretti, 1953). Finally, it may be mentioned that Wilson and Falkmer (1965) reported the presence of an insulin-like substance in *Eledone*.

The endocrinology of echinoderms is, in my opinion, surprising because of the large number of hormonal actions or hormone-like substances which can be demonstrated, and which are difficult to relate to the relatively simple anatomy and physiology of these animals. This is also true for the lower classes among the molluscs, but many gastropods and most cephalopods show a level of organization which suggests that, like their relatives the arthropods, they must rely to a considerable extent upon hormones for integration of their activities. However, the present survey shows that the number of hormones or hormone-like substances demonstrated so far are of the same order of magnitude in the two phyla, but it must be admitted that the scope for new discoveries, although large in both cases, still seems to be potentially greater in Mollusca.

Protochordata

Many attempts have been made to 'prove' the alleged kinship between Protochordata and Vertebrata through the demonstration of similarities on the functional level. Among these contributions endocrinological studies assume an important position. This work has been centred around two propositions, namely, that the neural complex in urochordates is a homologue of the pituitary, and that the endostyle, present in both subphyla, is a homologue of the thyroid gland.

As far as the first point is concerned, many authors have claimed the presence of oxytocin, vasopressin and melanocyte-stimulating hormone in extracts of the neural complex from various urochordates. It appears that the experimental approach employed in many cases has been questionable, and although some of the physiological effects reported earlier could be confirmed, the recent critical investigations by Dodd and Dodd (1966) seem to refute beyond doubt the supposed homology. The presence of gonadotrophic hormones cannot be substantiated either (Dodd, 1955; Hisaw, Botticelli and Hisaw, 1962).

The basis of the interest in the endostyle is the fact that during metamorphosis in the ammocoete larva of the lamprey this structure is converted in part to the thyroid gland, but it is engaged in the synthesis of iodotyrosines and iodothyronines even before this event (Barrington, 1968). It was thus natural to envisage that a similar activity takes place in the endostyles of protochordates.

Studies on iodine fixation in urochordates have indeed shown that this substance is incorporated into various tyrosine derivatives, including very small amounts of the two thyroid hormones, thyroxine (T_4) and triiodothyronine (T_3). Most of the binding occurs, however, in the tissue underlying the tunic, but some also takes place in specific regions of the endostyle (Roche, Salvatore and Rametta, 1962; Barrington, 1968).

The results obtained on *Branchiostoma* also show iodine binding elsewhere than in the endostyle, but the major part is bound in this organ to tyrosine, a substantial part of which is transformed into T_3 and T_4. The iodinated amino acids are in turn incorporated in a material, probably a glyco-protein, which is secreted into the pharynx (Barrington, 1965; 1968).

A biosynthetic pathway leading to the formation of thyroid hormones is thus unquestionably present in the protochordates, but this does not necessarily imply that they function as hormones. In actual fact, all attempts to demonstrate a physiological effect of these substances have been in vain and, as we shall see, the same holds for the lower vertebrates. The distribution of the iodinated amino acids in urochordates should be evaluated with particular reference to the observation that incorporation of iodine into amino acids, including T_3 and T_4, occurs in integumental sclero-proteins in various invertebrate taxa, for example Cnidaria (Roche, Fontaine and Leloup, 1963), Insecta and Mollusca, etc. (Berg, Gorbman and Kobayashi, 1959). It is of course correct that, as maintained by Roche *et al.* (1963), this fact does not necessarily mean that thyroid hormones occur in invertebrates, but the same seems to hold for urochordates and possibly even for cephalo-chordates.

Wilson and Falkmer (1965) have reported a protein with insulin-like properties in *Ciona*. No corticosteroids can be detected in *Branchiostoma* (Jones and Phillips, 1960).

Vertebrata

Endocrinological studies on vertebrates, including the cyclostomes, have revealed a great number of similarities both on the anatomical and the functional level, even though a gradual increase in the number of hormones and in the mechanisms of interaction have clearly occurred in phylogenetic evolution within the taxon. In the present context it is of particular interest to survey the conditions in cyclostomes.

In Hyperotreta (*Myxine*) a pituitary gland is present, distinctly subdivided into a neurohypo-physis and an adenohypophysis (Adam, 1963b; Fernholm, 1972). According to the latter author a hypothalamic regulation of the adenohypophysis is unlikely. The information about the hormones produced by the neurohypophysis is inconclusive, but it is possible that gonadotropins, adreno-corticotropin, thyrotropin and a melanocyte-stimulating hormone are present (cf. Fernholm and Olsson, 1969). Attempts to demonstrate prolactine in either *Myxine* or *Lampetra* were negative (Aler, Båge and Fernholm, 1971). The neurohypophysis contains octapeptides one of which has vasopressin-like properties (Adam, 1963b).

No thyroid gland but a large number of thyroid follicles are present, in which uptake of iodine and synthesis of thyroid hormones can be demonstrated (Gorbman, 1963).

The exocrine and endocrine parts of the pancreas are separated, the latter being located in an islet organ which displays similarities with that known from the higher vertebrates (Schirner, 1963). This author and Falkmer and Winbladh (1964) have demonstrated the presence and the function of insulin in *Myxine*. The corticosteroids cortisol and corticosterone are present in hagfish, but aldosterone cannot be detected (Jones, 1963; Idler, Sangalang and Weisbart, 1971). It has been possible to demonstrate the interconversion of various steroid hormones in the ovary of *Eptatretus* (Hirose *et al.*, 1975).

In *Lampetra* hypophysectomy leads to the suppression of the secretion of sexual hormones and the growth of the follicle cells. Some other events, including ovulation and spermiation, are retarded under these conditions. As in *Myxine*, the activity of the adenohypophysis is not, or to a slight extent only, under nervous control (Larsen, 1973).

Chromaffin tissue is found in *Petromyzon*, and since adrenaline and noradrenaline are present, it may be presumed that the hormonal action exerted by the adrenal medulla in higher vertebrates is also represented (Barrington, 1968). The two orders of Cyclostomata do not appear to be very different in this respect.

Score

The data compiled here are difficult to evaluate because, as emphasized above, a substance which exerts a specific hormonal action in one animal group may have a quite different function in other taxa. This situation is the more likely since the specificity of hormone action in many cases resides with the target cell rather than with the hormone. However, we shall conclude that osmoregulative neurosecretions, steroid sex hormones and insulin are present in Echinodermata, Mollusca and Vertebrata, implying that these agents cannot be used for comparative purposes. It is not without interest to note that production and secretion of a protein which possesses several of the properties of insulin is a plesiotypic character in the present context, and recent evidence even supports the contention that in the various invertebrates and in cyclostomes it assumes a function similar to that known from the higher vertebrates (Falkmer *et al.*, 1973; 1975). The branchial glands in cephalopods have been supposed to correspond to the adrenal cortex of vertebrates, but since nothing is known about the chemical nature of the hormones, this feature cannot be accepted as a point of similarity.

In discussing the work on the control of sexual maturity in *Octopus*, Wells (1960, pp. 104–5) observed: 'Clearly this control system has much in common both with the anterior pituitary-gonad control mechanism in vertebrates and with systems regulating the onset of sexual maturity in arthropods. The anterior pituitary, the optic glands, the X organ–sinus gland system of crustacea and the corpus allatum of insects are all regulated by efferent nerves arising in the highest centres of the central nervous system. All, moreover, serve in one way or another to delay the onset of sexual maturity until the animal is fully developed. . . . In detail, then, there is little similarity in the modes of operation of the mechanisms controlling the state of the gonad in cephalopods, vertebrates and arthropods. Clearly these systems have been evolved quite independently in the three groups. . . . In which case why the constant direct control by the highest centres of the central nervous system, parts that are otherwise concerned only with sensory integration and typically act only through intermediate centres?'

Since, in the present context, we should not make any *a priori* statements concerning the phylo-genetic origin of any character, we are forced to recognize that striking similarities obtain with respect to mechanisms controlling sexual maturation in the annelid superphylum and in Verte-brata. The agreement would probably be greater if the comparison did not involve the higher vertebrates, for the mechanisms prevailing for instance in cyclostomes, largely unknown, are likely to be much simpler. This viewpoint does not in any way exclude convergent evolution of the regulatory mechanisms discussed by Wells in the above quotation.

Table 3.25
Distribution of characters related to
hormonal regulations

Mollusca–Vertebrata
Control of sexual maturation
Neurohaemal organs
Transformed ganglia
Epithelial glands
Sertoli cells
Mollusca–Urochordata–Vertebrata
Iodine incorporation
Cephalochordata–Vertebrata
Iodine incorporation

Otherwise it seems that the endocrinological features of most importance for the comparison are histological ones, represented by the three types of glands, neurohaemal organs, transformed ganglia and epithelial glands. The differentiation pattern represented by the Sertoli cells may also be included here.

As regards Protochordata, the information gained so far does not suggest that they are highly developed from an endocrinological point of view. The only character of diagnostic value appears to be the incorporation of iodine. In urochordates this process occurs only to a minor extent in the endostyle, and it is therefore difficult to accept this feature as indicating specific affinity to Vertebrata, since similar observations have been made in various invertebrates, including the molluscs. No iodine incorporation has been reported in echinoderms, but since this activity in invertebrates seems to be associated with ectodermal scleroproteins, there would be no reason to expect positive results in this case. The iodine binding in cephalochordates is preferentially located in the endostyle, and may thus be accepted as indicating kinship with the vertebrates (Table 3.25).

Egg jelly coats

Eggs are usually covered by one or several layers of secreted material. In some animals the outer layer hardens to form a protective shell. This is usually the case in two of the groups under investigation, namely, Mollusca and Vertebrata, but in all three taxa a naked jelly coat may be present. Studies on the chemical nature of this 'jelly', 'mucin' or 'albumin' began in the last century, but significant progress in this field has been made only in the last three decades. The information obtained has led to the classification of these substances as either protein-containing glycosaminoglycans or glycoproteins.

Because of the importance of the jelly for the process of fertilization, the class Echinoidea has been extensively studied; in the other groups data come only from Gastropoda and Amphibia. The latter taxon is too far from the origin of Vertebrata to be of real interest in this context, but the results exhibit a number of striking similarities which may justify the present discussion.

No observations have been made on eggs of protochordates.

Echinodermata

The most typical feature of the jelly glycosaminoglycans in echinoderms is a very high content of sulphate. The monosaccharide components are mainly galactose and fucose. In 16 investigated

Table 3.26
Monosaccharides in egg jellies of Echinodermata (Hunt, 1970)

Echinoidea	
Arbacia lixula	Fucose, galactose
A. punctatula	Fucose, mannose, galactose, hexosamine
Echinus esculentus	Galactose
Echinarachnius parma	Fructose
Echinocardium cardatum	Fucose
Hemicentrotus pulcherimus	Fucose
Heliocidaris crassipina	Fucose
Lytechinus anamesus	Galactose
L. variegatus	Fucose, glucosamine, galactosamine
Paracentrotus lividus	Fucose, glucose, mannose, xylose, hexosamine
Pseudocentrotus depressus	Fucose
Sphaerechinus granularis	Glucose, mannose, fucose
Strongylocentrotus purpuratus	Galactose
S. pulcherimus	Fucose
S. droebachiensis	Fucose, galactose
Asteroidea	
Asterias amurensis	Fucose, galactose, glucosamine

species, of which 15 were from Echinoidea and one from Asteroidea, one or both of these substances were found in 15 cases (exception *Echinarachnius*), often as the sole, or otherwise as the main, substance. Hexosamines do occur, but not as a general feature, and hexuronic acids are never present (Table 3.26). Galactose and fucose occur as galactan sulphate and fucan sulphate, the latter probably as a branched molecule. The galactose from galactan sulphate in *Echinus* has the unusual L-configuration, but a small amount of the D-isomer may be present (Vasseur, 1952). Sialic acid has been demonstrated in several species of Echinoidea (Perlmann, Boström and Vestermark, 1959).

Mollusca

A galactan containing the D and L-forms of galactose in the ratio 6:1 has been isolated from the eggs and albumin glands of *Helix* (Bell and Baldwin, 1941). A similar galactan was observed in the albumin glands of *Biomphalaria*, the D–L ratio being 2:1 (Corrêa, Dmytraczenko and Duarte, 1967). Branched macromolecules were indicated in both instances. The presence of galactan was observed in the uterus and in the eggs of *Pila* (Meenakshi, 1954). Bayne (1966) has found that in the eggs of *Deroceras* (*Ariolimax*) the shell contains glucosamine, galactose and fucose, the jelly glucosamine and galactose. Galactans and polysaccharides containing both galactose and fucose have been demonstrated in the uterus or in eggs from several other species of Gastropoda (Hunt, 1970). All these gastropod glycosaminoglycans are unsulphated. Only in *Ariolimax* has a sulphated glycosaminoglycan been reported in the reproductive system. This substance contains galactose, fucose and glucosamine (Meenakshi and Scheer, 1968). Sialic acid has not been observed. However, the mucin in which the eggs of *Physa* are imbedded contains substantial amounts of sulphate and sialic acid. As well as a large amount of hexosamine it also contains five neutral sugars, of which galactose and xylose are the most important (Lunetta, 1971).

Vertebrata

A survey of analyses made on jellies from a number of amphibian species shows that the monosaccharides galactose and fucose have preserved their unique role (Table 3.27), but quantitative estimates indicate that the hexosamines assume a dominant position (Minganti, 1955). Sialic acid is present, possibly in rather large amounts (Lee, 1967); in *Rana temporaria* it amounts to 4·4 per cent on a molar basis (Höglund and Løvtrup, 1976).

In the investigations listed in the table, information about the sulphate content is either missing or else it is stated to be absent. In special studies it was established that the sulphate content is very low in three species of Salientia (Minganti and D'Anna, 1958; Bolognani *et al.*, 1966). Studies of ^{35}S-sulphate in the oviduct and in the jelly layers of the newt *Notophtalmus* (Humphries, 1970) and of *Rana* (Pereda, 1970) show that one or two of the innermost layers contain sulphate.

The presence of galactan has not been demonstrated, and L-galactose is conspicuously absent (Folkes *et al.*, 1950).

Table 3.27
Monosaccharides in egg jellies of Amphibia

Salientia	
Rana temporaria[1,2,3]	Galactosamine, glucosamine, galactose, fucose
R. esculenta[4]	Hexosamines, galactose, fucose
R. pipiens[5]	Galactosamine, glucosamine, galactose, fucose
R. japonica[6]	Hexosamines, galactose
Bufo vulgaris[4]	Hexosamines, galactose, fucose, mannose
Discoglossus pictus[4]	Fucose, mannose, glucose
Caudata	
Ambystoma mexicanum[4]	Fucose, mannose

1, Folkes *et al.* (1950); 2, James (1951); 3, Höglund and Løvtrup (1976); 4, Minganti (1955); 5, Lee (1967); 6, Hiyami (1949).

Score

The reported results are puzzling, insofar as a comparison between the three groups reveals an extraordinary conservatism with respect to the content of monosaccharides of egg jellies and yet within each group quite distinct changes have occurred. In the matter of the high content of sulphate Echinodermata and Mollusca seem to be related, but in the matter of the hexosamines a similar affinity is exhibited between Mollusca and Vertebrata. Sialic acid is present in all three taxa. The galactans containing the L-isomer of galactose form a character in common between the two invertebrate groups.

It is concluded that none of the characters discussed here is of sufficient weight to be used in the present context.

Cumulative score

Before we proceed to assess the cumulative score it may be mentioned that I am well aware that a number of objections will be raised against my selection and evaluation of the various characters, and also against their validity. Some of this criticism I shall try to meet in the section following, but certain points may be discussed at this stage.

In the first place, it may be said that the available data concerning all the selected characters are not comprehensive and precise enough to draw reliable conclusions. There is some truth in this contention, but this is a difficulty which is common in all classificatory work, including that based on morphological characters, and, as we have seen, it can be circumvented only by the expedient stated in our fourth premise, i.e. by using as many characters as possible.

As far as this is concerned it might be argued that I have stopped short of this goal. Against this it may be countered that the characters chosen in fact span a wide field of chemical, physiological and histological properties, and also that I have surveyed other topics without finding the available information comprehensive enough to be useful for the present purpose. In this connection I should like to point out that this situation is partly due to the fact that many investigations have been planned to 'prove' the current classification, so that comparative data are missing for the annelid superphylum. However, as far as the number of characters is concerned, it must be stressed that, as will be shown presently, anyone who accepts my premises will find those scrutinized above sufficient to reach a conclusion. And if the premises are not accepted, the continued comparison of characters will not strengthen any conclusion that might follow from this pursuit.

After these comments we may proceed to evaluate the cumulative score. In order to present this graphically we may, as concerns the relationship between Echinodermata, Mollusca and Vertebrata, imagine for each investigated property a virtual triangle EMV, at the corners of which the three taxa are situated, the sides representing an affiliation based upon the possession of the character in question. Since only properties shared by two of the three taxa are included, it follows that in each triangle only one side can be drawn. By summing up the various data recorded in Tables 3.4, 3.8, 3.9, 3.10, 3.11, 3.14, 3.16, 3.17, 3.18, 3.19, 3.20, 3.21, 3.22, 3.23, 3.24 and 3.25, we arrive at the result that the sides EM and EV are each represented by three characters and the side MV by 50 (Figure 3.17).

As the two protochordate taxa are each compared with all of the three other groups, a different graphical approach is dictated. Thus, we shall place the three major taxa at the corners of a virtual triangle and either Urochordata or Cephalochordata at the midpoint. Characters present in a protochordate taxon and in one of the other taxa will be represented by a diagonal uniting the midpoint with the appropriate corner. Following this approach, the outcome, taken from Tables 3.9, 3.10, 3.11, 3.14, 3.16, 3.17, 3.18, 3.19, 3.23, 3.24 and 3.25, will be two for the line UE, 16 for UM, 11 for UV, two for CE, 17 for CM and 15 for CV (Figure 3.18).

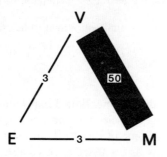

Figure 3.17. Cumulative score for Echinodermata (E), Mollusca (M), and Vertebrata (V). The numbers indicate characters in common and the widths of the connecting lines are proportional to these numbers

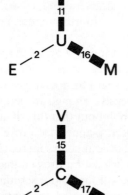

Figure 3.18. Cumulative score for Urochordata (U), and Cephalochordata (C). The numbers indicate characters in common and the widths of the connecting lines are proportional to these numbers

VERTEBRATA

In the above discussion of the method of phylogenetic classification derivable on the basis of the theory of cladism, it was shown that, because of the phenomenon of non-apotypy, it may be anticipated that all three pairs of any three arbitrarily chosen taxa will have characters in common. It was also concluded, however, that if phylogenetic classification is possible at all, then one of the pairs must stand out as having more characters in common than any of the other pairs. As is seen in Figure 3.17, this prediction is met with beyond expectation, the number of characters represented by the line MV being substantially higher than those corresponding to VE and EM.

It is of some interest to appraise the purportedly non-apotypic characters in the last two groups. The three features uniting Vertebrata and Echinodermata are presence of dermatan sulphate (Table 3.4), absence of chitin (Table 3.10) and adrenaline as a cardioaccelerator (Table 3.17). As discussed above, the first character is dubious, but if it is accepted, its occurrence in the two taxa must be an instance of convergence. The other two features may easily be explained by non-convergent apotypy, involving either losses or gains. In any case, I am convinced that nobody would venture far-reaching conclusions concerning the phylogenetic kinship between the taxa on the basis of these features.

The properties common to Echinodermata and Mollusca are calcium carbonate in the skeleton (Table 3.11), amino acid excretion (Table 3.20) and an integumental nerve plexus (Table 3.24). These characters are widespread in invertebrates, and are clearly plesiotypic ones which have been lost in Vertebrata. Hence, there is some justification for the proposition that, in agreement with the prediction of cladism, two of the three pairs of taxa possess only non-apotypic characters in common, those uniting Mollusca and Vertebrata thus being apotypic ones.

It was stated above that the result obtained is beyond expectation. This is strongly borne out when the data are used to calculate the value of t_1/t_2, the relative age of the primary and the secondary twin taxa. With $Q = 50/3 \sim 17$, we get $t_1/t_2 = 9$, and this value is evidently too large to be acceptable. In my opinion, the reason for this is that not all of the characters uniting Mollusca and Vertebrata fulfil the requirements stated above. I have suggested in Chapter 2 that apotypic convergence is much more likely than non-apotypic, because the acquisition of one character may be the prerequisite for the establishment of another arising through mutation. Therefore it is not at all unlikely that several characters are convergent. By this concession I expect to avert much criticism. More important is the other alternative, namely, that some characters are interdependent. I have mentioned one possible instance of this kind, the involvement of hyaluronate and chondroitin sulphate in the formation of cartilage. Many more cases may be represented in the various tables, but at our present stage of knowledge this question cannot be settled.

This circumstance implies that the characters used here are inferior to those obtained by molecular–biological methods as concerns the determination of the relative age of the taxa. On the other hand, since the present approach leads to an overestimation of the apotypic characters, it should lead to a safer establishment of the phylogenetic relationship.

As a mere supposition, I shall submit that only nine of the 50 apotypic characters are valid in the present context. This will lead to a value for t_1/t_2 of 2, probably the highest which can be accepted.

I shall furthermore postulate that, considering the nature of these characters, none of them can arise with the high frequency of the mutations observed in genetical studies, about 10^{-5}. It is probably modest to propose that their frequency corresponds to that of two consecutive, co-operating mutations, i.e. 10^{-10}. It will suffice in this context, however, since it establishes that, in the present basic classification, the secondary twin relationship between Vertebrata and Mollusca is more probable than is the one between the former and Echinodermata by the factor $10^{10(9-3)} = 10^{60}$.

Hence the basic classification between the three taxa must be represented by Figure 3.19, and our final conclusion may, with reference to Hennig's first theorem (T2.41), be phrased as follows:

The members of the taxa Mollusca and Vertebrata have ancestors in common that were not ancestors to the members of the taxon Echinodermata.

Since this classification differs from the one which might be anticipated from the current superphyletic classification, we may obviously further deduce:

Bateson–Grobben's superphyletic classification has been falsified through empirical observations and can no longer be upheld.

As we have already seen, this system never rested on a firm foundation, and I believe that the overwhelming evidence accumulated against it here must convince any biologist who admits that his trade is based on reasoning, rather than upon intuition and orthodoxy, that it must be abandoned.

Figure 3.19. Phylogenetic classification of the taxa Echinodermata, Mollusca and Vertebrata, as established on the basis of non-morphological characters. Echi, Echinodermata; Moll, Mollusca; Vert, Vertebrata

It is important to note, however, that the first conclusion does not imply that Mollusca are the unique invertebrate taxon whose recognition is the goal of phylogeneticists tracing the ancestry of Vertebrata. The reason for this is that the taxon Mollusca was arbitrarily chosen as a representative of the annelid superphylum. Clearly, it is necessary to continue the cladistic analysis, substituting other taxa for Echinodermata, to see if it is possible to find one which is still closer related to Vertebrata than Mollusca. Theoretically, it might even be necessary to pursue this search outside the annelid superphylum, but personally I believe that this would be unrewarding. However, as was announced in the introduction to the present section, we shall subject some arthropod taxa to a basic classification with Mollusca and Vertebrata, to see whether we can further narrow down which invertebrates are the closest kin to vertebrates.

I must mention here that, after the first writing of the present chapter was finished, a paper by Packard (1972) appeared in which the similarities between cephalopods and fishes, particularly on the level of functional 'adaptation', are discussed in great detail. Many of the points covered by Packard have also been dealt with by me but, over and above these, several further striking instances of 'convergence' are brought out. Some are morphological, and hence outside the scope of the present chapter, but functional features are also included, of which some have been overlooked by me while others have been reported too late to come to my attention. It might have been possible to include these here, an expedient which would, I believe, have strengthened my case. However, since it would not have changed my final conclusion, which, needless to say, deviates from that of Packard, I have refrained from making any changes in the manuscript.

PROTOCHORDATA

As regards Protochordata, the results summarized in Figure 3.18 indicate kinship with both Mollusca and Vertebrata, whereas the affiliation to Echinodermata is negligible. It therefore appears that, *vis-à-vis* Echinodermata, both protochordate taxa are secondary twins to either Mollusca or Vertebrata, or to both.

For reasons of expediency, the survey of the protochordates was not made according to cladistic rules, so that some of the characters represented in Figure 3.18 are common to Mollusca, Vertebrata and the protochordate taxon in question. For a proper evaluation of the phylogenetic relationships it is obviously necessary to omit all these plesiotypic characters.

If this is done, we arrive at the result that Urochordata have five characters in common with Mollusca and none with Vertebrata, whereas Cephalochordata have four characters in common with Mollusca and two with Vertebrata (Figure 3.20).*

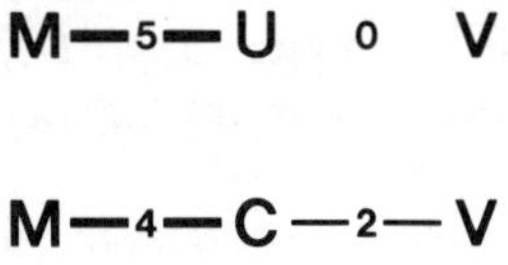

M—5—U   o   V

M—4—C—2—V

Figure 3.20. Final score for the phylogenetic relationship between Protochordata (U, Urochordata; C, Cephalochordata), Mollusca (M) and Vertebrata (V), respectively, on the basis of non-morphological characters. The numbers indicate characters in common and the widths of the connecting lines are proportional to these numbers

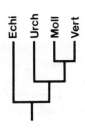

Figure 3.21. Phylogenetic classification of the taxa Echinodermata, Mollusca, Urochordata and Vertebrata as established on the basis of non-morphological characters. Echi, Echinodermata; Urch, Urochordata; Moll, Mollusca; Vert, Vertebrata. It will be appreciated that if Urochordata arose within the phylum Mollusca, a possibility which cannot be excluded on the basis of either the present investigation or morphological considerations, then a branch leading to Mollusca should originate from the line (taxon) including Urochordata and Mollusca plus Vertebrata

In view of the many characters common to Mollusca and Vertebrata, the former result seems reasonably unambiguous. Even without an analysis of the properties that unite Urochordata and Mollusca, it seems justified to infer that the former taxon is the primary twin to Mollusca and Vertebrata. Although the number of characters is rather small in this case, I shall nevertheless propose the following conclusion regarding the origin of urochordates (Figure 3.21):

The members of the taxa Urochordata and Mollusca plus Vertebrata have ancestors in common that were not ancestors to the members of the taxon Echinodermata, and the members of the taxa Mollusca and Vertebrata have ancestors in common that were not ancestors to the members of the taxon Urochordata.

The situation is less clear-cut with respect to Cephalochordata, particularly since there are certain characters that unite them with Vertebrata. As it happens, these are a tubular CNS (Table 3.24) and iodine incorporation in the endostyle (Table 3.25). As I have argued above, the former feature is an epigenetic consequence of the stretching effect of the notochord, hence this character is not independent. Since, in my opinion, convergence with respect to notochord formation is an acceptable proposition, the same evidently holds for a tubular CNS. As far as iodine incorporation is concerned, there are several alternatives to account for its occurrence in cephalochordates and vertebrates, but I believe a discussion of this question to be premature so long as we do not know more about iodine incorporation in molluscs. In any case, the characters associating the Cephalochordata with either Mollusca or Vertebrata are so few compared with

*I have recently learnt that one more feature, Reissner's fibre, unites the chordate taxa (Olsson, 1972). Since this character cannot upset my conclusions, I have decided not to make the appropriate changes in the text.

those uniting the latter that it may be permissible to state the following provisional conclusion:

The members of the taxa Cephalochordata and Mollusca plus Vertebrata have ancestors in common that were not ancestors to the members of the taxon Echinodermata, and the members of the taxa Mollusca and Vertebrata have ancestors in common that were not ancestors to the members of the taxon Cephalochordata.

The data presented above do not bear upon the relations between the two proto-chordate taxa. However, since these two subphyla have very few characters in common that are not shared by Mollusca or Vertebrata, I do not find much support for the supposed affinity between these two taxa. Without being able to substantiate this view, I shall therefore submit the following proposition:

The members of the taxa Urochordata and Protochordata have no ancestors in common that are not also ancestors to the members of some other invertebrate taxon.

'Arthropoda'

As already discussed, the proposal that the vertebrates arose from arthropod ancestors was advanced by Gaskell (1908) and Patten (1912), whose Arachnid theory was the subject of lively debates around the turn of the century. Both authors contended that the vertebrates originated in the distant past from animals belonging, presumably, to Merostomata, and that *Xiphosurus polyphemus* (Limulus) and a few other related species are the closest living relatives to the vertebrates.

Merostomata belong to the taxon Chelicerata, but Gaskell (1908) also demonstrated a certain affinity between Crustacea and Vertebrata. For this reason we must first con-sider the relation between Chelicerata and Crustacea. Various authors have maintained that both of these taxa have been derived from trilobites or trilobitomorphs. According to Størmer (1949) only the affiliation between the latter and Chelicerata is beyond doubt, and this author even suggests that the line in question has evolved directly from an annelid ancestor, independently from the Crustacea. However, new discoveries con-cerning the anatomy of the extinct trilobites have demonstrated the affinity between these and both of the two extant arthropod taxa (Cisne, 1974).

Therefore, if the latter are classified in a basic classification together with Mollusca, the two arthropod taxa should be the secondary twins (Figure 3.22). This proposition is supported by serological data (Leone, 1954).

A survey of the literature shows that relevant data are largely lacking for Chelicerata, so that it is impossible to use this taxon for a basic classification. On the presumption made above, I shall therefore deal only with the taxon comprising both Chelicerata and Crustacea, for which I shall adopt the name 'Arthropoda'. Although the two taxa are evidently quite closely related, it is by no means certain that 'Arthropoda' is a mono-phyletic taxon: this question can be answered only through cladistic analysis. If the

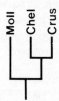

Figure 3.22. Proposed phylogenetic classification of the taxa Chelicerata, Crustacea and Mollusca. Chel, Chelicerata; Crus, Crustacea; Moll, Mollusca

answer is negative, I am committing a formal error by using the taxon 'Arthropoda' here. It should be noted that even if this taxon should turn out to be monophyletic, it does not follow that this is also true for Arthropoda: the opposite alternative is indicated by the demonstration that Uniramia (Onychophora, Myriapoda and Hexapoda) are, at the most, distantly related to the other members of the phylum (Manton, 1973; D. T. Anderson, 1973). But the phylogenetic relationship between the members of the phylum Arthropoda is of no relevance in the present context since, according to the rules of cladism, any set of three monophyletic taxa can be classified in a basic classification.

The problem posed in the present section may be solved by establishing a basic classification involving Mollusca, 'Arthropoda' and Vertebrata. We shall carry out this task in two steps by first trying to establish whether 'Arthropoda' is the primary twin taxon, and then, since that seems not to be the case, to ascertain to which of the other taxa it is a secondary twin.

PRIMARY TWIN RELATIONSHIP

The question to be settled in the present section is whether or not all the properties that were found to be common to Mollusca and Vertebrata are also found in 'Arthropoda'. Since so much more is known about Crustacea than about Chelicerata, we shall approach this question with reference to the former taxon. Although we do not know about the presence of several characters in Chelicerata, this procedure is in complete accord with the cladistic method, for those characters that are common to the other three taxa must be plesiotypic characters which may be expected even in Chelicerata.

It is obvious that in this classification we must allow for the possibility that information is missing about some characters, that losses or substitutions have occurred and, of course, that some instances of convergence obtain between the three pairs of taxa but these are the only acceptable reservations with respect to the implication of the word 'all' used above.

In Table 3.28 the outcome of this survey is presented. A few comments may be necessary. (7) This character is not well defined, but since partly or wholly uncalcified exoskeletons occur, the presence of 'horny' ectodermal skeletons is indicated. (9) Haemoglobin is found in the muscles and in other tissues of crustaceans, with the exception of malacostracans. (11) Great variation obtains with respect to gill morphology within Crustacea. (12) The heart has a single chamber, but a bulbus arteriosus is present. (13) The irregular three-dimensional network of myocardinal fibres is found only in decapods. (14) In molluscs the pericardium is the coelom, in crustaceans it is a blood sinus, forming part of the haemocoele, while in vertebrates the pericardium and the coelom are separate cavities, connected in hagfishes, but not in any other taxon. (21) At least two types of blood cells are found. (22–23) The hepatopancreas has been shown to carry out several of the functions exerted by the vertebrate liver and pancreas. (25–28) The antennal gland may be considered a specialized excretory organ, but gills, digestive glands, etc., are also involved in ionic regulation and in excretion. This situation corresponds to the one described above for the hagfishes. There is disagreement as to the extent to which the processes of filtration, absorption and secretion are involved. (29) 'At the moment, the occurrence of the urea cycle in Crustacea is far from demonstrated but cannot be completely ruled out' (Schoffeniels and Gilles, 1970, p. 218). (33) 5-HT affects certain chromatophores. This substance, or a closely related one, seems to occur in crustaceans, but it is not known whether it is concerned with the control of the chromatophores. (35) Chemoreception is well established, but the nature of the receptor organs is not. (40) Myelinated nerves are found in the crustaceans. (50) In malacostracans a male hormone is produced by an androgenic gland.

Table 3.28

Survey of the presence in Crustacea of the characters common to Mollusca and Vertebrata

(1) Chondroitin (not known)	(26) Filtration $(+)^{10}$
(2) Hyaluronate $+^1$	(27) Absorption $(+)^{10}$
(3) Keratan sulphate $+^1$	(28) Secretion $(+)^{10}$
(4) Epidermin $+^2$	(29) Ornithine cycle $(-)^{10}$
(5) Cartilage $-^3$	(30) Trimethylamine oxide $+^8$
(6) Calcified ectodermal skeleton $+^2$	(31) Rhodopsins $+^{11}$
(7) Horny ectodermal skeleton $(+)^2$	(32) Chromatophores, central control $+^{12}$
(8) Structure of dermis $+^4$	(33) Chromatophores, 5-HT $(+)^{13}$
(9) Myoglobin $+^5$	(34) Optic sense organs $+^{14}$
(10) Amino acid composition of haemoglobin (not known)	(35) Olfactory sense organs $+^{14}$
	(36) Static sense organs $+^{14}$
(11) Vascularized, ciliated gills $(+)^6$	(37) Proprioceptors $+^{14}$
(12) Chambered heart $(+)^7$	(38) Brain $+^{14}$
(13) Myocardial structure $(+)^7$	(39) Ganglia $+^{14}$
(14) Pericardium $+^7$	(40) Nerves $+^{14}$
(15) Accessory hearts $+^7$	(41) Oligodendroglia $+^{14}$
(16) Blood vessels $+^7$	(42) 'Astrocytes' $-^{14}$
(17) Valves in hearts and vessels $+^7$	(43) Multipolar heteropolar neurons $+^{14}$
(18) Elastic fibres in arteries $+^7$	(44) Neurons of varying size $+^{14}$
(19) Capillaries $+^7$	(45) Giant coordinating neurons $+^{14}$
(20) Nervous regulation of heart $+^7$	(46) Control of sexual maturation $+^{15,16}$
(21) Specialized blood cells $(+)^7$	(47) Neurohaemal organs $+^{17}$
(22) Liver tissue $+^8$	(48) Transformed ganglia $+^{17}$
(23) Pancreas tissue $+^8$	(49) Epithelial glands $+^{17}$
(24) Regulation of several ions $+^9$	(50) Sertoli cells $-^{15}$
(25) 'Kidney' $(+)^{10}$	

1, Rahemtulla and Løvtrup (1976); 2, Richards (1951); 3, Person and Philpott (1969); 4, Haeckel (1857); 5, Goodwin (1960); 6, Wolvekamp and Waterman (1960); 7, Maynard (1960); 8, Vonk (1960); 9, Robertson (1960); 10, Parry (1960); 11, Wald (1960); 12, Kleinholz (1961); 13, Welsh (1961); 14, Bullock and Horridge (1965); 15, Charniaux-Cotton (1960); 16, Wells (1960); 17, Highnam and Hill (1969).

It thus appears that the characters common to Mollusca and Vertebrata are to a large extent found in Crustacea also. That some are missing was to be anticipated, and may be explained as suggested above. Hence it seems justified to submit that most of the characters in Table 3.28 are plesiotypic characters in the basic classification involving the three taxa. This implies that 'Arthropoda' may be a secondary twin to either Mollusca or Vertebrata, and the resolution of this question is the subject of the following section.

SECONDARY TWIN RELATIONSHIP

In view of the reasonably exhaustive inquiry which preceded the compilation of the characters in Table 3.28, it is rather unlikely that further characters can be found to unite Mollusca and Vertebrata, and the primary task in the present section must therefore be to search for characters which may unite 'Arthropoda' and Vertebrata.

I shall begin with the phenomenon of metamery. Metameric segmentation of the coelom is much more than a matter of abstract form; it is a process which occurs in a material substrate in the course of embryogenesis. The fact that it exists in some animals but not in others must mean that only when the substrate possesses certain, for the time being unspecifiable, qualities does segmentation take place. By including metamerism in the list of characters in common between 'Arthropoda' and Vertebrata it is therefore implied that, among the substrate elements available for the construction of the bodies, one is common to these two taxa, while being absent in most Mollusca, or else a further factor is present in the latter which serves to suppress metameric segmentation.

Waterman (1961) has enumerated various physiological characters that are common to Crustacea and Vertebrata. Most of these have been dealt with in Table 3.28, but there are some remarkable similarities besides these (Table 3.29). Thus, in some crustaceans the blood is low in Mg^{2+} and high in Na^+, matching that found in Vertebrata and different from that obtaining in all other invertebrates, including molluscs (Robertson, 1960). Likewise, blood coagulation occurs in crustaceans (Florkin, 1960b). The mechanism is much simpler than the one in vertebrates, and this difference is usually strongly emphasized because the similarity is 'unexpected'. However, the very fact that it is based on a fibrinogen–fibrin transformation may be more than a striking coincidence. The growth regulation through a hormone produced by the eyestalk, which in many other ways functions as a pituitary (Passano, 1960), may also be noted as a character in common. Finally, it appears that the crustacean proprioceptors are closely similar to those found in vertebrates (Cohen and Dijkgraaf, 1961).

Waterman's work (1961) is based upon a very exhaustive review of the physiology of crustaceans. Therefore, since to my knowledge no dramatic new discoveries have been made in the intervening period, it seems unlikely that it will be easy to find more than these characters in common between the two taxa.

If Chelicerata and Crustacea are the secondary twins in a basic classification involving these taxa and Vertebrata, a question which I shall not attempt to settle, then, barring non-apotypy, the characters discussed so far should also be present in the former taxon.

Table 3.29
Characters common to 'Arthropoda'
and Vertebrata

Metamery
High Na^+, low Mg^{2+}
Blood coagulation
'Pituitary' growth regulation
Proprioceptors
Cartilage
Dermal skeleton
Two pairs of eyes
Median eye, upright simple retina
Lateral eye, inverted retina
Lateral eye, compound retina
Histology of retina

At present nothing is known about the blood ionic composition, blood coagulation does not occur (Grégoire and Tagnon, 1962), some kind of hormonal regulation of growth and moulting may be inferred, and proprioceptors are deniably present, but they have not been described.

In evaluating the characters which are common to Chelicerata and Vertebrata I shall rely upon the lists of characters which were mobilized by Gaskell (1908) and Patten (1912) in support of their theory. I shall not, however, include the various morphological characters which, once their view of the origin of Vertebrata is accepted, appear strongly suggestive, if not necessarily conclusive.

The first character to add to the list is cartilage which, although missing in Crustacea and therefore in Table 3.29, is found in Limulus (cf. Person and Philpott, 1969). Further, if comparison with extinct forms is allowed, we may also note certain similarities between the integumental skeleton of Limulus and the extinct ostracoderms, a point which was particularly emphasized by Patten. The possession of two pairs of eyes, one median and one lateral, distinguishes Chelicerata, and traces of the median eyes are found in the pineal eye or gland in Vertebrata.

Gaskell paid particular attention to the retina, showing that in Chelicerata, Crustacea and Vertebrata the median eye has an upright, simple retina. As concerns the lateral eye Chelicerata and Vertebrata have in common the fact that the retina is inverted, Crustacea and Vertebrata that it is compound. As far as the histology of the retina is concerned, similarities are again observable. It must be emphasized that Gaskell's studies did not concern the morphology of the eye. On this point obvious and well-known differences exist.

The compilation of characters from two distinct taxa carried out in Table 3.29 is not quite up to cladistic standards. Even so, the number of characters supporting the secondary twin relationship between 'Arthropoda' and Vertebrata is not sufficiently large to warrant any conclusions. However, I shall submit the following hypothesis, which may deserve further scrutiny:

The members of the taxa 'Arthropoda' and Vertebrata have ancestors in common that were not ancestors to the members of the taxon Mollusca.

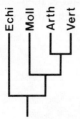

Figure 3.23. Phylogenetic classification of the taxa 'Arthropoda', Echinodermata, Mollusca and Vertebrata, as established on the basis of non-morphological characters. Arth, 'Arthropoda'; Echi, Echinodermata; Moll, Mollusca; Vert, Vertebrata

The proposed classification is shown in Figure 3.23. Should it be further corroborated, it clearly constitutes support of the ideas advanced by Gaskell (1908) and Patten (1912).

On the basis of the empirical evidence he had marshalled, Gaskell (1908, p. 497) felt entitled to assert that his theory 'strikes at the foundation of [the theory of parallel development]', i.e. of convergence. Indeed, what he anticipated was nothing but the rule of minimizing the cases of convergence. Since many readers will unquestionably be very sceptical towards the classification proposed here, I shall stress once more that this rule, based on the principle of parsimony, is the *sine qua non* in phylogenetics.

Assessment of the classification

Since the conclusions as to the ancestry of the vertebrates arrived at on the basis of the various chemical, physiological and histological–anatomical characters differ radically from the traditional view founded upon morphological features, it is necessary to ascertain whether any observations in the field of morphology, besides those discussed above, may refute or corroborate the outcome of our analysis. In addition, after I had written the present chapter, some biochemical data became available which may serve to test the proposed classification. Being a convinced Popperian, I shall not miss this opportunity to try to falsify the theory established concerning the origin of the vertebrates. Anticipating that this testing, in my opinion at least, does not lead to refutation of the conclusions, we shall finally discuss the implications of the latter with respect to the classification of some of the taxa which have been dealt with in the analysis.

MORPHOLOGICAL CHARACTERS

Morphological data which may be used for the testing of the theory are to be found partly in the fossil record and partly in the observations made by comparative morphologists and embryologists. These aspects will be discussed in separate subsections.

The palaeozoological record

The remarkable similarities in the external morphology of the extinct merostomes and ostracoderms were, undoubtedly, one of the factors which inspired Gaskell (1908) and Patten (1912) to put forward the Arachnid theory. They claimed that through acceptance of the latter the emergence of vertebrates would cease to be shrouded in mystery. The fact that the postulated ancestors were abundant in the Cambrian and in the Ordovician is certainly in favour of their theory, and so is the fact that, as appears from the preceding discussion, many of the functions and differentiation patterns which distinguish the vertebrate body must have been present in arthropods living at the time.

One stumbling-block which makes it difficult for palaeozoologists to accept the Arachnid theory is that bone has never been observed in any invertebrate. However, a closer scrutiny will show that in this respect the transition between 'Arthropoda' and Vertebrata may involve only slight changes, and if this is true then it can hardly be claimed that there is an abyss between invertebrates and vertebrates from the point of view of osteology.

Two features distinguish bone, namely, that the inorganic phase consists almost exclusively of calcium phosphate, and that it is a mesodermal differentiation pattern. As far as the first point is concerned, it should be noted that considerable quantities of calcium phosphate have been recorded in the carapaces of some trilobites (Størmer, 1949) and crustaceans (Richards, 1951). Thus, the differentiation pattern involving precipitation of calcium phosphate, rather than calcium carbonate, does occur in the line postulated to lead to the vertebrates, the only change required for bone formation, as far as the inorganic phase is concerned, being a quantitative one.

Secondly, as we have seen, cartilage, the matrix required for mesodermal calcification, is found in xiphosurans. And although the carapace of these animals is not calcified, it may not be too fantastic to postulate that there existed in the past animals possessing in their genome the information required on the one hand for the production of cartilage and on the other for the precipitation of inorganic calcium phosphate in their carapace. In these organisms a mutation involving a regulatory gene would suffice to ensure that the 'calcification gene' became activated in mesodermal cells also, and thus that bone formation would take place.

However this may be, it is a consequence of the Arachnid theory that the first vertebrates had a calcified exoskeleton, implying that in theory we should be able to follow their history back almost to the beginning. This does not mean, of course, that the earliest *known* vertebrates were actually the first existing ones.

I am not qualified to discuss in any detail the properties of the extinct vertebrate ancestors, but a few points may be mentioned. For one thing, with reference to Hyperotreta it is reasonable to presume that these animals lived in the sea. This question will be taken up later.

Secondly, we shall consider the problem of moulting. Patten (1912, p. 301) stated that 'in Limulus we see the beginning of a new type of exoskeleton, one that is subdermal and discontinuous, that need not be, and never is cast off after it once is formed'. This supposition has turned out to be wrong (Henriksen, 1931).

Owing to the fusion between endoskeleton and exoskeleton in the ostracoderms it is likewise assumed in this case that no form of moulting could occur and that the cephalic shield was formed by the fusion of smaller scales when the animal had reached its final size (Stensiö, 1958). It seems that this proposition is supported by certain changes occurring in the dermis of the metamorphosing ammocoete larva of Hyperoartii. In the latter, as in *Myxine*, the dermis is compact and hence of a type where the physical conditions for the formation of vascularized bone are lacking. As metamorphosis begins in the ammocoete larva, the collagenous bundles in the dermis are separated from each other, 'fibroblasts' becoming visible between them (Johnels, 1950). This indicates that the cells are engaged in a copious production of glycosaminoglycans, and the outcome of this

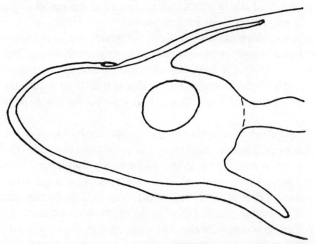

Figure 3.24. Anterior part of the head of *Petromyzon* (10 ×). In the outlined region the dermis is transformed during metamorphosis. (Reproduced from A. Johnels, *Acta Zool.*, **31** (1950), by permission of *Acta Zoologica* (Stockh.))

activity is that the dermis comes to consist of three layers, a superficial dense layer of collagenous bundles, the basement lamella, beneath which most of the chromatophores are located, next a thick layer of vascularized, loose connective tissue, and finally another collagenous layer. The process described here is confined to the head region, as shown in Figure 3.24. As emphasized by Johnels (1950), this supports the affinity between Hyperoartii and the extinct Cephalaspidomorphi, distinguished by the possession of a cephalic shield containing vascular bone while the rest of the body was covered by structures comparable to fish scales (Stensiö, 1958). It may be mentioned that a closely similar view was stated by Gaskell (1908).

The particular shape of the carapace in the arthropods is the outcome of an interaction between the exoskeleton and the body proper which can be traced back to very early ontogenetic stages. If, as proposed, the cephalic shield in the ostracoderms was formed by fusion of smaller elements at a late ontogenetic stage, then the latter cannot have been responsible for the shaping of the former, and since the same must hold for the soft parts of the body, it seems to follow that the endoskeleton is the only element that can be invoked in the determination of the weird forms of the head shields so typical of the ostracoderms.

Should this be true, then the similarity between the skeletons in these extinct animals would seem to be a matter of coincidence. Before this point is taken to be an argument against the Arachnid theory, it should be recalled that a morphogenetic mechanism of the kind outlined above has never been observed in any animal.

According to current notions the ostracoderms were jawless, bottom-dwelling food-strainers, altogether quite helpless animals when compared to the contemporaneous 'arthropods'. This image constitutes, so far as I can see, one of the most serious arguments against the rise of the vertebrates for, if the ostracoderms were so poorly endowed, it is very difficult to see how they could assert themselves, except through some kind of specialization which, purportedly, might bar the chances of further evolutionary advance. This reflection does not preclude that current views are correct but, if so, it appears that the survival of vertebrates must have involved an enormous amount of luck.

In the fossil record there is one element which is taken as supporting the origination of the vertebrates from the echinoderms, viz., Calcichordata (Jefferies, 1968). In Chapter 5 I shall make an attempt to relate some of the more important fossils to the classification of the living vertebrates, but I shall not do anything like that with respect to calcichordates. I admit that, wherever they are collocated in the dendrogram in Figure 3.23, it becomes necessary to grant a certain measure of convergence, and it is hardly possible to make a valid estimation of the classification which allows for the minimization of this phenomenon. I should like to emphasize, however, that the various 'chordate' characters that have been observed, notochord, post-anal tail, dorsal nerve cord, segmented ganglia and myotomes and a brain, are not independent. Rather, if the interpretation of the epigenesis of the 'chordate' body plan proposed above and elsewhere (Løvtrup, 1974) is correct, then all the remaining characters may probably be causally related to the elongation of the embryo arising through the stretching of the notochord.

It must be noted, however, that the difficulty pertains only to the relationship of Calcichordata to the other taxa dealt with here and has nothing to do with the phylogenetic classification of the latter. The reason for this is that, as we have concluded above, fossils should have no effect on the kinships established between the various taxa in the phylogenetic hierarchy.

Comparative aspects

That many striking similarities obtain on the functional level between the vertebrates and members of the annelid superphylum is, of course, a perfectly well-known matter of fact, even if the present compilation may perhaps serve to emphasize the truly remarkable degree of concordance.

On this basis, evidently, the Arachnid theory goes a very long way towards 'lightening the burden' of evolution, and one might therefore surmise that it would have been embraced with enthusiasm, and the fathers of the theory hailed as towering figures in biology. Yet what happened was that all the resemblances were discounted as instances of convergence and the theory almost unanimously disclaimed.

This attitude might be understandable if the opponents had a better substantiated way to account for the origin of vertebrates but, as we have seen above, this is hardly the case. The principal reason for their rejection of the theory was rather that as morphologists they were completely unable to envisage the transformation of the *adult* body of an 'arthropod' into the *adult* body of a vertebrate. As far as this point is concerned, the greatest obstacle, both as regards antagonists and protagonists, seems to lie in the fact that in the 'arthropods' the central nervous system is located ventrally, in the vertebrates dorsally. In order to account for this discrepancy, Gaskell (1908) and Patten (1912) both resorted to credulous formanalytical stratagems. Thus, the former presumed that the tubular intestine in an invertebrate ancestor became the neural tube in the vertebrates, a new alimentary canal being subsequently formed from the epidermis. Patten, on the other hand, simply proposed that the origination of vertebrates involved their forebears turning upside-down.

On the question of *how* the vertebrates arose, Gaskell and Patten thus were at odds and, in my opinion, they were both wrong. The flaw in their reasoning was that they attempted to derive one adult body plan from another. In spite of the fact that both of them stressed the importance of embryology, they did not realize that the clue to their problem lay in an understanding of the epigenetic mechanisms responsible for the transformation of the fertilized egg to the adult body. In order to explain the difference between the annelid superphylum and Vertebrata as regards the location of the central nervous system, we must begin by comparing the fate maps in the blastulae of the two groups (Figure 3.25). In drawing these fate maps I have committed an unorthodoxy by

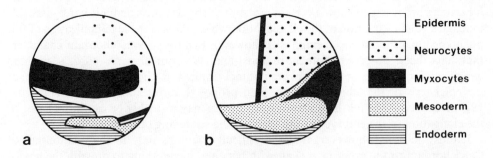

Figure 3.25. Comparison of the fate maps of the blastula in Polychaeta and Amphibia. a, *Podarke*; b, *Triton*. The fate maps are based on figures published by D. T. Anderson (1973) and W. Vogt (1929)

using the same code for the prototroch and telotroch in the annelid, and for the noto-chord and neural crest in the amphibian. The reason for this is that, in my interpretation (Løvtrup, 1974), these several primordia represent the same differentiation pattern, myxocytes, distinguished by the production of glycosaminoglycans. And from an epigenetic point of view it is probably more important to know about the chemical activities of the various cells than about the morphological structure they happen to contribute to. It must be remembered, of course, that not all of the neural crest cells end up as myxocytes. As is well known, some become neurocytes, others melanocytes.

Considering the time which has elapsed since these two taxa became separated, I think it is justified to claim that the similarity between the fate maps is striking and, although neither of them may be representative of the ancestors of 'Arthropoda' and Vertebrata, it is nonetheless possible to use them for a discussion of the epigenetic mechanism required to account for the postulated transformation between the two taxa.

Thus, it will be noticed that in both cases the primordium of the nervous system is located at the dorsal side, where it remains in the vertebrates but not in members of the annelid superphylum. The means by which the majority of the presumptive neurocytes leave their station in the latter case are well known. Two distinct mechanisms have been recorded: the cells may either spread over the surface of the embryo through epiauxesis (cf. Løvtrup, 1974), or they may detach themselves, enter the blastocoele and settle at the ventral surface of the latter. The second alternative may be anticipated in 'Arthropoda'.

If we compare this process with that occurring in Amphibia, the first and very remark-able thing to notice is that far from all neurocytes stay in the blastocoele wall. Many, notably the neural crest cells, but also some neural plate cells (cf. Løvtrup, 1974), enter the blastocoele where they form ganglia, etc. But most of the cells never detach them-selves, and they are the ones that subsequently contribute to the formation of the central nervous system. The reason for this particular behaviour is to be found in the fact that these cells are firmly bound to the surface coat,* as indicated by Holtfreter's observations on the properties and functions of the latter structure (1943). The process of gastrulation through epiauxesis occurring in many annelids and molluscs testifies to the absence of a syncytial surface coat. Yet the presence or absence of this structure does not constitute a distinction between Protostomia and Deuterostomia. The essential point is, that in the vertebrates the presumptive neurocytes to a large extent preserve their dorsal superficial location, in the annelid superphylum they do not. It is therefore possible to characterize the difference between the two groups through the presence and absence, respectively, of an agent serving to anchor most of the neurocytes at the embryonic surface. There is therefore no reason to reject the Arachnid theory on the basis of the explanations sub-mitted by Gaskell and Patten to account for the transformation from invertebrates to vertebrates; a simpler, and epigenetically much more likely, interpretation is possible.

Another distinct difference between the two taxa is the presence of a notochord in Vertebrata. This structure, which is probably the most important morphogenetic factor in the epigenesis of the vertebrates, must, in order to exert its proper function, be formed immediately after gastrulation. The differentiation patterns required for the construction of a notochord, myxocytes and collagenocytes, are ubiquitous in the annelid superphylum

---

*It seems that the existence of a particular surface coat may be questioned, but that does not detract from Holtfreter's observation that, in contrast to the interior ones, the external cells form a syncytium, possibly being kept together by cell junctions.

and, indeed, in most invertebrates. The fact that under these circumstances notochords are not more frequently formed, for instance in the annelid superphylum, shows very clearly that other factors are required to secure this outcome. I do not suggest that the conditions required for the creation of a notochord-possessing 'hopeful monster' are easily established in an 'arthropod' egg, but once this happened a process would be set up among the causal consequences of which might be a number of the features of the vertebrate body plan: stretching of part of the central nervous system, segmentation of the latter as well as of the mesoderm, etc., just as during normal embryogenesis. Should this happen in an egg in which a lot of information about useful differentiation patterns was encoded in the genome then, obviously, the chances of success would be greatly enhanced.

The views expressed here suggest that if the problem is approached from an epigenetic, rather than from a morphological, angle, then the transformation between 'Arthropoda' and Vertebrata becomes much less formidable than was formerly believed.

Much of the criticism directed against the Arachnid theory involved a plain rejection of the mechanisms of transformation invoked by Gaskell (1908) and Patten (1912). In particular, it was stressed that by proposing a conversion of the skin to gut, and of the gut to neural tube, Gaskell completely negated the germ-layer theory (e.g. MacBride, in Gaskell, 1910). This is, of course, completely correct; each of the tissues mentioned represents a differentiation pattern with a long phylogenetic history and, as we have seen, they are always formed at a circumscribed location in the embryo, the gut towards the vegetal pole, the others at the animal end and, usually, the presumptive nervous tissue at a dorsal location, the epidermis largely at the ventral side.

I have not made a close scrutiny of the literature, but so far as I can judge the objections raised against the transformation mechanisms proposed by Gaskell and Patten are both serious and valid; however, this does not hold for those directed against the submitted affiliation between 'Arthropoda' and Vertebrata. In a discussion on Gaskell's theory held in 1910 at the Linnean Society of London (Gaskell, 1910), Gadow pleaded for giving the Arachnid theory a chance. And the best chance one can give an empirical theory is to subject it to Popperian attempts at negating its predictions. Rather than trying to procure falsifying evidence, most of the opponents preferred to rely on pure reason by advancing formanalytical counterarguments.

Maybe the time is ripe to bring forward the Arachnid theory and subject it to relevant testing. We may not expect much assistance from morphologists in this endeavour, not necessarily because of unwillingness, but because, as emphasized above, morphology is of little help in superphyletic classification. From the present survey it appears that the known physiological and other functional characters may suffice to disprove the Protostomia–Deuterostomia theory, but they are not more than suggestive with respect to the decision about the taxon within the annelid superphylum which will eventually turn out to be the twin taxon of Vertebrata. Yet if morphologists and physiologists must thus acknowledge their impotence, we must rely on our molecular colleagues for assistance. As we shall see in the following section, their contribution to the solution of the problem of the ancestry of the vertebrates has already been initiated.

It follows from the preceding discussion that, in my opinion at least, evidence concerning the morphology of adult animals simply cannot be used for the purpose of falsifying the present theory but neither, of course, for its corroboration. The only morphological data which may potentially be marshalled in support of the conclusions arrived at above pertain to eggs. But here we encounter the difficulty that the discernible properties of these entities are so few, and the possible range of variation so slight, that they are of little help in the present context. The kind of information in the egg which exhibits the largest variability is the one contained in the genome, and this is the one which has been dealt with indirectly in the preceding survey.

If, then, comparative oology cannot, at present at least, be expected to either refute or corroborate the thesis about the origin of Vertebrata from 'Arthropoda', it may nonetheless be worth pointing out the striking similarity of the eggs of xiphosurans and hagfishes. In both cases the eggs are large, elongate and encapsulated, with discoidal cleavage (Dean, 1899; D. T. Anderson, 1973).

It is easy enough to dismiss this resemblance as due to convergence, but it is quite remarkable that the hagfish, which should be closer than any other vertebrate to the 'generalized forms' from which the latter supposedly sprang, bases its ontogenesis on such large eggs, which, furthermore, exhibit a number of quite advanced epigenetic mechanisms, for example yolk sac formation.

BIOCHEMICAL CHARACTERS

In the last few years the information obtained on the amino acid sequences in cytochrome c isolated from various organisms has demonstrated that this is an extremely conservative molecule, the composition of which has remained unchanged to such a degree that it is possible to use it for establishing the phylogenetic kinship between the five kingdoms Monera, Protista, Fungi, Plantae and Animalia (Dayhoff and Park, 1969).

On the basis of their results these authors have classed Insecta, Gastropoda and Vertebrata so that the latter are secondary twins (Figure 3.26). Studies on the amino acid sequence in haemoglobins published by Goodman, Moore and Matsuda (1975) have given the same result (Figure 3.27).

McLaughlin and Dayhoff (1973) point out that their findings falsify the Protostomia–Deuterostomia theory. Personally, I find that the molecular–biological records

Figure 3.26. Phylogenetic classification of Insecta, Mollusca and Vertebrata, as established on the basis of the amino acid sequence in cytochrome c. The figure on one of the vertical lines indicates the number of nucleotide replacements estimated to have occurred between the two bifurcations. Inse, Insecta; Moll, Mollusca; Vert, Vertebrata. Based on data published by McLaughlin and Dayhoff (1973)

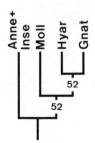

Figure 3.27. Phylogenetic classification of Annelida–Insecta, Mollusca, Hyperoartii and Gnathostomata on the basis of the amino acid sequence in haemoglobin. The figures on two of the vertical lines indicate the number of nucleotide replacements estimated to have occurred between the corresponding two bifurcations. Anne, Annelida; Gnat, Gnathostomata; Hyar, Hyperoartii; Inse, Insecta; Moll, Mollusca. Based on data published by Goodman *et al.* (1975)

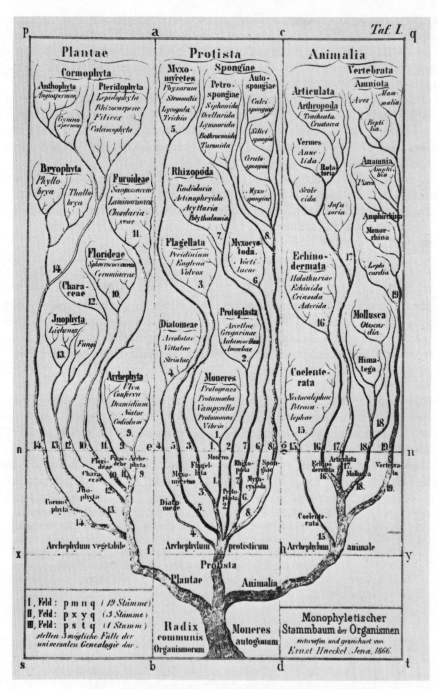

Figure 3.28. Phylogenetic tree for all living beings. From E. Haeckel (1866), *Generelle Morphologie der Organismen*, Georg Reiner

discussed here strongly support the classification at which I have arrived on the basis of non-morphological characters.

TAXONOMIC CONSEQUENCES

The conclusions stated above will, if they are accepted, have certain classificatory consequences. Thus, for one thing, the three groups of Chordata have been suggested to be independent monophyletic taxa which only form a monophyletic taxon together with members of one or more invertebrate taxa. If this assertion is correct it follows that the phylum Chordata, as well as the concept 'Protochordata', must be suppressed and replaced by the separate taxa (phyla) Tunicata, Cephalochordata and Vertebrata. The reason why I employ the name originally proposed by Lamarck for Urochordata is, of course, that the 'notochord' in the tunicate larva has turned the heads of zoologists for such a long time. I believe that this structure should be regarded as a curiosity, one of Nature's whims, which has little or no importance for classification.

Furthermore, it is quite clear that the present superphyletic classification should be abolished. The echinoderm superphylum simply does not correspond to any reality. Hence, the phylogenetic tree loses one of its branches and thus its likeness to an oak. In point of fact, it will be realized that if the cladistic theory is correct, then the days of phylogenetic trees are over, those of dichotomous dendrograms having arrived. Yet, as a tribute to past glory, I shall conclude by reproducing the creation of Haeckel (1866), which shows that, before the Protostomia–Deuterostomia theory had made its impact upon phylogenetic thought, Mollusca, if not Arthropoda, were considered more closely related to Vertebrata than were Echinodermata (Figure 3.28).

4

DIVERGENCE OF THE VERTEBRATES

The professed aim of the present chapter is to account for the course of the phylogenetic events which have led to the origination of the various members of the taxon
Vertebrata. Since the erection of a phylogenetic classification of a set of organisms
amounts to submitting a theory about their phylogeny, the goal may be reached by the
establishment of such a classification for the vertebrates.

In point of fact, the literature abounds with such classifications, submitted by evolutionists belonging to several biological disciplines, and since these people represent
such a tremendous wealth of knowledge, it may seem preposterous for me to embark
upon this problem. However, in justification I may mention firstly that the views propounded by the various authors are often incompatible and that the overall impression
one gets from the discourse is one of confusion. Secondly, one weighty argument in
favour of the present undertaking is that I have at my disposal a method permitting a
systematic approach to the task.

In the section following a classification will be established on the basis of a limited
number of selected morphological and, in some instances, non-morphological characters.
Subsequently, the suggested classification will be further tested through reference to
various characters, morphological as well as non-morphological, and compared with
already existing classifications, and finally, certain taxonomic consequences will be
outlined.

As claimed above, palaeozoological data ought not to be of direct concern for
phylogenetic classification. Nonetheless, as will be shown below, in some cases it is
impossible to carry through the present classification without reference to the fossil
record. I submit, however, that this situation reflects the lack of relevant data from living
organisms. However, fossil data are, of course, of paramount importance for the *testing*
of a phylogenetic theory. The present one will also be submitted to such a confrontation,
but this will be postponed until the last chapter of the book.

PHYLOGENETIC CLASSIFICATION OF VERTEBRATA

It was proposed basing the phylogenetic classification of Vertebrata on selected
morphological characters. The adoption of the qualitative approach in the present

context is dictated by necessity. In some of the taxa involved so few characters are known that it would be impossible to employ the quantitative method used in the preceding chapter. Even so, the dictum of numerical taxonomy will be followed insofar as a substantial number of characters will be used for the establishment and assessment of the classification.

It will be appreciated that in the present context the fifth of the basic premises is not strictly applicable, for the classification of Vertebrata is not a superphyletic one. Thus, since not only non-morphological but also morphological characters may be used, the proposed choice is justified.

As for the standard characters to be used, it is proposed to follow, as far as it is possible and necessary, the list presented in Table 4.1. This is seen to comprise quite common characters, most of which have been used at some occasion for the purpose of establishing phylogenetic kinships. Also, I shall employ only features which may be found in well-known textbooks, notably those of Goodrich (1958) and Romer (1962), except when for certain reasons it may be necessary to rely on more detailed sources.

Table 4.1
List of characters used in the phylogenetic
classification of Vertebrata

Scales
Axial skeleton
Median and caudal fins
Paired fins or limbs
Jaw suspension and visceral skeleton
Skull
Teeth
Gills
Air sacs and lungs
Digestive tract
Kidney and urogenital tract
Heart
Sensory organs
Nervous system and brain
Endocrine organs
Reproduction

We may now turn to a discussion of some properties of the classification. For one thing, since it is intended to classify the taxon Vertebrata, the classification must evidently be inclusive, i.e. the terminal taxa represented in the dendrogram must together comprise all members of the taxon. Secondly, it is also necessary to make a decision with respect to the degree of resolution of the classification. Clearly, for the present purpose the analysis may well be confined to taxa of relatively superior rank. However, when trying to submit exactly what this statement implies, we face a terminological embarrassment because we cannot set down our case without reference to Linnean taxa. Surely, a resolution on the class level would be a most satisfactory achievement. However, it is almost unanimously agreed that some, at least, of the vertebrate classes are not monophyletic, and consequently they cannot be embodied in a phylogenetic classification. Hence, if we accept the class as a starting point, we must state as a programme that the analysis should be carried beyond this level as far as it is necessary to get a resolution into truly monophyletic taxa. This implies that the classes which meet this demand may be classified as such, while the remaining ones must be further resolved. In the present context I shall presume that the two endothermic vertebrate classes, Aves and Mammalia, are monophyletic. Admittedly, agreement on this point is limited to the former taxon only, since the latter is often claimed to be 'polyphyletic' (cf. Simpson, 1959). I have already expressed reservations as to the use of this word, and shall here only point out that this view is based on fossil data; hence, as far as I can judge, it may be of no relevance to the living representatives of the class. The only remaining very large vertebrate taxon of high rank, Teleostei, has also been proposed by some authors to be 'polyphyletic' (cf. the discussion by Patterson, 1967). I am not in a position to evaluate the validity of this contention, but here also I shall treat the taxon as being monophyletic. As far as the other taxa which will be subject to classification are concerned, it may be advisable to postpone their enumeration until they are encountered in the proper context.

The following discussion will be subdivided into five main sections, dealing with Cyclostomata, Pisces, Amphibia and Amniota, respectively, and with the classification of the whole taxon.

Cyclostomata

I have already commented on the purported 'polyphyletic' origination of Cyclostomata, pointing out that the only classifications which do not require entirely unacceptable convergence are those three according to which, in a basic classification involving Gnathostomata and the two orders of Cyclostomata, either the latter, or one of them together with the former, constitute the set of secondary twins (Figure 2.7a and 2.7b).

The members of the two agnathous taxa possess many characters in common which, as we know, may be either plesiotypic or apotypic. In order to decide between these alternatives we must, according to the rule of the qualitative approach, try to discover whether one of these taxa, but not the other, has characters in common with Gnathostomata. Should this be the case, we may propose that the former cyclostome taxon and Gnathostomata are the secondary twins.

Turning to the first item in the table, scales, it appears that these, as well as any other kind of calcification except otoliths (Carlström, 1963), are absent in the recent cyclostomes but present in all lower gnathostomes. This circumstance suggests *prima facie* that absence of calcification is a plesiotypic character. However, according to the ideas advocated by Stensiö (1958) and Jarvik (1968b), hagfishes and lampreys are descended from separate fossil forms possessing calcified skeletons. If this is true, the absence of calcification must be due to convergent teleotypy; thus, in either case this character cannot be used for phylogenetic classification.

The main constituent of the axial skeleton is an unconstricted notochord, and since this structure recurs in some of the lower gnathostomes it may be plesiotypic. The same presumably holds for the absence of ribs. However, in Hyperoartii one encounters cartilaginous arcualia (Goodrich, 1958), and these structures are found in one gnathostomous taxon, viz., Actinistia (*Latimeria*), according to Millot and Anthony (1958a).

The median and caudal fins in the two agnathous orders exhibit many, presumably plesiotypic, characters in common, but through the possession of radial muscles (Goodrich, 1958) Hyperoartii show affinity to Gnathostomata.

Five features, paired fins, jaw suspension and visceral skeleton, skull and air sacs, are clearly plesiotypic in the cyclostomes; except for the visceral skeleton and the skull, the plesiotypy pertains to the absence of the character in question. According to Jarvik (1965), the rasping tongues in the two orders of Cyclostomata are so different that they must have been acquired independently. Should this plausible suggestion be true, it follows that this feature cannot be used for classificatory purposes. The location of the gills internal to the visceral arcs unites the recent cyclostomes with the extinct ostracoderms (Stensiö, 1968); this must be a plesiotypic character.

In the digestive tract Hyperoartii exhibit one character otherwise widespread among fishes, a spiral intestine (Romer, 1962).

The structure of the kidney in Hyperotreta is unique and very 'primitive', while in Hyperoartii the gnathostomous condition is approached. At least three features unite the members of this order with the higher vertebrates, viz., the absence of the pronephros in the adult, the morphology of the mesonephros (Goodrich, 1958) and the presence of collecting tubules (Hickman and Trump, 1969).

In having accessory hearts (Fänge, Bloom and Östlund, 1963; Johansen, 1963), the hagfishes are also unique among the vertebrates. This may be a teleotypic or a plesiotypic character, but in either case the absence of accessory hearts is a feature common to the other taxa.

The sensory organs are poorly developed in cyclostomes. I have found two features which may be used in the present context. One is that there are two, rather than one, semicircular canals in Hyperoartii, suggesting an approach towards the condition in Gnathostomata. The second is that in the retinal receptors of the lamprey one finds synaptic ribbons comparable with those found in all gnathostomous vertebrates. These are absent in hagfishes, where spherical synaptic bodies are found (Holmberg and Öhman, 1975). The brain and the nervous system are 'primitive' in the cyclostomes, presumably a consequence of plesiotypy. I have not found any features which definitely unite either of the two taxa with Gnathostomata, except that in Hyperotreta, but not in Hyperoartii, the dorsal and ventral nerves are united in a compound spinal nerve (Goodrich, 1958).

As regards the endocrine system it may be observed that, according to Fernholm and Olsson (1969, p. 351), the adenohypophysis in *Myxine* 'is unique in being composed mainly of isolated connective tissue-enclosed cell groups and in having a very poor vascularization . . . . Except for [specialized adenohypophysial tissue in the caudal part of the gland] it is not possible to divide the adenohypophysis into histologically different parts as in all other vertebrate groups.' Further arguments supporting this point may be found in Fernholm (1972).

With respect to eggs and fertilization I shall here only mention the remarkable fact, already commented on, that in Hyperotreta, presumably the oldest vertebrate taxon, one finds very large eggs with a hardened egg envelope and discoidal cleavage. According to the view advanced in Chapter 3, this feature may be explained as a plesiotypic character, pointing towards the invertebrate stock from which the vertebrate arose.

The result of the preceding survey shows that 10 characters unite Hyperoartii and Gnathostomata (Table 4.2), while only one is common to Hyperotreta and Gnathostomata. Therefore, according to the established rule, it is necessary to conclude, at least provisionally, that the first pair of taxa are the secondary twins and Hyperotreta the primary one (Figure 4.1).

It may be appropriate, with reference to Figure 4.1, to discuss a principle adopted in

Table 4.2
Characters common to Hyperoartii and Gnathostomata

Arcualia
Radial muscles
Spiral intestine
No persistent pronephros
Morphology of mesonephros
Collecting tubules
No accessory hearts
More than one semicircular canal
Synaptic ribbons in retinal receptors
Histology of adenohypophysis
Blood volume < 10 per cent
Hyperosmoregulation
Properties of insuline
Perfected immune response
Nervous regulation of heart
Chondroitin 6-sulphate in cartilage
Absence of a unique dermatan sulphate
Amino acid composition of collagen
Peptide pattern and amino acid composition of haemoglobin

Figure 4.1. Phylogenetic classification of the cyclostomes in relation to Gnathostomata. Gnat, Gnathostomata; Hyar, Hyperoartii; Hytr, Hyperotreta

drawing phylogenetic dendrograms, namely, that if it is possible among the three taxa in a basic classification to single out one as being 'advanced' or 'dominant', then it should be placed to the extreme right. Thus, in the present case there is no doubt that Gnathostromata should assume this position. If the two agnathous orders had been the secondary twins, then obviously the whole dendrogram should be turned round to allow Gnathostomata to be on the right. As we shall discuss later, ambiguities may occasionally arise in the application of this rule.

No doubt it will be felt that the characters compiled for the purpose of deciding the present question are quite unsatisfactory. Yet from my scrutiny of the literature it appears to be difficult, if not necessarily impossible, to find further morphological characters of relevance for the present classification. However, since the phylogenetic affinities between the cyclostome orders and Gnathostomata have been the subject of so much debate, it is desirable to corroborate the suggested basic classification through further empirical evidence. For the reason stated, it is necessary to rely on physiological and biochemical characters for additional support.

Of physiological characters I have found six demonstrating affinity between Hyperoartii and Gnathostomata and none in common between the latter taxon and Hyperotreta. Among these are first that the blood volume is less than 10 per cent

(Holmes and Donaldson, 1969) and that these animals can regulate the blood tonicity above that of the surrounding medium (Conte, 1969). The evidence discussed in Chapter 3 indicates that the endocrine system, and particularly the regulating influence of the pituitary, is less developed in Hyperotreta than in Hyperoartii. This fact must be correlated with the histology of the gland, a character which we have already listed. The only other feature of relevance in this context is the physiological properties of insulin; the hagfish hormone appears to be distinctly different from that found in other vertebrates (Epple, 1969).

A number of studies have shown that in lampreys the immune response is well developed, while negative results were obtained with hagfishes (Good, 1969; Linna *et al.*, 1970). In *Eptatretus* it has been possible, however, to demonstrate allograft rejection and antibody production to soluble antigens and to microorganisms (Hildemann and Thoenes, 1969; Thoenes and Hildermann, 1970; Acton, Weinheimer, Hildemann and Evans, 1969). These responses seem to be matched by similar ones described in members of the annelid superphylum (Cooper, 1968; Evans *et al.*, 1968; Seaman and Robert, 1968). It may therefore be justified to conclude that the immune response observed in members of Hyperotreta is a plesiotypic character.

To these four characters we may add nervous regulation of the heart, as discussed in Chapter 3.

A few comments may be called for on the question of osmoregulation. It has been generally accepted for a long time that the vertebrates originated in fresh water (cf. H. W. Smith, 1932; Romer, 1955a). If this is true, then the condition in Hyperotreta would be teleotypic and the character in the other taxa plesiotypic. This view has been contested by several authors (e.g. W. Gross, 1933; 1950; Robertson, 1957; White, 1958; Spjeldnaes, 1965), who are of the opinion that the vertebrates arose in the sea and that the hagfishes are descendants of these first vertebrates. In this case hyperosmoregulation must be an apotypic or a convergent character.

Among the biochemical characters uniting Hyperoartii and the higher vertebrates may be mentioned the presence of chondroitin 6-sulphate in cartilage; hagfishes seem to be unique among vertebrates in having chondroitin 4-sulphate, a feature otherwise found in invertebrates (Mathews, 1967). As discussed in Chapter 3, the hagfishes are distinguished from all other vertebrates through the chemical composition of the glycosaminoglycans found in the notochord and the skin (Anno *et al.*, 1971; Seno *et al.*, 1972).

We have already observed in the preceding chapter that the overall amino acid composition of collagen in Hyperoartii is somewhat closer to that of the other vertebrates than is that in Hyperotreta (Figure 3.13); in particular, the lysine content in the latter case is much lower than in the remaining vertebrates (Pikkarainen, 1968). Finally, it should also be recalled that the peptide pattern and the amino acid composition of haemoglobin in Hyperoartii is quite similar to that of the higher vertebrates, and distinctly different from that found in Hyperotreta (Figure 3.15).

Since, in spite of my endeavours, I have been unable to find any non-morphological characters indicating affinity between Hyperotreta and Gnathostomata, I believe that the classification proposed in Figure 4.1 is reasonably well corroborated by the non-morphological features mentioned. Indeed, striking new discoveries are required before the 19 properties listed in Table 4.2. can be acknowledged as being non-apotypic rather than apotypic characters.

Pisces

Occasionally, the word 'fishes' is used to designate members of the three taxa Cyclostomata, Chondrichthyes and Osteichthyes, but Pisces, purportedly encompassing the last two taxa (Goodrich, 1958), is no longer accepted as a separate taxon. Nevertheless, since the gnathostomous, non-amphibian and non-amniote vertebrates unquestionably possess a number of characters in common, it would certainly, in the present context, be unwarranted to embark upon the classification acknowledging a subdivision of these animals in two classes. Hence, the present section will be concerned with both Chondrichthyes and Osteichthyes, for which, following Goodrich, the name 'Pisces' has been adopted. Whether or not this expedient will be acceptable in the phylogenetic classification of these animals is, of course, a point which can be settled only when we have come to the end of our task.

The taxa which have to be classified in the present section are, in alphabetical order: Actinistia, Chondrichthyes, Chondrostei, Cladistia (Brachiopterygii), Dipnoi and Neopterygii. According to the principles of cladism this task must be completed through a succession of basic classifications. For the first of these I propose to choose three taxa of 'primitive' fishes, namely, Actinistia, Chondrostei and Dipnoi. The remaining ones will then be classified in the following order: Neopterygii, Chondrichthyes and Cladistia.

ACTINISTIA, CHONDROSTEI AND DIPNOI

The present basic classification holds in my opinion the clue to the classification of the whole group. Unquestionably, it also presents the most intricate problems.

Indeed, it seems so difficult to find characters which can be rated with confidence as apotypic that it may be justifiable to deviate from the cladistic procedure and begin by discussing some current views on the classification of the three groups. Many authors contend that an affinity between Dipnoi and the extinct Rhipidistia is borne out by the fossil record, and since the rhipidistians through some important features, cosmoid scales, lobed paired fins, intracranial joint, etc., are united with the extinct actinistians and their living representative, *Latimeria*, a special subclass, Sarcopterygii, has been erected by Romer (1955b) and Thomson (1969) to accommodate the living and extinct Dipnoi and 'Crossopterygii' (Rhipidistia and Actinistia). Jarvik (1964; 1968a) has cogently argued against this that the name 'Sarcopterygii' applies equally well to Chondrichthyes and Cladistia, and further that the recent studies of the morphology of *Latimeria* demonstrate that this animal is only distantly related to Rhipidistia; so slight in his opinion is the kinship that he recommends the suppression of the name 'Crossopterygii'.*

This argument represents in a nutshell the issue raised by Hennig's theory of phylogenetic classification. Evidently, the stand taken by Romer and Thomson implies that, in the present context, Actinistia and Dipnoi are the secondary twins and Chondrostei the primary one. In opposing this view, Jarvik is clearly of the opinion that Chondrostei are one of the set of secondary twins and one of the 'crossopterygian' taxa the other one. In doing so, he alleges that the 'sarcopterygian' characters are plesiotypic ones

---

*I believe that the classification which ensues from my analysis amply supports Jarvik's opinion. Still, I find that in the following discussion it is convenient to apply this name, in a completely informal way, to the group of fishes distinguished by the possession of the characters mentioned above.

which appear to be apotypic only because, in Chondrostei, they have been replaced by teleotypic actinopterygian ones.

Before we begin our survey of the standard characters it should be pointed out that some of the apparent 'primitiveness' of the living animals dealt with here is due to loss, so that it will be necessary in the present section to rely in part on fossil evidence pertaining to characters possessed by extinct members of the three taxa.

In the three taxa the scales are either more or less reduced or absent. The cosmoid scale (cf. Figure 4.9) distinguishes most of the extinct 'crossopterygians' (Goodrich, 1907). It is therefore natural that attempts have been made to refer the scales of *Latimeria* to this class (Roux, 1942; Bernhauser, 1961). In fact the scale comprises two components, the scale and a number of denticles which ornament the exposed part of the scale (M. M. Smith, Hobdell and Miller, 1972). The scale proper closely resembles the type found in some recent teleosts, consisting of a very thin, densely calcified, non-vascular and acellular outer layer and a basal, uncalcified portion composed of alternating laminae of parallel bundles of collagen fibres. The denticles, which are secondarily attached to the surface of the scale (C. L. Smith *et al.*, 1975), are closely similar to the selachian placoid scale (Figure 4.2). Clearly, a lot of imagination and goodwill is required to find traces of the cosmoid scale in those of *Latimeria*, and even naming the thin mineralized layer 'isopedine' (M. M. Smith *et al.*, 1972) seems rather questionable.

The scales of fossil dipnoans are related to the cosmoid scale (Goodrich, 1907), but those found in recent forms are not too different from those of *Latimeria* (Kerr, 1955). The reduction of calcification has obviously occurred convergently in these two taxa.

If it is assumed that Actinopterygii, i.e. Chondrostei + Neopterygii, are a monophyletic taxon, then it may be inferred that the common ancestors of these two taxa were members of Palaeonisciformes, the oldest known fishes with actinopterygian affinities. In these the scales were already palaeoniscoid. Among the typical features of this scale type are that the mineralized tissue is deposited in concentric layers around the scale (concentric growth) and that the material at the distal side (ganoin) is different from that deposited at the inner side (bone). That this type of scale represents a teleotypic character is supported by the observation that in the early palaeoniscids one may find scales with a vascular layer (Lehman, 1966a). The scales are thus of no value in the present classification.

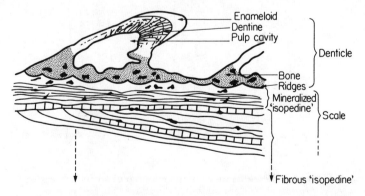

Figure 4.2. Histology of the scale of *Latimeria*. (Redrawn from M. M. Smith, M. H. Hobdell and W. A. Miller, *J. Zool. London*, **167** (1972), by permission of The Zoological Society of London)

All three taxa have unconstricted notochords in the living forms, and in *Latimeria* there are no centra, only cartilaginous, or slightly ossified, arcualia, forming neural and haemal arches according to the usual pattern (Millot and Anthony, 1958b). In certain Devonian dipnoans holospondylous centra have been found, consisting of bone or calcified cartilage, with considerable constriction of the notochord (Jarvik, 1952; Schultze, 1970). In the palaeoniscoid ancestors of Chondrostei a calcified axial skeleton was present also, but the notochord was persistent and unconstricted, exactly as in the living forms (Lehman, 1966a). When we consider that although the centra were ossified the notochord was unconstricted in Rhipidistia (Lehman, 1966b), it appears that the notochordal constriction in Devonian Dipnoi is a teleotypic character, lack of constriction being a plesiotypic feature.

According to Lehman (1966a, b) the calcification in Palaeonisciformes and Dipnoi comprises the notochordal sheath, the external skeletogenous tissue and the bases of the arcualia (chordacentric, autocentric and arcocentric calcification). In the fossil actinistians the notochord was never ossified in this way. In the living forms, where little calcification occurs, it has been observed that 'mesoblastic' cells invade the noto-

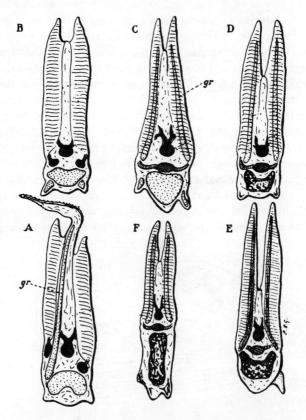

Figure 4.3. Sections through the gill-arch of various fishes. A, *Mustelus*; B, *Ceratodus*; C, *Acipenser*; D, *Lepisosteus*; E, *Salmo*; F, *Polypterus*. (Reproduced from E. S. Goodrich (1909), in *A Treatise on Zoology*, Vol. 9.1, by permission of A. & C. Black)

chordal sheath in chondrosteans and dipnoans (Goodrich, 1958), but not in *Latimeria* (Millot and Anthony, 1958a).

The typical 'crossopterygian' feature, the possession of two dorsal fins, the endo-skeleton of the anterior one being quite simple whereas that of the posterior one, as well as of the anal fin, is complex and similar to that found in the paired fins, is found in all actinistians, including *Latimeria*, but also in early dipnoans. Fusion of the median fins with the caudal in later lungfishes is evidently a teleotypic character.

Originally, the tail was heterocercal in all three groups (Thomson, 1969; Schultze, 1973), and this feature is still found in Chondrostei. The symmetrical tails found in the other taxa, diphycercal in Actinistia and gephyrocercal in Dipnoi, are clearly teleotypic characters.

Both *Latimeria* and Dipnoi have lobed paired fins, the basic structure of which is the 'archipterygium', at least in the pectoral fins. This sets them apart, together with the extinct Porolepiformes and Pleuracanthodii, from all other fish with 'lobed' fins, i.e. Chondrichthyes, Cladistia and Osteolepiformes. Chondrostei have the typical actinop-terygian fins, in which the endoskeleton is confined to the skin base. As indicated by the name, the archipterygian fin appears to be a plesiotypic character.

The skull in *Latimeria* has the 'crossopterygian' intracranial joint, and this feature is absent in both of the other taxa. This lack might conceivably be an apotypic character, but the fact that Chondrostei and Dipnoi exhibit many dissimilarities concerning skull bones, dermal bones and palate suggests that several other important modifications took place independently in the two lines, possibly in association with the events leading to the loss of the intracranial joint.

Excepting secondary losses, Dipnoi from their first origin, or at least from an early stage in the history of the taxon, to the present day have been distinguished by the possession of tooth plates, situated on the prearticular in the lower jaw and on the vomers and the entopterygoid in the palate (Lehman, 1966b; Denison, 1974). A scrutiny of Peyer's monograph (1968) suggests that the various features related to the dentition in the three groups are either plesiotypic or teleotypic characters.

Supporting gill-rays are absent in Dipnoi, present in both of the other taxa (Good-rich, 1958; Millot and Anthony, 1958b; cf. Figure 4.3). The loss in Dipnoi seems to be a teleotypic character.

The most characteristic feature in the digestive tract of the three groups is probably that they have a spiral intestine. The pancreas in *Latimeria* and in Dipnoi is of the compact type, while that of Chondrostei is of the 'diffuse' actinopterygian type (Epple, 1969), once more presumably a teleotypic character.

The data published so far on the anatomy of the kidney and urogenital tracts in *Latimeria* by Millot and Anthony (1958a) are not very detailed, but it appears that these organs exhibit a number of unique features; whether the latter are plesiotypic or teleotypic is impossible to decide. In contrast, the organization of the male urogenital tract is exactly the same in *Acipenser* and in *Neoceratodus* (Figure 4.4).

The heart of *Latimeria* is almost linear (Millot and Anthony, 1958a), corresponding quite closely to the hypothetical primitive condition envisaged by Goodrich (1958). In this respect the animal is unique among Gnathostomata, including the two other taxa under scrutiny, and the curvature of the heart in the latter may be considered a character in common to these. Otherwise the heart in all three groups incorporates a well-developed conus arteriosus and is surrounded by a sturdy pericardium, features which are acknowledged as plesiotypic. The living Dipnoi have a partial separation of the pulmonary from the remaining venous blood, a situation similar to that encountered in Amphibia, but 'to some extent the resemblance may be due to convergence; for *Ceratodus*, the most primitive of living Dipnoi, has a less specialized heart than *Lepidosiren* or *Protopterus*' (Goodrich, 1958, p. 553).

Millot and Anthony (1958a, p. 2572, author's translation) write the following about the brain of *Latimeria*: 'Its position is unique among the vertebrates ... it is in fact localized in the most posterior, occipital part of the cranial cavity . . . . The volume of the brain is extremely low, even for a fish. Weighing less than three grams in a male of 40 kg, the nervous centres occupy less than one hundredth of the cranial cavity, otherwise full of fat. It was already known that in primitive fishes the cerebral mass by no means fill all of the intracranial space, but such disproportion in size between brain and endocranium is without known equivalence.' The last observation may possibly be accounted for by allometric growth; *Latimeria* is, after all, a very large fish. Yet the fact remains that a small brain apparently is a plesiotypic feature, and it is therefore justifiable to suggest that brain development in the other forms is an apotypic character.

The structure of the forebrain has been found to be a valuable diagnostic character in comparative morphology. This does not hold, unfortunately, in the present basic classification, for, according to the most recent studies, 'the forebrain of *Neoceratodus* has some features in common with that of the actinopterygians ..., but ... the telencephalon of the crossopterygian *Latimeria* approaches the actinopterygian condition even more closely' (Nieuwenhuys and Hickey, 1965, p. 442). On the basis of this statement it seems necessary to conclude that the inverted forebrain in the other lung-

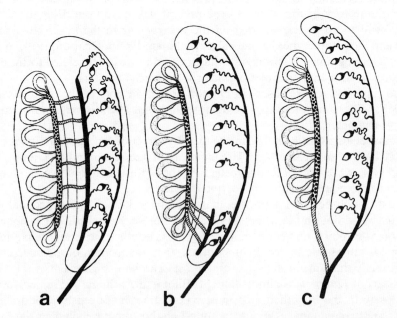

Figure 4.4. Diagrams of the male urogenital system in some fishes. a, *Acipenser, Neoceratodus, Lepisosteus*; b, *Lepidosiren*; c, *Polypterus*. (Reproduced from A. J. P. Van den Broek (1933), in *Handbuch der vergleichenden Anatomie der Wirbeltiere*, Vol. 6, by permission of Urban & Schwarzenberg)

fishes, *Lepidosiren* and *Protopterus*, in Selachii and in Tetrapoda is a convergent, teleotypic character.

In the cerebellum one feature is common to Actinistia and Dipnoi, viz., the absence of a valvula, a structure which is present in Chondrostei (Millot and Anthony, 1958a; Larsell, 1967). Mauthner's fibres are absent in a *Latimeria* (Nieuwenhuys, 1964); their presence in the other two groups may possibly be an apotypic character. A jugal canal is present in dipnoans and in extinct actinistians (Goodrich, 1958), and also in *Latimeria* (Millot and Anthony, 1958a). The absence of this branch of the lateral-line system is a general actinopterygian feature, which may be interpreted as a teleotypic character. A remarkable feature is that the cross-section of the spinal cord in *Latimeria* is flattened, a feature otherwise found only in cyclostomes (Nieuwenhuys, 1964). The more rounded cross-section found in the two other taxa may be an apotypic character (cf. Figure 4.7).

No information of comparative importance exists concerning the endocrine organs of *Latimeria*. The eggs of *Latimeria* are very large, measuring up 90 mm in diameter (Anthony and Millot, 1972). According to these authors the mature egg is protected only by a thin ovular envelope. Although there is no copulatory organ in the male (Millot and Anthony, 1958a), internal fertilization would seem essential in these vulnerable eggs, and viviparity was, indeed, suggested by J. L. B. Smith (1940). This supposition has now been confirmed (C. L. Smith *et al.*, 1975). The rather small eggs in the other taxa are fertilized externally. Dipnoan eggs are covered by a layer of jelly, while those of chondrosteans are distinguished by the hardened egg membrane (chorion) characteristic of Actinopterygii. The particular conditions found in *Latimeria* are clearly teleotypic, and their absence in the other taxa presumably constitutes a plesiotypic character.

It would be very gratifying if to the present list of morphological characters it was possible to add an equally long list of physiological and chemical characters. However, in this field *Latimeria* remains almost completely *terra incognita*. Some data have been obtained, however, and, except for one, they will be discussed in the next section. The remaining character is the physiological properties of haemoglobin which, together with that found in Elasmobranchii and Chondrostei, form a group of primitive haemoglobins in Gnathostomata. However, apparently the substance found in *Latimeria* is intermediate between those found in the other two groups (Wood, Johansen and Weber, 1972).

As might be anticipated, Table 4.3 demonstrates that some characters are common to each of the three pairs of taxa. However, those uniting Actinistia and Chondrostei have all been rated as plesiotypic. If this characterization is correct, it follows from the rule of the qualitative approach that Dipnoi is one of the secondary twins. Of the two sets of characters uniting this taxon with the others, one should be plesiotypic and the other apotypic. As appears from the table, all but one of those common to Actinistia and Dipnoi have been assessed as plesiotypic, while only one of the nine common to Chondrostei and Dipnoi has this rating. If these evaluations are accepted, there can thus be no doubt that the latter taxa constitute the set of secondary twin taxa, while Actinistia is the primary twin (Figure 4.5). Since it is impossible at this stage to decide which of the two secondary twins is the dominant one, the two taxa have been arranged alphabetically in the figure.

As I said in the introduction to the present section, the basic classification of these

Table 4.3

Characters common to Actinistia, Chondrostei and Dipnoi

Actinistia–Chondrostei	Actinistia–Dipnoi
(p) Unconstricted notochord	(p) 'Cosmoid scales'
(p) Dentition	Two dorsal fins
(p) Gill rays	(p) Archipterygial fins
(p) Haemoglobin	(p) Absence of modified dentine
	(p) Dorsal opening of 'air sac'
	(p) Compact pancreas
	(p) Jugal canal
	(p) Absence of valvula

Chondrostei–Dipnoi

Calcification of axial skeleton
Invasion of notochordal sheath
Male genital system
Curvature of heart
Development of brain
Mauthner's fibres
Shape of spinal cord
(p) Small eggs

(p) signifies that the character in question is classed as plesiotypic.

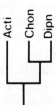

Figure 4.5. Phylogenetic classification of Actinistia, Dipnoi and Chondrostei. Acti, Actinistia; Chon, Chondrostei; Dipn, Dipnoi

primitive fish is the most troublesome, but also the most important, in the phylogenetic classification of Pisces. The primary reason for this difficulty is undoubtedly that we are concerned with taxa which have been separated for hundreds of million years, during the course of which extensive teleotypic changes have occurred, and that consequently the chances for the occurrence of non-apotypy must be great. Another handicap is, of course, that the information available about *Latimeria* is still rather scanty in many respects. Granting this, it may nevertheless be argued that my classification is wanting in objectivity. In defence of it I can only say that I have investigated the consequences of the two other possible classifications and both of them appear to increase the degree of non-apotypy compared to the one proposed here. This, I believe, will become apparent from the remainder of the present chapter.

NEOPTERYGII

Classification of Neopterygii is probably the simplest of all in the present group for, as we have seen, the chondrosteans possess so many actinopterygian characters that there can be no doubt that Chondrostei and Neopterygii are secondary twins, together forming the monophyletic taxon Actinopterygii.

However, since in the following section we shall be discussing characters pertaining to the three subtaxa of Neopterygii, namely, Amiidae, Lepisosteidae and Teleostei, it may be expedient to establish a classification of these three groups. By classing the first two taxa together in the taxon Holostei, some taxonomists have obviously made them secondary twins. Yet it is a remarkable fact that each of the taxa has characters in common with Teleostei which are conspicuously absent in the other (Goodrich, 1958). This has occasionally led to attempts to place either one of the taxa as intermediate between the other two neopterygian taxa. Thus, as shown by Patterson (1973), the skeleton of *Amia* resembles that of the teleosts in a number of features which are probably apotypic (Table 4.4). Accepting this, Amiidae and Teleostei are the secondary twins, Lepisosteidae the primary twin (Figure 4.6).

Table 4.4
Characters common to Amiidae and Teleostei (C. Patterson, 1973)

Median, unpaired neural spines above the vertebral column
Mobile maxilla, free from the cheek and swinging about a peg-like internal head
Supramaxilla
Interopercular
Loss of quadratojugal as an independent element
Uncinate processes on the epibranchials
Large posterior myodome
Large post-temporal fossa
Intercalar which spreads extensively over the surface of the otic region of the braincase

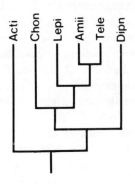

Figure 4.6. Phylogenetic classification of Neopterygii in relation to the basic classification in Figure 4.5. Acti, Actinistia; Amii, Amiidae; Chon, Chondrostei; Dipn, Dipnoi; Lepi, Lepisosteidae; Tele, Teleostei

CHONDRICHTHYES

Holocephali and Selachii are usually united in the taxon Chondrichthyes. In point of fact, if this classification is accepted, then the separate discussion of these taxa is not required according to the programme stated above, but it is done here because it allows discussion of some interesting questions.

Although the members of Holocephali exhibit several peculiar features, there is still a considerable number of characters which support the common origin of all Chondrichthyes. Among these I would like to mention the claspers as a model of an apotypic character. Whether one believes, in the Darwinian tradition, that these copulatory organs arose in along series of infinitesimal steps, each of which conveyed selective advantage to its bearer, or whether one believes that they arose through one lucky stroke of fortune, functional at the outset, even if not necessarily in their ultimate form, it is evident that the odds against the repeated occurrence of this phenomenon are so small that convergence should be invoked only if very compelling evidence prescribes this interpretation.

This argument appears to lead to a conclusion which I would like to avoid by any means, namely, that the Chondrichthyans descended from placoderms. The reason for this is that Miles (1967) has recorded the presence of claspers on the pelvic fins of the Devonian placoderm *Rhamphodopsis*. Yet, as we shall discuss in the next chapter, some early members of Chondrichthyes did not have claspers. Hence, assessment of the claspers in these two instances as an apotypic character would imply that some placoderm taxon is a twin to the one including the clasper-bearing chondrichthyans, or some equally unacceptable alternative 'explanations'. So far as I can see, the discovery of Miles (1967) forces upon us the conclusion that claspers have arisen convergently twice, but it does not follow from this that they must have done so three times.

When we face the problem of allocating Chondrichthyes to their proper place in the dendrogram it is impossible to ignore the very impressive list of characters which, in the opinion of Jarvik (1968a), indicate an affinity between Holocephali and Dipnoi. It must be emphasized, however, that we are here dealing with animals belonging to the most superior taxa in the phylum which, in spite of various teleotypic modifications, may have preserved a substantial number of the characters possessed by the ancestors of the living Gnathostomata. In other words, it may be relatively easy (and for other taxa than these two), to find characters in common, but there is a considerable risk that these traits are plesiotypic ones. It is not always simple to decide which characters are of this kind, but it seems that at least some of those listed by Jarvik occur also in *Latimeria*, and should this be so, I would not hesitate to rate them as plesiotypic. One feature, the tooth plates, also found in some primitive members of Selachii, does not belong to this category and must either be an apotypic or a convergent teleotypic character. If Denison (1974) is right in his assertion that the oldest lungfishes were devoid of tooth plates, then their acceptance as an apotypic character implies that Holocephali have originated within the taxon Dipnoi.

If Jarvik's proposition is rejected it follows that Chondrichthyes cannot be the secondary twin to Dipnoi. It is rather obvious that the same holds with respect to Actinopterygii, and we have therefore reduced the present classification to the basic one involving the taxa Actinistia, Chondrichthyes and Actinopterygii + Dipnoi.

Before we approach this problem it should be mentioned that the greatest obstacle to the classification of Chondrichthyes, at least along the classical morphological approach, lies in the fact that no ossification occurs in the members of this taxon, implying the absence of the various bones, particularly the dermal bones of the head, which are so important in classification. This has the consequence that we must try to settle the problem facing us without reference to osteology. It should also be emphasized that wherever bones are involved in the formation of other anatomical features, the latter will also be missing in Chondrichthyes. As long as we know so little about the epigenetic mechanisms involved in the creation of the animal body, it is difficult to predict the likely effects of the lack of ossification. However, as a possibility I would like to mention that in many fishes the growth of the bony operculum is unquestionably causally involved in the formation of the common opening for the gills. And if this is true, then the serial arrangement of the open gill slits in selachians may be explained by the fact that dermal ossification is absent in these animals.

It is almost a dogma in vertebrate phylogenetics that Chondrichthyes are widely different from Osteichthyes, the bony fishes, thus implying that they should be the primary twin in the present classification. In order to demonstrate the criteria on which this view is based, I shall rely on one authority (Romer, 1966), who presents the following list: (1) claspers, (2) large yolky eggs, (3) absence of lungs, (4) marine habitat from the earliest origin and (5) absence of bone.

We have already discussed the claspers and concluded that they must be a teleotypic character, acquired within the taxon. The eggs of *Latimeria* were not known when Romer wrote his book, otherwise he could not have used this character, which in fact unites Actin and Chondrichthyes.

As regards the lungs, one should certainly consider 'that 300 million years of continuous existence in the sea, where the auxiliary respiratory function may not be needed, is plenty of time for them to have been lost' (Thomson, 1971, p. 152). It must be recalled, however, that the lungs may have served a purpose besides gas exchange, namely, to lower the specific gravity (Thomson, 1971). All the known early vertebrates were heavily armoured with bony plates, and for them a reduction of the specific gravity may have been of great importance for locomotion. If Chondrichthyes descended from bony, lung-possessing ancestors, then the fortuitous loss of the lungs may have occurred without dire consequences. This explanation presupposes, as indicated in the above quotation, that the lungs were not necessary for respiration.

In fact, it appears that chondrichthyans are not without expedients for reducing the specific gravity, for 'it should not be overlooked that many sharks are actually neutrally buoyant due to the accumulation of large volumes of oils in the liver and viscera' (Thomson, 1971, p. 150). Under these circumstances it becomes understandable that they could well do without lungs. Incidentally, a large fatty liver is also found in *Latimeria* (Millot and Anthony, 1958).

It is true that, with some few exceptions, the chondrichthyans have been marine throughout their history. Yet, in the Devonian, when they first made their appearance in the fossil record, there were also many 'crossopterygians' and dipnoans living in the sea (Thomson, 1969; 1971), so this 'character' does not seem to carry any weight. Thomson discusses the idea put forward by H. W. Smith (1953) that these early marine fishes used urea for their osmoregulation. Should this be the case, it constitutes a character in common with the chondrichthyans which is still preserved in *Latimeria* (Pickford and Grant, 1967).

'In past times it was generally assumed that the absence of bone in the sharks was a primitive condition and that the sharks represented an evolutionary stage antecedent to that of the bony fishes. This assumption appears, at the present day, to be a highly improbable one. Bone, as we have seen, appears in groups much lower down the evolutionary scale; and if we believe the sharks to be primitive in this regard, we must believe, rather improbably, that bone was evolved a number of times by the vertebrates' (Romer, 1966, pp. 37–8). If this quotation, which probably represents a majority view today, is true, the loss of ossification is a teleotypic character in Chondrichthyes. It is, of course, of value for the distinction between this taxon and Osteichthyes, but it sheds no light on phylogenetic relationships.

After this discussion on the basis of the list compiled by Romer (1966), we thus arrive at the result that three characters are most likely teleotypic (claspers, absence of lungs and absence of bone), while three characters are common to Chondrichthyes and *Latimeria* (large yolky eggs, fatty liver and osmoregulation by uraemia). This result is hardly convincing support for the traditional view on the classification of Chondrichthyes.

If we follow the cladistic procedure, then the classification of Chondrichthyes as the primary twin would have the consequences that this taxon should possess at most some of the characters common to the other two taxa. The characters with which we are concerned here are those in Table 4.3 that were classified as plesiotypic. Because of teleotypy, the presence of all these characters cannot be made a requirement for Chondrichthyes to be a secondary twin, but, in point of fact, the only ones missing are an unconstricted notochord, small eggs and teeth and scales, and as far as the last two items

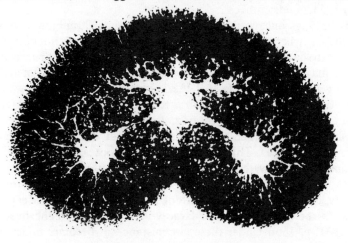

(a)

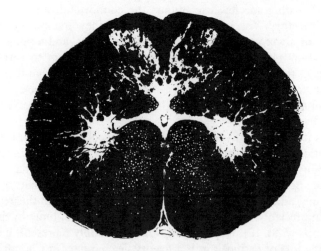

(b)

Figure 4.7. Cross-section of the spinal cord. a, *Latimeria*; b, *Acanthias*. (Reproduced from R. Nieuwenhuys (1964), in *Organization of the Spinal Cord*, by permission of ASP Biological and Medical Press (Elsevier division))

are concerned, the non-osseous constituents—enameloid and dentine—are present, whereas the osseous ones are absent, as might be expected. In this context it should be noted that if, by mutation, the dermal ossification was suppressed in *Latimeria*, then the scales would be reduced to the denticles, or placoid scales, found in contemporaneous selachians.

This still does not exclude Chondrichthyes from the primary twin position, but it certainly shows that further inquiry is required before the present basic classification is definitely settled.

There are not many characters in common between Chondrichthyes and Actinopterygii + Dipnoi. However, it has been observed that in both taxa the notochord is invaded by mesoblastic cells (Goodrich, 1958), and the heart and the brain are also better developed than in *Latimeria*.

Turning now to the question of whether it is possible to find any characters in common between Actinistia and Chondrichthyes I may point to the remarkable fact that in several studies on the soft anatomy of *Latimeria* the authors have stressed more or less striking, but certainly unanticipated, similarities between this animal and members of Chondrichthyes. I am unable to estimate the validity of these claims personally, and I must therefore enumerate them without further comments.

The nodular gland in *Latimeria* is the homologue of the digitiform rectal gland in the selachians (Millot and Anthony, 1958a); this organ is missing in the holocephalians. The structure of the eye in *Latimeria* is very similar to that of the chondrichthyans (Millot and Carasso, 1955). Other observed similarities pertain to the central nervous system; thus Nieuwenhuys (1964, p. 32) reported that 'cross sections through the cord of *Latimeria* . . . reveals the surprising fact that the shape of the grey matter in this form shows a striking resemblance to that found in sharks' (Figure 4.7), and that Mauthner's fibres are absent. A brain feature common to the two taxa is the fact that the first pair of nerves is contained in a hollow stalk, the cavity of which is a continuation of the telencephalic vesicle (Millot and Anthony, 1958a).

In its very simple structure, the thymus exhibits distinct similarities with that of Selachii, and the same holds for the thyroid gland (Millot and Anthony, 1956).

To these characters we must add the three discussed above, the very large eggs, the fatty liver and isosmoregulation through uraemia.

The ten characters discussed above are listed in Table 4.5. For the reader brought

Table 4.5
Characters common to Actinistia and
Chondrichthyes

Fatty liver
Rectal gland
Structure of eye
Shape of grey matter in spinal cord
Absence of Mauthner's fibres
First pair of nerves in telencephalic vesicle
Structure of thymus
Structure of thyroid gland
Very large eggs
Osmoregulation through uremia

up with the notion that Chondrichthyes and Osteichthyes are two separate and quite distinct taxa this list may not be very impressive. But when it is recalled that no serious support can be found for any of the other classificatory alternatives, then I think that for the time being there is only one possible choice for the classification of Chondrichthyes, namely, as the secondary twin to Actinistia (Figure 4.8.).

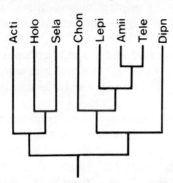

Figure 4.8. Phylogenetic classification of Chondrichthyes in relation to the classification in Figure 4.6. Acti, Actinistia; Amii, Amiidae; Chon, Chondrostei; Dipn, Dipnoi; Holo, Holocephali; Lepi, Lepisosteidae; Sela, Selachii; Tele, Teleostei

The classification proposed here implies that one of the initial steps in the evolution of the chondrichthyans involved the loss of ossification and that, consequently, they must have originated from some kind of bony fish. This course of events appears to be corroborated by the fact that extinct sharks show similarities with the latter which are absent in living members of the taxon (Romer, 1937).

To conclude the present section, I would like to quote one expression of the suspicion that the traditional attitude to the relationship between Chondrichthyes and 'Osteichthyes' may be wrong: 'One is greatly tempted to bestow upon the Acanthodii the honor of a position close to or within the direct ancestry of the Osteichthyes or Elasmobranchii. [The fact that one cannot choose between these two possibilities may mean that the two major groups are less distinct than supposed]' (Thomson, 1971, p. 161; Thomson's brackets).

CLADISTIA

The classification of Cladistia (Brachiopterygii) has met with great difficulties and is still a matter of dispute today. In the classical opinion, advocated by Huxley and Dollo (cf. Goodrich, 1928), these fishes were considered to be related to the 'Crossopterygii'. This is not very remarkable for, so far as I know, there is no group of fishes, except possibly Chondrichthyes, which have not been supposed to be affiliated with these extinct fishes. However, as emphasized above, classification with respect to extinct forms may or may not be of importance, but the essential point is to establish the relationships between the taxa of living animals.

Goodrich (1907; 1928) made a step in this direction by suggesting that, since Cladistia exhibit various palaeoniscoid features, they must be affiliated to Actinopterygii. In the present classification this would presumably imply that the two taxa are secondary twins. This view has gained quite wide acceptance, and has been forcefully advocated by Daget (1950; 1958). The only alternative to this classification seems to be the one proposed by Jarvik (1968a), according to which Cladistia are distinct from all other fishes, implying presumably that they are the twin taxon to the one including all other gnathostomes. This view might have been acceptable before *Latimeria* was known, but acknowledging the 'primitiveness' of this animal compared to all other fishes, it is rather unlikely that Cladistia are a still older taxon. That proposition simply involves too much convergence to be likely. Since there is no evidence whatsoever indicating affinity between this taxon and Actinistia + Chondrichthyes, it appears that the present task may be confined to establishing a basic classification of the taxa Actinopterygii (Chondrostei), Cladistia and Dipnoi.

The character which originally led Goodrich to suspect affinity between Cladistia and Actinopterygii is the kind of scales found in the former. Goodrich (1907) distinguished two types of ganoid scale, the palaeoniscoid and the lepidosteoid. The former, found in certain members of Palaeonisciformes, is composed of three layers, lamellated bone, cosmine and an outer lamellated layer, present only where the scale has been in contact with the epidermis, thus indicating that this tissue is involved in its formation. Around the edges of the scale the two lamellated layers are continuous, indicating that 'concentric layers of new substance are continually being deposited over the whole surface' (Goodrich, 1907, p. 757). This scale type represents an intermediate form between the cosmoid scale, found in extinct 'Crossopterygii' and Dipnoi, which it is like through the presence of cosmine, and the lepidosteoid scale, which it resembles in the absence of vascular bone (Figure 4.9).

According to Goodrich (1907; 1928), the scale of *Polypterus* grows concentrically and contains the three elements of the palaeoniscoid scale, and he thus classified it as being of this type (Figure 4.10a). Daget (1950) has reinvestigated this question and has come to the conclusion that the scale structure pictured by Goodrich deviates from reality on certain points, notably that denticles are normally absent and that the layer of ganoine (enameloid) is very thin (Figure 4.10b). This does not, however, shake the foundation of Goodrich's argument, for in this way it turns out to be extremely similar to the scale of the palaeoniscoid *Cheirolepis* (Figure 4.10c). Goodrich's theory about the absence of vascular bone in the scale of *Polypterus* is not convincingly borne out by Daget's illustration (cf. Figure 4.10a and 4.10b). In fact, Daget (1950) seems to be in doubt about how to classify the 'vascular' layer of calcified tissue but, settling for 'trabecular dentine', he remarks that, in contrast to real dentine, the former contains osteoblasts.

Considering that in most other lines of fish, including Dipnoi, the cosmoid scale was either retained with extensive modifications or lost completely, it seems that the scale type found in Cladistia constitutes a very strong support for the views held by Goodrich.

Incidentally, if we compare Figures 4.9 and 4.10 it appears that, although the notion about 'concentric growth' may be upheld as far as the palaeoniscoid scale is concerned, this simile is not very convincing when the layers are as thin as in *Polypterus* and in *Cheirolepis*. However, there is nonetheless a striking difference between the cosmoid and the palaeoniscoid scale which forms the basis of Goodrich's distinction, namely, that in the former case only one layer of the outer shiny material is deposited, whereas in the latter several successive layers are present. This kind of growth may represent a particular instance of Holmgren's principle of delamination (Jarvik, 1959).

The axial skeleton is largely ossified in Cladistia, as in Neopterygii. According to Remane (1936), the amphicoelous, holospondylous vertebrae of these animals are 'almost completely' like those of Teleostei, but they are not, on the other hand, so closely similar to those of the other neopterygian taxa. Since an unconstricted notochord

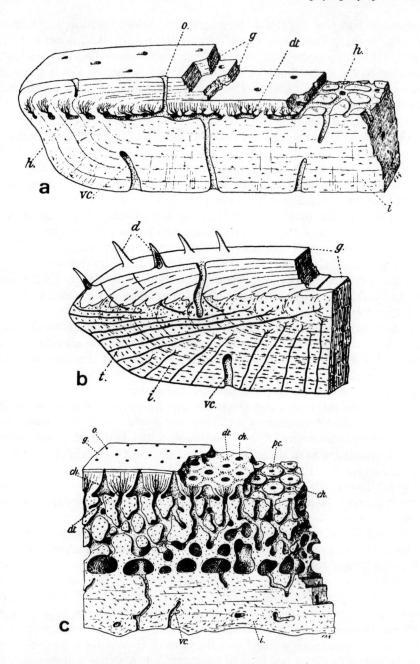

Figure 4.9. Histology of various types of fish scale. a, palaeoniscoid scale from *Eurynotus*; b, lepisosteoid scale from *Lepisosteus*; c, cosmoid scale from *Megalichthys*. ch, chamber of cosmine (dentine) layer; d, denticles; dt, canaliculi in cosmine layer; i, isopedine (bone) layer; o, outer opening of ventral canals; pc, pulp cavity; t, tubules with branching inner ends; vc, vertical canals. (Reproduced from E. S. Goodrich (1909), in *A Treatise on Zoology*, Vol. 9.1, by permission of A. & C. Black)

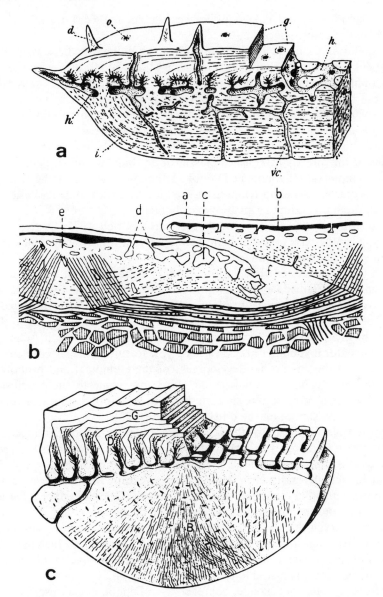

Figure 4.10. Histology of the scale of *Polypterus*. a, Goodrich's conception. For the meaning of the reference letters, see caption to Figure 4.9. (Reproduced from E. S. Goodrich (1909), in *A Treatise on Zoology*, Vol. 9.1, by permission of A. & C. Black.) b, Daget's conception. a, epidermis; b, ganoine; c, cavity in the unexposed part of the scale; d, openings of vertical canals; e, trabecular dentine; f, isopedine. (Reproduced from J. Daget (1958), in *Traité de Zoologie*, Vol. 13.3, by permission of Masson S. A.) c, scale of the palaeoniscoid *Cheirolepis*. B, isopedine; D, dentine; G, ganoine. (Reproduced from T. Ørvig, after Goodrich and Aldinger (1967), in *Structural and Chemical Organization of Teeth*, Vol. 1, by permission of Academic Press)

is found in Chondrostei and Palaeonisciformes (Lehman, 1966a), we may be dealing here with a case of convergence.

The median fins in Cladistia are so different from anything found elsewhere that they must represent a teleotypic character. Goodrich (1928, p. 90) argues, rather embarrassingly I think, that 'the series of dorsal finlets of the *Polypterini* . . . may well have arisen by the subdivision of a single elongated dorsal fin such as occurs in many Actinopterygians'. The caudal fin is symmetrical, rather similar to that of *Amia*, but once more we are certainly dealing with a teleotypic character.

The paired fins in Cladistia are lobed, a feature which we may rate as plesiotypic on the basis of the already established classification, but the endoskeleton is not the plesiotypic archipterygium found in Dipnoi. Although it is not of the tribasal type found in Palaeonisciformes and Rhipidistia, we may still, with some hesitation, accept the loss of the archipterygium as an apotypic character.

The jaw suspension in Cladistia is methyostylic (Goodrich, 1928; Daget, 1950), and since this type is otherwise found only in Actinopterygii, it may with some confidence be regarded as an apotypic character.

Goodrich (1928) suggested that various details in the bones of the head and the skull were similar in Cladistia and Actinopterygii. But the evaluation of such characters evidently involves a certain measure of subjectivity. At least, this is indicated by the fact that all of them are rejected by Jarvik (1968a). Yet the suggestion by this author that the absence of myodomes in Cladistia argues against an affinity with Actinopterygii is hardly tenable, for myodomes may be a teleotypic feature in the latter taxon.

On the basis of the embryonic development of the ethmoidal and orbitotemporal regions of the cartilaginous skull, Bertmar (1968) has come to the conclusion that *Polypterus* assumes a position intermediate between Actinopterygii on the one hand, and Chondrichthyes–Dipnoi on the other (cf. Figure 4.22).

According to Peyer (1968), the teeth of Cladistia contain 'modified dentine', a character which is otherwise found only in Actinopterygii. Two rows of gill-rays are present in *Polypterus*, as in Actinopterygii, while none are found in Dipnoi (Figure 4.3). There is one efferent branchial vessel, as in Actinopterygii, except for Chondrostei, where a situation approaching that in the Selachii is encountered (Goodrich, 1958). The anterior branchial arteries and the jugular and cerebral veins of *Polypterus* are more similar to those of Actinopterygii than to those of Dipnoi and Chondrichthyes (Bertmar, 1968).

Lungs with a ventral opening are present in Cladistia, a feature in which they resemble the Dipnoi. But since we have already accepted this as a plesiotypic character, it cannot be used as an argument against affinity with Actinopterygii.

The pancreas in Cladistia is neither compact, as in Dipnoi, nor diffuse, as in Actinopterygii, but rather forms a hepatopancreas together with the liver (Daget, 1950). Whether this is a teleotypic trait or represents a transition from one form to another is impossible to decide.

According to Goodrich, the urogenital systems in *Polypterus* and Actinopterygii 'are built on the same plan' (1928, p. 91). As far as the male urogenital system is concerned, it appears from Figure 4.4 that in this respect *Acipenser* is constructed like Dipnoi, while the pattern found in *Polypterus* is quite unique (Van den Broek, 1933). In the female urogenital tract a striking similarity obtains between *Amia* and *Polypterus*, but the importance of this character is somewhat reduced by the fact that the condition in Chondrostei resembles that in Dipnoi.

No jugal canal is present in Cladistia and Chondrostei (Goodrich, 1958). The presence of a large solid otolith was suggested by Goodrich (1928) as a diagnostic feature in the present context. This type is indeed characteristic of Neopterygii, but since it has subsequently been found in the extinct rhipidistian *Megalichthys* (Romer, 1937) and in *Latimeria* (Millot and Anthony, 1958a), this is obviously a plesiotypic character.

Among the features rallied in support of his classification, Goodrich (1928) also mentions the structure of the brain, particularly the telencephalon, which is slightly everted in *Polypterus*, as in Chondrostei and 'Holostei'. However, as discussed above, this feature is also found in *Latimeria*, and it must therefore be a plesiotypic character, and the inversion in Dipnoi a teleotypic one.

In the cerebellum of *Polypterus* we find a corpus cerebelli, auricle and valvula, just as in Actinopterygii (Larsell, 1967). It has often been noted that the type of larva found in *Polypterus* and Dipnoi exhibits a number of striking similarities, but since this feature is shared with the Amphibia, it may be regarded as plesiotypic.

Only one physiological character seems available for the present classification, viz., that the neurohypophysial principle of Actinopterygii (ichthyotocin) is found in *Polypterus* (Perks, 1969).

Table 4.6
Characters common to
Actinopterygii and Cladistia

Scales
No archipterygium
Methyostyly
Chondrocranium
Modified dentine
Two gill-rays
One efferent vessel
Certain cephalic vessels
No jugal canal
Valvula
Ichthyotocin

Our list (Table 4.6) thus comprises 11 characters, of which only a few can confidently be rated as apotypic. Yet, in spite of the rather meagre evidence which can be mobilized in favour of Goodrich's view, I fail to see that any other classification could possibly reduce further the incidence of non-apotypy (Figure 4.11). This viewpoint coincides with the ones taken recently by Gardiner (1973) and Schaeffer (1973).

Among recent authors Daget is the one who has studied most exhaustively the problem of the phylogeny of Cladistia (1950; 1958). Unfortunately, his works were written in a precladistic era, and as a consequence the enormous list of characters he accumulated is a thorough mixture of plesiotypic, apotypic and teleotypic characters. However, Daget arrives at the following conclusion: 'Polypterini cannot be maintained among the crossopterygians nor even be considered an aberrant or distant group evolved from them. All that has been presented above concerning their morphological and anatomical organization proves on the contrary that they are affiliated to Actinopterygii' and 'All considered, they [Cladistia] differ less from a typical palaeoniscoid than does a sturgeon or a siluroid' (1950, pp. 160 and 164, author's translation).

It is interesting to contemplate how obsolete traditional phylogenetic reasoning has become with the advent of cladism for, according to the classification in Figure 4.11, which incorporates Daget's views on the affinities of Cladistia, this taxon is, indeed, an aberrant group phylogenetically related to the 'crossopterygians'.

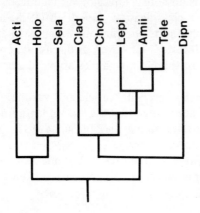

Figure 4.11. Phylogenetic classification of Pisces. Acti, Actinistia; Amii, Amiidae; Chon, Chondrostei; Clad, Cladistia; Dipn, Dipnoi; Holo, Holocephali; Lepi, Lepisosteidae; Sela, Selachii; Tele, Teleostei

Amphibia

The class of living Amphibia comprises only three orders, Caudata, Salientia and Gymnophiona so, *prima facie*, their classification ought to be a simple matter, but the very fact that this problem is still subject to comprehensive discussions (Eaton, 1959; Szarski, 1962; Parsons and Williams, 1963) shows that this is far from being the case. Before we enter upon an analysis of the question, it may be appropriate to stress a conspicuous difference between the more traditional and the present approach to the classification of the Amphibia. According to Parsons and Williams (1963), five alternatives obtain for the affinity between the three orders, namely, they may be quite separate, they may form either one of three different combinations of two and one, or they may be united in one group. According to the cladistic method the three taxa must together constitute a basic classification; hence the first and the last of the above alternatives are without relevance.

The great difficulty involved in the present classification is to find characters which can be rated as apotypic with assurance. There are, indeed, some characters in common between the taxa, but very often there is reason to suspect that they are plesiotypic.

Owing to this state of affairs we shall begin by trying to find the apotypic characters distinguishing the whole class, i.e. the class characters. If we refer to an authority on Amphibia we find: 'Amphibia may . . . be defined as cold-blooded vertebrates having a smooth or rough skin rich in glands which keep it moist; if scales are present, they are hidden in the skin' (Noble, 1954, p. 1). In this definition Noble was unquestionably overcautious; in order to include Gymnophiona he did not call these animals tetrapods, but since the loss of limbs has occurred repeatedly among the tetrapods in association with an elongation of the body, such as found in Gymnophiona, there seems to be no objection to defining Amphibia as tetrapods. One more feature seems to be common to all three groups, that they have pedicellate teeth (Parsons and Williams, 1963). The fact that they are ureotelic is sometimes quoted, but this seems to be a plesiotypic character. Hence there may be three characters in common for the whole class, viz., skin, teeth and limbs. These are hardly enough to establish a monophyletic taxon, especially if, as is usually the case, limbs are regarded as plesiotypic for Tetrapoda.

Among the three orders, Gymnophiona stand apart, being 'more primitive than any other modern Amphibia' (Noble, 1954, p. 12). An idea of the implication of Noble's word 'primitive' may be obtained from the following quotation: 'Thus, in the stapes, as in many other features of the skull [of Gymnophiona] we must look to the reptiles for more primitive conditions than exist in modern Amphibia' (Noble, 1954, p. 223). From this statement it appears that the 'non-primitive' characters in Caudata and Salientia must be either apotypic ones common to the two taxa or teleotypic ones which have arisen independently.

Further support for the isolation of Gymnophiona from the other Amphibia is found in the following quotations: 'The mesonephric tubules of caecilians exhibit a type of modification not found in other Amphibia but one which was taken up by the more advanced vertebrates . . . . Caecilians thus show the first step in the origin of the true ureter of higher vertebrates' (Noble, 1954, p. 272) and 'In this way the blastopore [of embryos of Gymnophiona] becomes surrounded by blastodisc while the latter still remains on the upper surface of a partly divided egg. This is a very important step in the direction of reptilian development' (Noble, 1954, pp. 22–3). Furthermore, it may be observed that the choanal tube in Gymnophiona and Amniota is formed from a naso-buccal groove, while in Caudata and Salientia it is developed from a choanal and a gut process (Bertmar, 1969). And the four characters mentioned here do not exhaust those which affiliate Gymnophiona with Reptilia.

The isolated position of Gymnophiona may be accounted for if it is made the primary twin in a basic classification involving the three amphibian orders. But this classification must be reconciled with their possession of the advanced characters quoted above. Therefore, to account both for their isolation from the other amphibians and their having properties in common with the reptiles, I propose that these are apotypic characters, uniting Gymnophiona and Amniota. Depending upon the rating of the characters which may be common to the other amphibian orders, we arrive at two alternative classifications: if the characters are apotypic, Caudata and Salientia form a pair of secondary twins (Figure 4.12); if they are plesiotypic or convergently teleotypic, one of the two taxa is the secondary twin to Gymnophiona + Amniota, the other one the primary twin. Two options are possible for the latter classification but, as will appear from the following discussion, only the one making Caudata the primary twin needs serious consideration (Figure 4.13).

If we want to find support for the classification in Figure 4.12, we must look for characters in common between Caudata and Salientia. Eaton (1959) claims that the list given in Table 4.7 supports his contention that a closer affinity obtains between these two taxa than between either of them and Gymnophiona. This list is not very extensive, but the similarities between Caudata and various primitive members of Salientia listed under (2), (4) and (6) appear to be weighty.

The dendrograms in Figures 4.12 and 4.13 allow for the origination of Amphibia and Amniota from a common, tetrapod ancestor. This theory probably represents the most widely accepted view, but dissident opinions have been voiced. Thus, on the basis of a number of unmistakable similarities between Dipnoi and Caudata, Holmgren (1933) and Säve-Söderbergh (1934) postulated that the latter originated from the former. Jarvik (1942; 1972) has found that the restored snout region of Porolepiformes resembles a

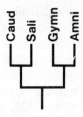

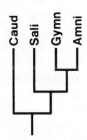

Figure 4.12. Alternative phylogenetic classification of Caudata, Gymnophiona and Salientia in relation to Amniota. Amni, Amniota; Caud, Caudata; Gymn, Gymnophiona; Sali, Salientia

Figure 4.13. Alternative phylogenetic classification of Caudata, Gymnophiona and Salientia in relation to Amniota. Amni, Amniota; Caud, Caudata; Gymn, Gymnophiona; Sali, Salientia

Table 4.7
Characters common to Caudata and Salientia (Eaton, 1959, p. 177)

A similar reduction of dermal bones of the skull and expansion of palatal vacuities
Movable basipterygoid articulation in primitive members of both orders
An operculum formed in the otic capsule, with opercularis muscle
Many details of cranial development, cranial muscles, and thigh muscles, especially between *Ascaphus* and the Urodela
Essentially similar manner of vertebral development, quite consistent with derivation of both orders from Temnospondyli
Presence in the larva of *Leiopelma* of a salamanderlike gular fold, four limbs, and no suggestion of modification from a tadpole

caudate model, being different from that restored in Osteolepiformes, which has affinities with a model based on all other tetrapods. The suggested affiliations between the two pairs of taxa have purportedly been further corroborated by investigations covering a number of other morphological features. From his findings Jarvik has drawn the conclusion that Caudata originated from Porolepiformes and all the other tetrapods from Osteolepiformes. These two hypotheses obviously imply the dual origin of the tetrapod limb, and Jarvik (1965), as well as Holmgren (1933), advanced various arguments in support of this possibility. From an epigenetic point of view there is nothing objectionable in this contention, for when morphological evolution has reached the stage when the acquisition of limbs is epigenetically possible, then the chances of convergence, although slight, are yet within reasonable bounds. The great problem is that the same argument applies equally well to the snout and to other anatomical features. It must be stressed that the dendrogram in Figure 4.13 is not a necessary condition for the acceptance of Jarvik's theory about the origin of the Caudata, for the latter is compatible with the classification in Figure 4.12 but then, of course, it is necessary to postulate that the tetrapod limb has originated convergently three times.

In my opinion we reach an impasse with respect to the two alternatives discussed and, considering the effort which has so far been devoted to this problem, I question whether it can be settled on the basis of morphological characters. Unfortunately, none of the available physiological or biochemical data are of much use either. It should be possible to approach the problem through molecular–biological methods which allow for the determination of the age of the taxa. As appears from Figures 4.12 and 4.13, the time separating Caudata and Salientia from Amniota (and Gymnophiona) should be the same in the former and different in the second instance. Experiments of this kind have been made by Salthe and Kaplan (1966) who, using antigens to muscle lactate dehydrogenase, could show that the two amphibian orders are closer to each other than to Reptilia and that the 'distances' between the latter and each of the two other taxa are approximately identical. Without knowing the precision of the method it is difficult to estimate the validity of these results but, since I have to make a choice between the two classifications discussed, I use them as a means to support Figure 4.12, rather than accepting a trifurcation. In any case, it appears that Amphibia is not by itself a monophyletic taxon, since it comprises either two (Figure 4.12) or three (Figure 4.13) separate monophyletic taxa. The taxon Caudata + Salientia will in the sequel be designated by 'Amphibia'.

Before we finish the present section it may be rewarding to reflect briefly upon the possible consequences of the phenomenon of neoteny observed in some caudates. It is a remarkable fact that neoteny is largely confined to the otherwise most 'primitive' members of the taxon, suggesting that it does not refer to the amphibians, but rather to their ancestors. This is to say that Caudata, and possibly Salientia, may be neotenous descendants of 'non-amphibian' forms. Several of the features peculiar to the two taxa, for example the reduced ossification and the various epigenetic consequences of this phenomenon, among which we may surely include the loss of scales, are explicable in this way.

Of particular interest is the fact that this may also be true for the pedicellate teeth, probably the one feature which has formed the strongest argument in favour of a close affinity between the three amphibian orders, because it is otherwise found only in very distantly related taxa (Parsons and Williams, 1962). Yet, if we make the presumption that their common ancestors had a deciduous dentition of pedicellate teeth which was eventually replaced by labyrinthine teeth, then the amphibian teeth are a plesiotypic feature and hence of no consequence for a decision about the kinship between the three taxa. Support for this contention may be found in the observation of a deciduous foetal dentition in certain viviparous Gymnophiona, in which the teeth, in contrast to those observed in the adults, are of the pedicellate type (Parker, 1956). The reader should refer to de Ricqlès (1975), who rejects the interpretation of amphibian neoteny given here.

Amniota

As the name implies, the members of this taxon are united primarily on the basis of an embryonic character, the amnion. This, and the various other membranes forming around the embryos in this taxon, is unique. Acceptance of this feature as an apotypic and diagnostic character for the monophyletic taxon Amniota is therefore amply vindicated; needless to say, the classification is supported by several other traits.

It is usual to distinguish three separate classes within Amniota, namely, Aves, Mammalia and Reptilia, and it is generally held that of these the first two are monophyletic but that since, purportedly, Aves originated within Reptilia, the latter cannot be so. This implies that, in the present context, the phylogenetic classification must comprise the reptilian orders plus the classes Aves and Mammalia.

For classification within Amniota, of living animals as well as of fossils, the number of temporal openings has traditionally been used. The four main groups that may be distinguished in this way are Anapsida, with no opening, Synapsida and Euryapsida with one, and Diapsida with two. To the first group belongs Chelonia, to the last all remaining living reptiles and Aves.

Mammalia are assumed to be descendants of the synapsid mammal-like reptiles, while Euryapsida are extinct.

In employing this character for the phylogenetic classification of Amniota, comparative morphologists seem to adhere to the postulate of Darwin, according to which evolution proceeds gradually, although seemingly in large steps in this case, and to the rule of Dollo, which states that evolution is irreversible. Hence, since the ancestors of Amniota had no temporal openings, it has been proposed that Chelonia are the oldest order, that subsequently animals arose with one temporal opening, among which were the ancestors of Mammalia, and finally all the Diapsida. Who says that mathematics is of no use in biology?

The number of temporal openings is undoubtedly an important feature which may well be apotypic for certain taxa and is, of course, of particular value in work with fossils. It is, however, a single character and, unless the classification can be supported by others, the possibility of convergence must always be kept in mind as a source of error. It may even have been a factual source of error, for apparently the present classification involves a number of inconsistencies (Bellairs, 1970).

This classification may or may not be correct; that is not so vital. But the danger is that, once it is accepted, specialists will often be inclined to dismiss contradictory characters as being convergent.

In the present section it will be possible to follow the cladistic procedure, and we shall begin by trying to establish the affinities between Aves, Crocodylia and Squamata. Subsequently we shall deal in turn with Chelonia and Rhyncocephalia, thus finishing the taxon Aves + reptiles. For the preliminary classification we shall use the standard characters with some necessary modifications; however, in view of the serious criticism that has been levelled against the use of the temporal openings, we shall forget these for the moment. Whether or not they are of importance will be known once the classification is finished. In the last subsection we shall deal with the classification of Mammalia.

AVES, CROCODYLIA AND SQUAMATA

Birds are distinguished by a great number of teleotypic characters, among which are some that, for the majority of the members of the taxon, contribute to their ability to fly. They do, however, possess many characters in common with reptiles, which leaves no doubt about their origin. It is not easy to find characters in common between Aves and Squamata, but I have found a few. Thus, the type of scutes found in birds is of the same type as that found in Squamata (Boas, 1931), and 'the non-vascular bone of many small birds closely resembles the non-vascular bone tissue of snakes and lizards, but the primary and secondary vascular bone of larger birds is unlike corresponding structural patterns found in fossil or recent reptiles' (Enlow and Brown, 1958, p. 226).

In agreement with the generally accepted affinity between Aves and Crocodylia, it is much easier to find characters that unite these groups. It should be mentioned that in the osteology of extinct birds and crocodiles many similarities have been described, particularly, in the skull (Goodrich, 1958; Romer, 1966). A single penis is found in various flightless birds and in Crocodylia, in contrast to the hemipenes found in Squamata. The ovaries in Squamata are of the saccular type, but in Crocodylia they are of the compact type (Weichert, 1965), also found in Aves. A four-chambered heart is found in Aves and Crocodylia, a feature which has often been used to support the affinity between these taxa (Goodrich, 1958).

With respect to the sensory organs, it appears that the anatomy of the ear of birds and crocodiles exhibits great similarities (Baird, 1970). Likewise, Jacobson's organ is absent or reduced (Parsons, 1970). The histology of the adrenal gland in Aves and Crocodylia is remarkably alike, and quite distinct from that found in Squamata (Gabe, 1970).

The eggs of birds and crocodiles, in contrast to those of Squamata, are truly 'cleidoic', i.e. through the presence of a lime-impregnated shell they are cut off from exchange of fluids with their surroundings, the supply of water being ensured by the presence of albumen. Finally, the young of Squamata have an egg tooth, while those of the other two taxa have a horny caruncle.

It is unlikely that all of the characters listed in Table 4.8 are apotypic, and consequently it may seem that nine characters is a rather small number for deciding the present classification. However, since the latter is one about which I think there will be no dispute, not least because it is so well supported by palaeozoological observations, it may be a waste of effort to look for further characters, and the classification of the three taxa should consequently be as shown in Figure 4.14. By placing Aves to the extreme right, their position as a 'dominant' taxon is emphasized.

Table 4.8
Characters common to Aves and Crocodylia

Single penis
Solid ovary
Four-chambered heart
Anatomy of ear
Absence or reduction of Jacobson's organ
Histology of adrenal tissue
Lime-impregnated shell
Albumen
Horny caruncle

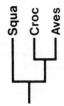

Figure 4.14. Phylogenetic classification of Aves, Crocodylia and
Squamata. Croc, Crocodylia; Squa, Squamata

CHELONIA

Owing to lack of temporal openings Chelonia should, according to prevailing notions, be the twin to the taxa Squamata and Aves + Crocodylia. But when we begin to delve into this question it appears that a number of characters unite Chelonia and Crocodylia. Thus, the type of scutes is the same (Boas, 1931) and, furthermore, 'living and fossil turtles and crocodiles have complex, predominantly primary bone tissues which conform, in basic structural plan, to the pattern found in most other archosaurs' (Enlow and Brown, 1958, p. 225). We also find a single penis and a solid ovary in this taxon (Weichert, 1965). Further, the histology of the adrenal tissue appears to be intermediate between that in *Sphenodon* and that in Aves and Crodylia (Gabe, 1970). In the egg we find a lime-impregnated shell and albumen, and the embryo has a horny caruncle.

It is possible to enumerate more morphological characters common to Chelonia and Crocodylia, but I shall refrain from doing so and mention only one more character, the properties of the blood proteins. Concerning this point Dessauer (1970, pp. 51–52) states: 'Serology demonstrates the fairly close relationship of some lizards and snakes, indicates a very remote affinity between turtles and crocodiles, and shows the wide divergence of the Squamata from the Testudines and Crocodilia'.

This quotation, together with the other characters mentioned, clearly supports the proposal that Chelonia cannot assume the systematic position mentioned above, but must be classed together with Aves and Crocodylia. From Tables 4.8 and 4.9 it appears that the properties common to Crocodylia and to either of the other two taxa are found in all three of them, and hence must be rated as plesiotypic. Since, as discussed in the preceding section, Crocodylia and Aves have characters in common which are absent in Chelonia, it seems necessary to adopt the classification shown in Figure 4.15, in which Aves and Crocodylia are made the secondary twins.

Table 4.9
Characters common to
Chelonia and Crocodylia

Type of scutes
Histology of bone tissue
Single penis
Solid ovary
Lime-impregnated shell
Albumen
Horny caruncle
Blood proteins

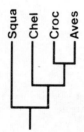

Figure 4.15. Phylogenetic classification of Chelonia in relation to the basic classification in Figure 4.12. Chel, Chelonia; Croc, Crocodylia; Squa, Squamata

Needless to say, this classification differs radically from the one established on the basis of the temporal openings, according to which the anapsid skull is a plesiotypic character, thus placing the Chelonia at the apex of the hierarchy of Amniota. The present classification implies that, although this particular feature has been lost in Chelonia, their ancestors had a diapsid skull. I must admit, however, that I cannot see anything objectionable in the presumption that the heavy ossification which distinguishes this order might have interfered epigenetically with the formation of the skull, with the result that the temporal openings were obliterated.

RHYNCOCEPHALIA

There are two characteristics which distinguish the members of the sole surviving species, *Sphenodon punctata*, of the taxon Rhyncocephalia: they have many properties in common with Squamata, suggesting a rather close affinity, and they are very 'primitive' animals. It is, accordingly, customary to classify them as the primary twin to Sauria and Serpentes, the two taxa included in Squamata.

The correctness of this classification evidently depends upon whether *Sphenodon* is 'primitive' in the sense 'old' or in the sense 'teleotypically retarded'. In the former case the classification is right, in the second case it may or may not be so. To settle this question we must try to establish whether it is possible to find characters in common between *Sphenodon* and the taxon composed of Chelonia + Crocodylia + Aves.

Some morphological characters of this kind are the uncinate process on the ribs (absent in Chelonia) (Weichert, 1965), the solid ovary, the cleidoic egg and the horny caruncle in the young. No male copulatory organ is present in *Sphenodon* (Table 4.10).

Some people may feel that I put too much weight on the properties of the egg in the present context. There can be no doubt, however, that the proper functioning of a cleidoic egg must require so many adjustments that the likelihood of this character

Table 4.10
Characters common to
Aves + Chelonia + Crocodylia and
Rhyncocephalia

Uncinate processes on ribs
Solid ovary
Lime-impregnated shell
Albumen
Horny caruncle

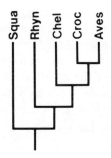

Figure 4.16. Phylogenetic classification of Rhyncocephalia in relation to the classification in Figure 4.13. Chel, Chelonia; Croc, Crocodylia; Rhyn, Rhyncocephalia; Squa, Squamata

arising convergently several times is exceedingly small. Consequently, I believe that the cleidoic egg is a more significant character than many morphological ones.

In any case, the properties listed above suggest that in the present basic classification Squamata are the primary twin (Figure 4.16).

MAMMALIA

There is no doubt that any basic classification comprising Mammalia and two of the taxa in reptiles + Aves would give the result that Mammalia are the primary twin. This outcome would agree with current notions, based mainly on palaeozoological evidence. The latter shows with all the necessary clarity that there is an early reptilian line which, dating back to the lower Carboniferous, leads from Pelycosauria over Therapsida to the early Mammalia. This proposition is supported not only by various features in the skull and the dentition (Romer, 1966), but also by such an apparently insignificant trait as the absence of the hooked fifth metatarsal otherwise found in living reptiles and birds (Goodrich, 1916).*

The conclusion that Mammalia represent a separate evolutionary line therefore seems unavoidable, and that they must consequently be classified as shown in Figure 4.17. It will be noticed that in this figure I have placed Mammalia to the extreme right. The reason for this is that, even allowing for the resourcefulness of the birds, this taxon must be rated as 'dominant' relative to the remaining Amniota.

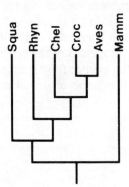

Figure 4.17. Phylogenetic classification of Amniota. Chel, Chelonia; Croc, Crocodylia; Mamm, Mammalia; Rhyn, Rhyncocephalia; Squa, Squamata

At this stage it may be worth mentioning two characters which have been used above in the classification of the other amniotes, the penis and the egg. The penis in Mammalia is single, as in Chelonia, Crocodylia and Aves, and it appears that various attempts have been made to derive the type found in Monotremata from that in these taxa, but according to Gerhardt (1933) the differences are too many to make these endeavours convincing. The eggs of Monotremata contain albumen, and lime impregnation may occur, but they also exhibit some affinity to those of Squamata, for instance through the fact that, apparently, they exchange water with their surroundings and hence are not truly cleidoic (Raynaud, 1969). The significance of these features may be questioned, but they support the adopted classification. Furthermore, in the young of Monotremata we find both an egg tooth and a horny caruncle (Young, 1962).

*Shortly before the present manuscript went to the printer, a paper was published by Robinson (1975), showing that Goodrich's generalization cannot be upheld.

In the introduction to the present section I mentioned that a basic classification of Mammalia together with taxa from reptiles + Aves would give the result which is here accepted as the correct one. Yet there are two reasons why this is not quite satisfactory from a cladistic point of view.

The first is that most biologists would probably be inclined to regard Mammalia as a dominant taxon, and as such they ought to be one of the secondary twins. This discrepancy is relatively easy to explain for, as we shall see in the next chapter, there is good support for the view that the ancestors of the mammals were non-dominant animals which, owing to the caprices of fortune, managed the almost impossible, to reclaim dominance once lost.

The other reason is that most of the characters setting Mammalia apart would be mammalian, and hence teleotypic, ones, while those uniting the other taxa would be reptilian characters, and hence, presumably, plesiotypic ones. The latter point is not easy to settle, since we do not know how many new reptilian characters were acquired during the time the mammalian ancestors were losing theirs. At any rate, it is evident that the phylogenetic classification of Mammalia would be very difficult without the assistance of fossil data.

I shall finish by mentioning that a number of similarities exist between Crocodylia and Mammalia. These are the subdivision of the vertebral column into five distinct regions, a secondary palate and thecodont teeth. In addition, we have, of course, the four-chambered heart and a complete diaphragm, the presence of a cochlea in the ear and of a true cerebral cortex. All these features must, according to the proposed classification, be accepted as instances of convergence.

Vertebrata

All that remains to do in this section is to assemble the four dendrograms representing the phylogenetic classification of the cyclostomes, fishes, 'Amphibia' and Amniota into one, comprising the whole phylum. And this task is facilitated by the fact that the first taxon was classified with respect to Gnathostomata and hence to the fishes, and 'Amphibia' with respect to Amniota. The resulting two dendrograms must be united, however, and here it appears that only two alternatives can be seriously contemplated. The first is that the various similarities which can be demonstrated between Dipnoi and Caudata are rated as plesiotypic and that, consequently, Cladistia + Actinopterygii and Tetrapoda are secondary twins. The other possibility, a more conservative stand and the one adopted here, is that the characters in question are apotypic and that thus Dipnoi and Tetrapoda are the secondary twins (Figure 4.18).

It is not possible to mobilize much support for this choice; indeed, I am inclined to believe that, with the characters available at the moment, including those surveyed in the following section, the probabilities for the two alternative classifications of Dipnoi are about fifty-fifty, a circumstance which, of course, might warrant the introduction of a trichotomy in the dendrogram.

Among the $3 \times 5 \times \ldots \times 37$ possible dendrograms representing the phylogeny of the 20 vertebrate taxa with which we operate in the present context, we have thus finally arrived at a single one. It is obvious that the overwhelming majority of classifications were excluded in advance, but it is certain that Figure 4.18 does not represent the only acceptable one. I am aware that the classification of a few taxa will be strongly contested.

In rejoinder, I can only point out that the dendrogram is an empirical theory which can and, if possible, should be falsified. However, considering the past disputes over the classification of these taxa I doubt whether this can be done on the basis of existing knowledge. New data, morphological or non-morphological as the case may be, are unquestionably needed before this issue can be settled.

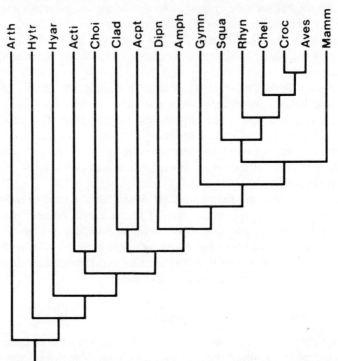

Figure 4.18. Phylogenetic classification of Vertebrata. Acpt, Actinopterygii; Acti, Actinistia; Amph, Caudata + Salientia; Arth, 'Arthropoda'; Chel, Chelonia; Choi, Chondrichthyes; Clad, Cladistia; Croc, Crocodylia; Dipn, Dipnoi; Gymn, Gymnophiona; Hyar, Hyperoartii; Hytr, Hyperotreta; Mamm, Mammalia; Rhyn, Rhyncocephalia; Squa, Squamata

ASSESSMENT OF THE CLASSIFICATION

The main purpose of the present section is to test the proposed classification through confrontation with further empirical evidence. It must be acknowledged that 'testing' is used here in a sense that does not quite correspond to the one implied by Popper (1969), for most of the data are well-known facts. This does not mean that they were all known to me when I made the classification, indeed, some were not, but I must admit that I made frequent references to one set of data, those concerning lens antigens, to check if my classification was proceeding in the right direction, i.e. whether the number of non-apotypies required by the classification went down when a change was introduced.

In the following four subsections we shall test the classification with respect to properties which may be classified as morphological, physiological, biochemical and genomic characters, respectively. The characters used are not the outcome of a systematic selection. However, by and large they may be distinguished by one or both of the following features, they can be traced in the major taxa of the phylum and they have been used previously for classificatory purposes. I have not, however, aimed in the slightest extent at completeness; there are many more characters available in the literature for those who want to falsify my classification. If strict adherence to the cladistic rules were adopted, testing should involve estimates of the degree of non-apotypy involved in this and alternative classifications. As far as I can see, it is not possible to employ this expedient for the time being; all that can be done is to establish whether or not the actually observed distribution of the characters makes sense on the basis of the proposed classification.

In the last two subsections we shall compare the present classification with previous ones and discuss some taxonomic consequences of the suggested classification of Vertebrata.

Morphological characters

Traditionally the anatomy and histology of the hard tissues have had great impact in phylogenetics, primarily because scales, teeth and skeletons are the only parts preserved in fossils.

One point which has been borne out by the fossil record is that, often enough, transitional stages are missing between the ossification patterns distinguishing separate, though undoubtedly somehow related, groups. In line with the Darwinian tradition, according to which evolution occurs in small steps, observations of this kind are usually blamed on the incompleteness of the fossil record. Yet, recalling that the process of calcification involves the interplay between a number, probably limited, of parameters, it may be envisaged that if one or more of the latter are changed, possibly only quantitatively, then the resulting pattern of ossification may well be radically different from the one found before the change.

This statement, which to some degree has been advanced by Devillers and Corsin (1968), is a conjecture and must remain so as long as our knowledge about mechanisms of calcification is as fragmentary as it is at present. It is remarkable, however, that this state of affairs can hardly be referred to the lack of factual observations, for of these there are plenty in the literature; what is missing may rather be a comprehensive theory of calcification, allowing for predictions about modifications in ossification patterns following changes in the various parameters involved in the calcification process.

This suspicion as regards the reliability of hard tissues for classificatory purposes is confirmed when we look into the phylogeny of the axial skeleton. It seems well substantiated that aspondyly occurred in the early gnathostomes. This condition is represented today by *Latimeria*, Dipnoi and Chondrostei. Vertebrae are found in Chondrichthyes, Cladistia, Neopterygii and Tetrapoda, but also in some extinct members of Dipnoi. It is evident that, whichever classification is adopted for Vertebrata, convergence must have occurred several times with respect to the calcification of the axial skeleton, involving both gains and losses.

There is, however, one feature which must be discussed in the present context, the otoliths. Carlström (1963) has studied the distribution of these in the vertebrates and found that they vary with respect to chemical composition, crystal form and size. Two different salts occur, calcium phosphate as apatite in Hyperotreta and Hyperoartii, and calcium carbonate in the remaining taxa. The most common crystal form of the latter is aragonite, found in all but the endothermic animals, Mammalia and Aves. In these only calcite occurs, but this crystal form is also present, together with aragonite, in some members of Squamata and in Chelonia. Calcite is further found in *Latimeria* and in one shark. An unusual, and unstable, crystal form, vaterite, is present in Cladistia, Chondrostei, Lepisosteidae and Amiidae.

The otoliths are either large statoliths (ear stones) or masses of minute statoconia (ear dust). The former type is found in *Latimeria*, and in Cladistia + Actinopterygii. In those cases where both statoliths and statoconia occur the former are made up of aragonite, the latter of vaterite.

The data are presented in Figure 4.19. It is seen that they fit the present classification quite well. The absence of statoliths must be accounted for in three cases, in Chondrichthyes, Dipnoi and Tetrapoda. Since the two latter taxa are made secondary twins, only two instances of loss are required.

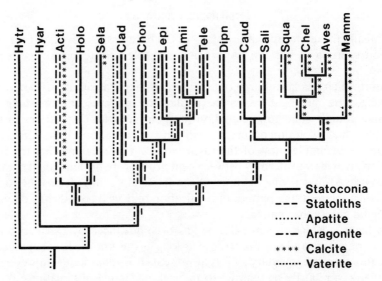

Figure 4.19. Distribution of otoliths in Vertebrata. Acti, Actinistia; Amii, Amiidae; Caud, Caudata; Chel, Chelonia; Chon, Chondrostei; Clad, Cladistia; Dipn, Dipnoi; Holo, Holocephali; Hyar, Hyperoartii; Hytr, Hyperotreta; Lepi, Lepisosteidae; Mamm, Mammalia; Sali, Salientia; Sela, Selachii; Squa, Squamata; Tele, Teleostei

A feature which has been much used in considerations of fish phylogeny is the form of the tail fin. The original type is presumed to be the protocercal caudal fin found in the cyclostomes; the heterocercal tail is an apotypic character for Gnathostomata. This tail, more or less modified, is still found in Chondrichthyes, Chondrostei, Lepisosteidae and Amiidae (Figure 4.20), but the other four terminal taxa in the figure are each distinguished by a special type of caudal fin which, according to the present classification, must have arisen teleotypically within the taxon. This conclusion is in complete agreement with current notions, for the four types of tail fin are so different that no closer relation between any pair of them can be presumed to exist.

Another character which has frequently been discussed in phylogenetic contexts is the jaw suspension. According to de Beer (1937) the primitive condition may have been autodiastyly, found in various extinct fishes. Modifications of this jaw suspension have occurred teleotypically in all the extant vertebrates except the members of Gymnophiona + Amniota (Figure 4.21). One modification, amphistyly, is found in a selachian, *Heptanchus*, and also in *Latimeria* (Millot and Anthony, 1958a). Another modification, autosystyly, is found in Dipnoi and in 'Amphibia', a condition which according to the proposed classification must be due to convergence. It may be noted that in *Acanthodes* the jaw suspension is amphistylic (Miles, 1968). If de Beer's (1937) view is correct, then my classification implies that this jaw suspension was convergently acquired by the extinct acanthodians. This conclusion coincides with the opinion of Miles (1968).

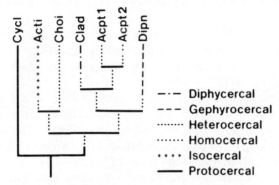

Figure 4.20. Distribution of various types of tail fin in cyclostomes and fishes. Acpt, Actinopterygii; Acti, Actinistia; Choi, Chondrichthyes; Clad, Cladistia; Cycl, cyclostomes; Dipn, Dipnoi. It should be noted that 'Cycl' and 'Acpt 1' are not monophyletic taxa, but 'Acpt 2', which stands for Teleostei, is

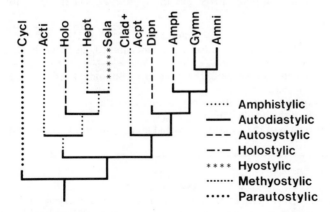

Figure 4.21. Distribution of various types of jaw suspension in Gnathostomata. Acpt, Actinopterygii; Acti, Actinistia; Amni, Amniota; Amph, Caudata + Salientia; Clad, Cladistia; Cycl, cyclostomes; Dipn, Dipnoi; Gymn, Gymnophiona; Hept, *Heptanchus*; Holo, Holocephali; Sela, Selachii

In his work on the vertebrate skull, de Beer (1937) compiled lists of characters which distinguish the chondrocranium of various groups. The lists which summarize the differences between Chondrichthyes and Actinopterygii are not interesting in the present context, since the proposition that these two taxa are very distinct groups is common to all classifications, including the present one. Of more significance perhaps, are the characters listed in Table 4.11, which serve to emphasize the affinity between Chondrostei and Cladistia.

The particular position of Dipnoi relative to the other fishes can be seen in Table 4.12. Since none of the characters in the list are present in 'Amphibia', we may probably rate all the features as plesiotypic gnathostomous ones which have been lost in Tetrapoda. Furthermore, the first has been lost in Actinopterygii, the second in Chondrichthyes and the third in Selachii.

Among the characters which, according to de Beer, are common to Dipnoi, Caudata and Salientia, the only one not dealt with elsewhere in the present work is 'the associated chondrification of anterior parachordals and trabeculae as "Balkenplatten", flanking the tip of the notochord' (1937, p. 459).

Against this must be weighed a considerable number of characters common to Caudata and Salientia but absent in Dipnoi (Table 4.13). This is an impressive list that might well be claimed to support the classification in Figure 4.12, but it is possible to assemble an equally long list of differences between the two taxa which may or may not support the alternative classification in Figure 4.13 (Table 4.14). It is thus seen that here, as with so many other characters, an impasse is reached with respect to the affiliation between the 'amphibian' orders.

The relations between the various reptilian orders, and between these and the birds, were also analysed by de Beer (1937), on the basis of various chondrocranial characters. The affinity between Squamata and Rhyncocephalia is indicated by the characters listed in Table 4.15. Besides the egg caruncle, a feature already dealt with, two more characters unite Rhyncocephalia, Aves, Chelonia and Crocodylia, namely, 'persistence of continuity between columella auris and the stylohyal, via the pars interhyalis' and

Table 4.11
Chondrocranial characters common to Chondrostei and Cladistia (G. R. de Beer, 1937)

Enclosure in the cartilaginous rostrum of the transverse ethmoidal lateral line canal
Similarity in form of the nasal capsules
Atypical relation of the internal carotid artery to the trabecula
Similarity in relations of the jugular canal
Forking of the dorsal ends of the anterior branchial arches into infra- and suprapharyngo-
    branchials

Table 4.12
Chondrocranial similarities between Dipnoi, Selachii and Actinopterygii (G. R. de Beer, 1937)

Nasal capsules have no floor (as in Selachii)
Glossopharyngeal nerve passes through the auditory capsule (as in Actinopterygii)
Auditory capsule has no medial wall (as in Holocephali and Actinopterygii)
Abducens nerve leaves the skull posterior to the pila antotica
Chondrocranial roof is complete (in *Neoceratodus*)

Table 4.13
Chondrocranial similarities between Caudata and Salientia (G. R. de Beer, 1937)

Shortness of occipital region of the skull
Chondrification of the medial wall of the auditory capsule
Formation of fenestra ovalis and foramen perilymphaticum
Existence of a columella and an operculum in the fenestra ovalis
Formation of the occipito-atlantal joint as a result of splitting of an invertebral cartilage, the anterior portion forming the paired condyles
Position of the trabecular horns, ventral to the external nostrils
Larval hypobranchial skeleton, with fusion of the dorsal ends of the ceratobranchials
Number and position of the bones, especially the paired bones

Table 4.14
Chondrocranial differences between Caudata and Salientia (G. R. de Beer, 1937)

Suprarostral and infrarostral cartilages of Salientia
Complete transformation of the quadrate at metamorphosis in Salientia
Usual absence of a nasal septum and presence of separate medial walls to the nasal capsules in Caudata
Form of the hypobranchial skeleton after metamorphosis
Existence of a basal connection between the quadrate in Caudata, and a pseudobasal connection in Salientia
Fusion of the parachordals in the hindmost region dorsal to the notochord in Salientia and ventral to it in Caudata
Absence of a basicranial fenestra in Salientia

Table 4.15
Chondrocranial similarities between Rhyncocephalia and Squamata (G. R. de Beer, 1937)

Supraparachordal course of notochord
Fusion of the pleurocentrum of the proatlas vertebra with the tip of the odontoid process
Formation of a median hypocentral occipital condyle

Table 4.16
Chondrocranial similarities between Aves, Chelonia and Crocodylia (G. R. de Beer, 1937)

Intraparachordal course of the notochord
Fusion of the pleurocentrum of the proatlas vertebra with the basal plate
Median pleurocentral occipital condyle

Table 4.17
Chondrocranial similarities between Aves and Crocodylia (G. R. de Beer, 1937)

Presence of a median prenasal process
Formation of infrapolar processes
Formation of subcapsular processes or metotic cartilages
Fragmentation of the parasphenoid into a median rostrum and paired basitemporals
Pneumaticity of the bones of the skull

'connection between the processus dorsalis of the columella auris and the extracolumella by means of the laterohyal, thus enclosing "Huxley's" foramen' (de Beer, 1937, p. 463).

The characters uniting Aves, Chelonia and Crocodylia are shown in Table 4.16, those shared by Aves and Crocodylia in Table 4.17. Only one character is common to Chelonia and Crocodylia and absent in Aves: 'connection of the columella auris via the pars interhyalis with Meckel's cartilage' (de Beer, 1937, p. 462).

The various characters described by de Beer roundly support the present classification, except for the last one mentioned. It may be that this feature has been lost or modified teleotypically in Aves.

I have reported here the characters which are directly concerned with the classification established above. They are not very many, but in view of the fact that they are characters dealing solely with the chondrocranium they may still carry some weight.

Bertmar (1968) has surveyed the embryonic chondrocranium in various fishes and amphibians. Figure 4.22 consists of diagrams of representative chondrocrania arranged in accordance with the present classification. I may be biased, but it seems to me that these diagrams show Cladistia + Actinopterygii as constituting a reasonably homogeneous group, and the amphibian chondrocrania as possessing features which may be related to the condition found in Dipnoi. Since the classification of Selachii is not disputed, it may suffice to acknowledge that the chondrocranium in this taxon appears to exhibit plesiotypic features from which those of the other taxa are derivable.

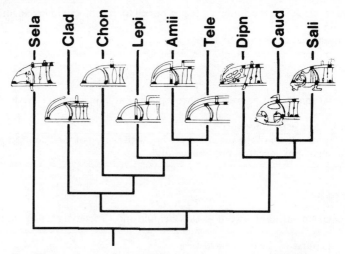

Figure 4.22. Diagrams of the ethmoidal and orbito-temporal regions of the cartilaginous skull in some fishes and amphibians. (Reproduced from G. Bertmar (1968), in *Current Problems in Lower Vertebrate Phylogeny*, by permission of the author and Almqvist & Wiksell)

One further feature which has been discussed in phylogenetic contexts is the anatomy of the nose. In Gnathostomata three different nasal structures may be distinguished, viz., achoanate, choanate and 'pseudochoanate'. From the distribution shown in Figure 4.23 it follows that lack of choana is a plesiotypic character, that true choanae have arisen twice, in the ancestors of the tetrapods and in some teleosts (Bertmar, 1969), and that the 'pseudochoana' is also a convergent character which has arisen independently three times.

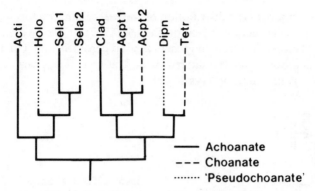

Figure 4.23. Distribution of choanae in Gnathostomata. Acpt, Actinopterygii; Acti, Actinistia; Clad, Cladistia; Dipn, Dipnoi; Holo, Holocephalia; Sela, Selachii; Tetr, Tetrapoda. It should be noted that while 'Sela 1' and 'Acpt 1' are not monophyletic taxa, 'Sela 2' and 'Acpt 2' may or may not be

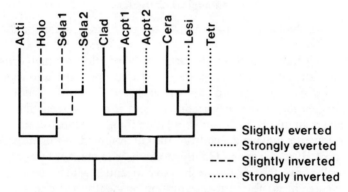

Figure 4.24. Distribution of telencephalic shape in Gnathostomata. Acpt, Actinopterygii; Acti, Actinistia; Cera, Ceratodidae; Clad, Cladistia; Holo, Holocephali; Lesi, Lepidosirenidae; Sela, Selachii; Tetr, Tetrapoda. It should be noted that 'Sela 1' and 'Acpt 1', and possibly 'Sela 2', are not monophyletic taxa, but 'Acpt 2', which stands for Teleostei, is

Of characters related to the soft anatomy we shall only discuss the telencephalon and the cerebellum here. Holmgren (1922), in his discussion of the phylogeny of the telencephalon, came to the conclusion that the slightly inverted forebrain in Holocephali was the original form, supposedly also present in 'Crossopterygii', from which all the other types could be derived. And, indeed, this proposition did fit very well before the shape of the telencephalon in *Latimeria* was known. However, as the latter was found to be slightly everted, and reminiscent of that found in *Neoceratodus* and in Actinopterygii (Nieuwenhuys, 1963), it seems necessary to postulate that this form is the original one, and that the strong inversion found in some Selachii, Lepidosirenidae and Tetrapoda has arisen convergently (Figure 4.24). Incidentally, it appears from this figure that we cannot deduce whether the telencephalon of the rhipidistian ancestors of Tetrapoda was slightly everted or slightly or strongly inverted.

Figure 4.25 summarizes the distribution of the cerebellar structures which have been dealt with in an earlier section, based on Larsell (1967). It can be seen that the corpus cerebelli arose before the branching between Hyperoartii and Gnathostomata, the auricle in the line leading to the latter taxon. The valvula appears to be a teleotypic character in Cladistia + Actinopterygii.

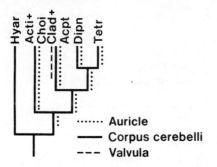

Auricle
Corpus cerebelli
Valvula

Figure 4.25. Distribution of cerebellar structures in Hyperoartii and Gnathostomata. Acpt, Actinopterygii; Acti, Actinistia; Choi, Chondrichthyes; Clad, Cladistia; Dipn, Dipnoi; Hyar, Hyperoartii; Tetr, Tetrapoda

Physiological characters

In this section we shall take up two interrelated problems, osmoregulation and urea production.

Osmoregulation is found in all vertebrates, but the problem we shall consider here is the one facing aquatic animals whose surfaces are freely permeable to water and ions; hence the amniotes, even the aquatic ones, are excluded. Figure 4.26 shows the various kinds of osmoregulation found in vertebrates, and it can be seen that the simplest, also found in invertebrates, is the isosmoregulation found in Hyperotreta. The first very important step towards independence from the environment in this respect was made by the ancestors of Hyperoartii and Gnathostomata, who managed to keep the tonicity of the blood well above that of the fresh-water localities which they inhabited. This achievement, so significant for further evolution, evidently had one great drawback, exclusion of the vast expanses of the oceans from colonization. Among living fishes we encounter two different expedients which have been used to overcome this difficulty. One is the preservation of hyperosmoregulation, accomplished through retention of urea, as found in *Latimeria* and Chondrichthyes; the other is the hyposmoregulation found in the lampreys and in some actinopterygians, in which blood tonicity is preserved through active regulatory mechanisms.

It is seen from Figure 4.26 that the classification advocated here requires hyposmoregulation to have arisen at least twice, in Hyperoartii and in the ancestors to actinopterygians, a proposal which is supported by the fact that distinct differences are observable in the mechanism of hyposmoregulation in the two taxa (Hickman and Trump, 1969). This proposition implies, however, that it has been lost several times in the course of the evolution of Actinopterygii. Various kinds of evidence seem to argue against this proposition, so it seems more likely that it has been acquired more than once by members of this taxon. Two instances are suggested in Figure 4.26, but that is possibly an underestimate.

Whether or not hyposmoregulation was also possessed by the various 'crossopterygians' which are known to have invaded the sea is impossible to decide. In one line, including, we may presume, all marine actinistians, urea retention was adopted, and it is possible that it also occurred in other taxa (Thomson, 1969), but in that case we would, presumably, be dealing with instances of convergence.

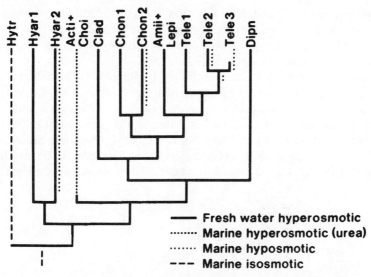

Figure 4.26. Distribution of osmoregulation in cyclostomes and fishes. Acti, Actinistia; Amii, Amiidae; Choi, Chondrichthyes; Chon, Chondrostei; Clad, Cladistia; Dipn, Dipnoi; Hyar, Hyperoartii; Hytr, Hyperotreta; Lepi, Lepisosteidae; Tele, Teleostei. It should be noted that 'Amii + Lepi' is not a monophyletic taxon, while 'Hyar 1', 'Hyar 2', 'Chon 1', 'Chon 2', 'Tele 1', 'Tele 2' and 'Tele 3' may or may not be

If the marine 'crossopterygians' belonged to the line leading to Actinistia and Chondrichthyes, they might possess eggs capable of developing in the sea. Otherwise they would have to return to fresh water for spawning, for from the phylogenetic classification it must be presumed that the eggs and larvae of Cladistia, Dipnoi and 'Amphibia', which exhibit so many similarities, represent those of their common ancestors and hence, presumably, also those of the rhipidistians which gave rise to the tetrapods. It seems very unlikely that eggs and larvae of this type could ever develop and survive in salt water.

We teach our students that urea production is primarily a means for detoxification of ammonia, and that this expedient is not necessary in animals living in ambient water. This does not square with urea production in Chondrichthyes and Dipnoi, but then, of course, the former 'need' urea for osmoregulation and the latter for ammonia detoxification during aestivation. Considering that urea production is a very complex process, involving many enzymes, this teleological explanation, with its concomitant implication of convergence, must make enormous demands on natural selection, so great, indeed, that even believers in the almightiness of this agent ought to feel some embarrassment.

As we have observed above, the ornithine cycle is an expedient which the vertebrates may have inherited from their invertebrate ancestors. If this possibility is accepted, it has, according to the present classification, been lost through convergence a number of times, thus, presumably, in both cyclostome orders and at least once in Actinopterygii (or Cladistia + Actinopterygii). As far as this proposition is concerned, it should be mentioned that arginase can be demonstrated in the teleost liver (Cohen and Brown, 1960).

One great advantage of the traditional classification of Amniota, which is preserved in the present one, is that, by making Mammalia the twin taxon to all other Amniota, it is easy to explain the preservation of ureotelism in the former taxon and its purported substitution by uricotelism in reptiles and Aves.

The idea behind the story of uricotelism (Needham, 1931) is intimately associated with the cleidoic egg which, cut off from exchange of fluid with the surroundings, may be poisoned by the urea accumulating during development. We have already seen above that the egg of Squamata is not 'cleidoic' and, indeed, they secrete most of their nitrogen waste in the form of ammonia and urea (H. Clark, 1953; Clark and Sisken, 1956); the same holds for the alligator embryo (Clark, Sisken and Shannon, 1957). In Chelonia it is possible to find forms that are almost uricotelic, and others which are similar to those

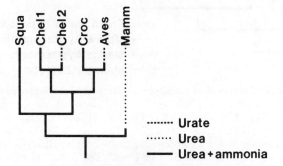

Figure 4.27. Distribution of end-products of embryonic nitrogen metabolism in Amniota. Chel, Chelonia; Croc, crocodylia; Mamm, Mammalia; Squa, Squamata. It should be noted that 'Chel 1' and 'Chel 2' may or may not be monophyletic taxa

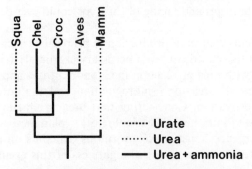

Figure 4.28. Distribution of end-products of adult nitrogen metabolism in Amniota. Chel, Chelonia; Croc, Crocodylia; Mamm, Mammalia; Squa, Squamata

just described (Bellairs, 1970). However, even in birds the situation does not seem to be so clear-cut as originally envisaged, for substantial amounts of ammonia and urea are certainly produced (Clark and Fischer, 1957). The data shown in Figure 4.27 suggest that, so far as is known, 'cleidoic' uricotelism only occurs in Aves and in some Chelonia, and that this situation must have arisen through convergence.

The end-products of adult metabolism are shown in Figure 4.28. It appears that the uricotelism in Squamata and Aves is convergent. The secretion of ammonia, together with urea, in Chelonia and Crocodylia may possibly be referred to their aquatic habitat.

Biochemical characters

In the present section we shall deal with four different biochemical characters, lens proteins, bile salts, pituitary peptide hormones and the amino acid sequence in cytochrome c.

The importance of the work on the distribution of lens proteins in vertebrates published by Manski and his collaborators (1967a, b, c) rests on two facts: first, that many different taxa have been studied, namely, 11 out of the 20 terminal taxa in Figure 4.18, and second, that many immunologically different lens proteins were found, so that the results represent many different taxonomic characters.

Altogether, six different lens proteins were found in Hyperoartii, out of which four, AG 1, 2, 3 and 4, are preserved in all investigated taxa, while one, AG 5, is present in Dipnoi and Cladistia, and one, AG 6, only in the latter taxon (Figure 4.29). As indicated in the figure, these two proteins, which date back to the time before the separation between Hyperoartii and Gnathostomata, have each been lost three times in the course of subsequent evolution.

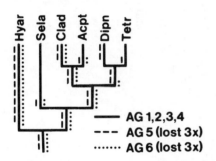

Figure 4.29. Distribution of a set of six lens proteins which originated before the separation between Hyperoartii and Gnathostomata. Acpt, Actinopterygii; Clad, Cladistia; Dipn, Dipnoi; Hyar, Hyperoartii; Sela, Selachii; Tetr, Tetrapoda. It should be noted that the classification is exclusive. This and the following five figures are based on data published by Manski *et al.* (1967a, b, c)

Nine different lens proteins were acquired by the ancestors of Gnathostomata. Two proteins, PL 1 and 4, are present in all investigated taxa. One protein, PL 6, has been lost in Amniota, while one, CH 1, is present in Selachii and Tetrapoda but absent in Cladistia + Actinopterygii and Dipnoi (Figure 4.30). Two further proteins, GN 1 and 4, are present in Selachii, Cladistia, Lepisosteidae and Amiidae, suggesting losses in Teleostei and Dipnoi + Tetrapoda. The last group of three proteins, GN 2, 3 and 5, is present in all taxa but Dipnoi; however, within Amniota one or two of the proteins are missing in Crocodylia, Mammalia and Aves, while all three are present in Squamata and Chelonia (Figure 4.31).

At the next branching three further proteins were acquired, AC 1, 2 and 3. Of these the first is found in all taxa except Dipnoi, while the second has been lost four times, and the third at least twice (Figure 4.32). The exact number of losses is unknown, since the fate of these proteins has not been established in Amniota.

In the taxon Dipnoi + Tetrapoda another three proteins, DI 1, 2 and 3, have been acquired. D 1 is found in all taxa, D 2 in all but Aves and Mammalia, while D 3 is absent from Amniota (Figure 4.33).

Certain lens proteins are exclusive to the taxon Cladistia + Actinopterygii. Thus, two proteins, PA 1 and 2, are found in all taxa, one, PA 3, in all but Teleostei, and one, PA 4, only in Cladistia and Amiidae. Finally, it may be noted that two proteins, TE 1 and 2, occur only within the taxon Teleostei (Figure 4.34).

As appears from Figures 4.29–4.34, the number of losses, and hence the degree of non-apotypy, required to account for the actual distribution of the lens proteins is remarkably slight when interpreted with reference to the present classification. Whether or not the latter represents the minimum number of convergences is difficult to say, but

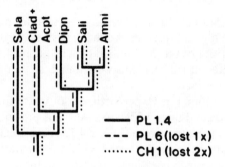

Figure 4.30. Distribution of a set of four lens proteins which originated in early members of the taxon Gnathostomata. Acpt, Actinopterygii; Amni, Amniota; Clad, Cladistia; Dipn, Dipnoi; Sali, Salientia; Sela, Selachii. It should be noted that the classification is exclusive, and also that Squamata have not been investigated with respect to the presence of CH 1

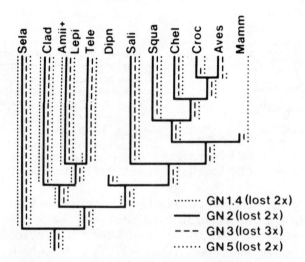

Figure 4.31. Distribution of another set of five lens proteins which originated in early members of Gnathostomata. Amii, Amiidae; Chel, Chelonia; Clad, Cladistia; Croc, Crocodylia; Dipn, Dipnoi; Lepi, Lepisosteidae; Mamm, Mammalia; Sali, Salientia; Sela, Selachii; Squa, Squamata; Tele, Teleostei. It should be noted that 'Amii + Lepi' is not a monophyletic taxon and that the classification is exclusive

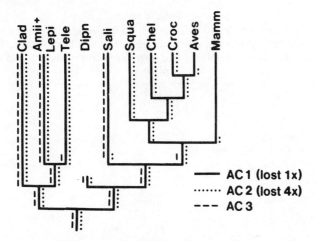

Figure 4.32. Distribution of a set of three lens proteins which originated after the Actinistia + Chondrichthyes had become separated from the remaining Gnathostomata. Amii, Amiidae; Chel, Chelonia; Clad, Cladistia; Croc, Crocodylia; Dipn, Dipnoi; Lepi, Lepisosteidae; Mamm, Mammalia; Sali, Salientia; Squa, Squamata; Tele, Teleostei. It should be noted that 'Amii + Lepi' is not a monophyletic taxon and that the classification is exclusive

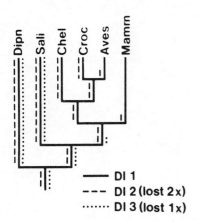

Figure 4.33. Distribution of a set of three lens proteins which originated after Cladistia + Actinopterygii had become separated from Dipnoi + Tetrapoda. Chel, Chelonia; Croc, Crocodylia; Dipn, Dipnoi; Mamm, Mammalia; Sali, Salientia. It should be pointed out that AC 3 cannot be determined in Amniota, and it is therefore impossible to decide whether or not it has been lost in this taxon

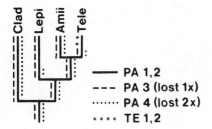

Figure 4.34. Distribution of a set of four lens proteins which originated after Cladistia + Actinopterygii had become separated from Dipnoi + Tetrapoda, and a set of two lens proteins which originated after Teleostei became separated from the remaining Actinopterygii. Amii, Amiidae; Clad, Cladistia; Lepi, Lepisosteoidae; Tele, Teleostei. It should be noted that the classification is exclusive

Table 4.18

Bile salts in Vertebrata. α: –––; β: ——. From (8) no distinction is made between 5α- and 5β-compounds. The numbers above the arrows indicate the number of changes involved in the transition from one molecule to another. The asterisks indicate where the changes have taken place

(1) Myxinol

(2) Hypothetical intermediate

(4) Latimerol

(5) 5a-Cyprinol

(7) Chimaerol

(6) Scymnol

(3) Petromyzonol

(9) Ranol

(8) Bufol

(10) 3a,7a,12a,-Trihydroxy-
coprostanic acid

(12) Cholic acid

(11) 3a,7a,12a,22ξ-Tetrahydroxy-
coprostanic acid

it deserves mention that there is one classification which in this respect is equivalent to the one proposed, namely, the one which makes Dipnoi the twin to Cladistia + Actinopterygii and Tetrapoda. With this classification the losses of AC 1, 2 and 3 in Dipnoi are avoided, but instead, DI 1, 2 and 3 must have been lost in Cladistia + Actinopterygii. Although the lens proteins thus leave us in the lurch as concerns this most difficult problem, the classification of Dipnoi, I think it is fair to claim that they constitute a reasonably satisfactory corroboration of the classification proposed here.

The story of the phylogeny of the bile salts, as told by Haslewood (1967; 1968), makes fascinating reading. In Table 4.18 are shown the most important substances and one hypothetical intermediate (2), and it is indicated that, except for the transition (1) → (2), which requires three, all the others imply only one or two modifications of the molecule. The distribution of the substances is shown in Figure 4.35. Among the interesting observations to be made is that the bile salt found in *Latimeria*, latimerol, is chemically very close to myxinol, the bile salt found in Hyperotreta, thus once more stressing the early origin of this animal. The occurrence of cholic acid in Selachii, Actinopterygii and Tetrapoda, but not in Actinistia, Holocephali and Dipnoi, must imply losses in the latter taxa or independent gains in the former. The presence of taurine conjugation in Actinopterygii and Tetrapoda may be a plesiotypic character lost in Dipnoi. The presence of bufol in Dipnoi and Amphibia may be a plesiotypic character lost in Amniota.

The main conclusion arrived at by Haslewood (1967, p. 72) is: 'In steroid bile salts

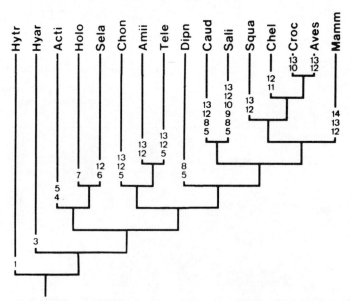

Figure 4.35. Distribution of bile salts in Vertebrata. The numbers refer to the substances shown in Table 4.18, except for (13) and (14) which stand for taurine and glycine conjugates, respectively. No distinction has been made between α- and β-compounds. Acti, Actinistia; Amii, Amiidae; Caud, Caudata; Chel, Chelonia; Chon, Chondrostei; Croc, Crocodylia; Dipn, Dipnoi; Holo, Holocephali; Hyar, Hyperoartii; Hytr, Hyperotreta; Mamm, Mammalia; Sali, Salientia; Sela, Selachii; Squa, Squamata; Tele, Teleostei. It should be noted that the classification is exclusive

there is a progression from substances containing the entire C_{27} skeleton of cholesterol to C_{24} bile acids, and also from C_{27} (and C_{26}) bile alcohols conjugated with sulphate to C_{27} acids as taurine conjugates and then to C_{24} acids, finally conjugated with both glycine and taurine. In summary, the evolutionary course has evidently been: C_{27} (and C_{26}) alcohols (sulphates) → C_{27} acids (taurine conjugates) → C_{24} acids (glycine conjugates)'. If this is true, then the C_{24} acid–taurine conjugates in Actinopterygii and Tetrapoda have probably originated through convergence. The particular state of affairs in Chelonia and Crocodylia must then imply teleotypic losses.

As they stand, the results on bile salts are too variable to be of use in deciding between various classificatory alternatives, but, by and large, they make good sense seen against the background of the present classification.

The amino acid composition and sequence of the several neurohypophysial hormones have been determined in various vertebrate taxa (Perks, 1969; Acher, Chauvet and

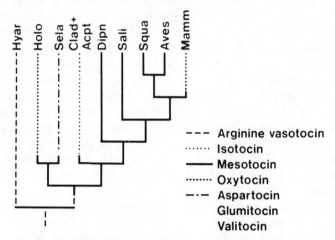

─ ─ ─ Arginine vasotocin
········ Isotocin
───── Mesotocin
········ Oxytocin
─·─ Aspartocin
Glumitocin
Valitocin

Figure 4.36. Distribution of neurohypophysial hormones in Vertebrata. Acpt, Actinopterygii; Clad, Cladistia; Dipn, Dipnoi; Holo, Holocephali; Hyar, Hyperoartii; Mamm, Mammalia; Sali, Salientia; Sela, Selachii; Squa, Squamata. It should be noted that the classification is exclusive

Table 4.19
Neurohypophysial hormones in Vertebrata

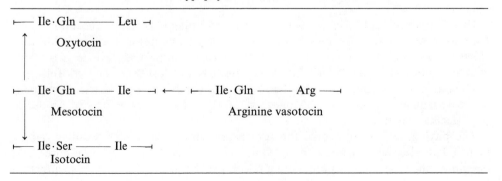

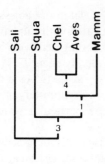

Figure 4.37. Phylogenetic classification of Chelonia, Mammalia, Salientia and Squamata, as established on the basis of the amino acid sequence in cytochrome c. The figures on the vertical lines indicate the number of nucleotide substitutions estimated to have occurred between the respective bifurcations. Chel, Chelonia; Mamm, Mammalia; Sali, Salientia; Squa, Squamata. Based on data published by Margoliash (1973)

Chauvet, 1972). From Figure 4.36, in which their distribution is shown, it appears that the remarkable conservatism displayed with respect to the chemistry of these hormones can be accounted for on the basis of one-step mutations (Table 4.19), with the result that convergence obtains between Holocephali and Mammalia. It is evident that the changes to which the hormones have been subjected in the course of evolution are too few to make the observed distribution of classificatory significance.

Some results obtained by Margoliash (1973) from analyses of amino acid sequences in cytochrome c are shown in Figure 4.37. This author is concerned about the fact that Chelonia are closer to Aves than are Squamata. As will be appreciated, I do not find this result disturbing, for it is a corroboration of my classification. However, a more serious point is that Squamata branch off before Mammalia. Since this must be wrong if the current views, as adapted by me, are correct, we may concur with Margoliash (1973) when he states that his phylogenetic tree is not perfect. It should be noted, however, that according to Penny (1974) several classifications of Squamata are compatible with Margoliash's data. In any case, this result only serves to show that, since non-apotypy can never be excluded, no satisfactory classification can ever be based on a single character. It may be mentioned that the classification of Hyperoartii, Selachii and Teleostei arrived at by Margoliash (1973) does not fit with the one proposed here either. Better agreement is observed with regard to the results obtained on the basis of amino acid sequence determinations in haemoglobins (cf. Dickerson, 1971).

Genomic characters

In the present section we shall deal with certain genomic characters which have been used by various authors, in particular Ohno (1970), for inferences about phylogenetic affinities.

One distinctive feature in the genome of many animals is the presence of microchromosomes. Figure 4.38 shows the distribution of these bodies as reported by Matthey (1949; 1954) and Ohno (1970). Evidently, microchromosomes are a plesiotypic gnathostomous character which has been convergently lost six times, in Teleostei, in Dipnoi, in some Caudata, in some Salientia, in Crocodylia and in Theria (marsupials and true mammals), but not in Monotremata. As a consequence, it is of no use in phylogenetic classification.

The phylogenetic relevance of the sex chromosomes and of the sex-determining system has also been dealt with by Ohno (1970). As appears from Figure 4.39, heteromorphous sex chromosomes are a teleotypic character in certain subtaxa within

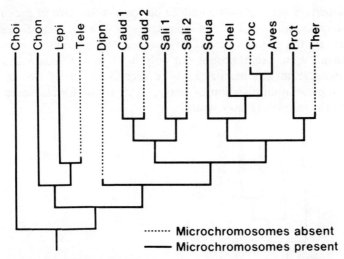

Figure 4.38. Distribution of microchromosomes in Vertebrata. Caud, Caudata; Chel, Chelonia; Choi, Chondrichthyes; Chon, Chondrostei; Croc, Crocodylia; Dipn, Dipnoi; Lepi, Lepisosteidae; Prot, Prototheria; Sali, Salientia; Squa, Squamata; Tele, Teleostei; Ther, Theria. It should be noted that 'Caud 1', 'Caud 2', 'Sali 1' and 'Sali 2' may or may not be monophyletic taxa and that the classification is exclusive

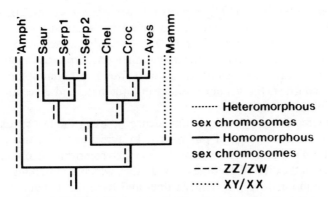

Figure 4.39. Distribution of sex chromosome characteristics in Tetrapoda. 'Amph', Caudata + Salientia; Chel, Chelonia; Croc, Crocodylia; Mamm, Mammalia; Saur, Sauria; Serp, Serpentes. It should be noted that 'Serp 1' and 'Serp 2' may or may not be monophyletic taxa and that the classification is exclusive

Serpentes, in Aves and in Mammalia. Among the tetrapods the XY/XX sex-determining system is found only in mammals, while the ZZ/ZW system occurs in all the remaining taxa we have information about. This, unfortunately, is not the case for Chelonia and Crocodylia.

Another character of interest in the present context is the size of the genome, as represented by the DNA content of a normal diploid nucleus. According to Ohno (1970), the ancestral vertebrate genome was about 20 per cent of the size of the mammalian one, a value which is still found in some members of Teleostei and Salientia. Most frequently, however, the genomic size is larger, even though values exceeding those found in Mammalia have been encountered only in Dipnoi, Caudata and Salientia. The observed increase in genome size may be accounted for by two mechanisms, polyploidy and tandem multiplication of genes; of these the former seems to be the most efficient for the purpose (Figure 4.40).

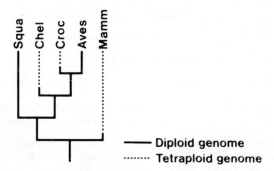

Figure 4.40. Distribution of genome size in Amniota. Chel, Chelonia; Croc, Crocodylia; Mamm, Mammalia; Squa, Squamata. It should be noted that the classification is exclusive

In the other amniotes Ohno (1970) reports two genome sizes. One, corresponding to 40–60 per cent of the mammalian value, is found in Squamata and Aves, and one twice as large (80–90 per cent) in Chelonia and Crocodylia. According to Ohno, the latter genomes and, we may presume, the mammalian one, represent a tetraploidization of the original one; this size incidentally is also found in Gymnophiona (Goin and Goin, 1967). In contrast, the genomes found in Squamata and Aves are supposed to represent the diploid condition.

Ohno (1970) stresses that genetical equivalence between the sex chromosomes is a precondition for the occurrence of polyploidy. In agreement with this contention, in those taxa, Mammalia and Aves, where distinct heteromorphism obtains, a characteristic constancy of the diploid genome is observable, which is in striking contrast to the great variation found in Teleostei (Hinegardner and Rosen, 1968) and in Caudata and Salientia (Goin, Goin and Bachmann, 1968). It is true that in Amniota as a whole a reasonable degree of constancy exists but, as we have seen, there are nonetheless three taxa in which tetraploidy occurs, namely, Mammalia, Chelonia and Crocodylia. On the basis of the present classification this state of affairs can be accounted for only by three separate events of genome duplication. This interpretation, which implies that heteromorphism of the sex chromosomes cannot occur in the phylogenetic line leading from Gymnophiona to Aves, is in accord with the fact that, as discussed above, this character is teleotypic in the taxa where it occurs.

In his work on the phylogenetic implications of the various genomic properties, Ohno (1970) has shown that, in its size and in the absence of microchromosomes, the crocodilian genome is very similar to that of the mammals, and he seems to infer that this similarity may indicate affiliation between these taxa. Disregarding the fact that the whole argument collapses with the occurrence of microsomes in Monotremata, Ohno seems to repeat the mistake of the classical approach, i.e. basing a classification on a single property. In the present case it would certainly be rather misguided to postulate that no empirical data argue against a close affiliation between Crocodylia and Mammalia.

Comparison with other classifications

From the preceding section it appears that the proposed phylogenetic classification of Vertebrata is supported by various empirical observations, even though it must certainly be tested with many more before it can be accepted as a serious alternative to other classifications. In any case, it may be useful to establish the extent to which it deviates from earlier views.

As concerns cyclostomes it seems that, by dividing these into two independent monophyletic taxa, the present classification concurs with the claim that the two included orders are 'diphyletic' (Stensiö, 1958; Jarvik, 1968b).

Chondrichthyes are usually considered to represent a more 'primitive' (i.e. older) evolutionary stage than Osteichthyes, and by making them a primary twin to the pair of secondary twins comprising all other fishes (except Actinistia), the proposed classification completely corresponds to this view. Actinistia definitely occupy a position at variance with current notions, according to which they belong to Osteichthyes. Yet the unsuspected primitiveness of *Latimeria*, the several features in common with Chondrichthyes and Hyperoartii, support the classification proposed here.

One main point which comes out of the present classification is that the three osteichthyan taxa, Actinistia, Cladistia + Actinopterygii and Dipnoi, are in quite separate lines, even if the two last taxa, being secondary twins, are closer to each other than to Actinistia.

In this context I shall present two recent views on this question. First: 'I can find little unquestionable evidence for believing that any two of the major groups of osteichthyans [i.e. Actinopterygii, "Crossopterygii" and Dipnoi] are more closely related to each other than either is to the third group' (Schaeffer, 1968, p. 221). This claim, of course, runs counter to the cladistic rule according to which, out of three taxa, two must (almost) always be more closely related to each other than to the third one. However, if 'Crossopterygii' are regarded as encompassing Actinistia, Porolepiformes and Osteolepiformes, and if the latter are accepted as ancestors to Tetrapoda, then the classification suggested here will lead to the classification shown in Figure 4.41, in which 'Crossopterygii' are represented at the lowest and highest taxonomic rank. This circumstance certainly lends credibility to Schaeffer's opinion.

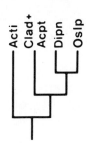

Figure 4.41. The relationship between Actinopterygii, 'Crossopterygii' and Dipnoi. Acpt, Actinopterygii; Acti, Actinistia; Clad, Cladistia; Dipn, Dipnoi; Oslp, Osteolepiformes

Jarvik took the following view (1968a, p. 242): 'it is difficult to find significant resemblances indicating close relationship of Dipnoi to any of the teleostome groups, e.g. Actinistia and Actinopterygii'. According to my classification, Actinopterygii and Dipnoi are secondary twins, Actinistia the primary twin, in a basic classification. Yet Actinopterygii are known to have undergone many teleotypic changes, and the same may have happened in Dipnoi; it is therefore not so surprising that it is difficult to find anything but plesiotypic vertebrate and gnathostomous characters in common between the groups.

In the endless debate about the phylogenetic status of Cladistia, kinship with 'Crossopterygii', Actinopterygii and, less frequently, Dipnoi, has been proposed. The suggested position of this taxon, as demonstrated through the exclusive classification in Figure 4.42, turns out to be the most perfect compromise imaginable, even though it settles the issue somewhat in favour of Goodrich's view (1928).

Jarvik (1968a) has produced an impressive list of 24 characters which, in his opinion, indicate a close relationship between Holocephali and Dipnoi. This situation is not at all incompatible with the present classification. For if we establish a basic, exclusive classification comprising these two taxa and Actinistia, then Dipnoi will be the primary twin, the other two taxa the pair of secondary twins (Figure 4.43). We may therefore expect to find the following groups of characters in both Holocephali and Dipnoi: (1) plesiotypic characters present in *Latimeria*, (2) plesiotypic characters lost in *Latimeria* and (3) convergent characters. I shall not try to allot each of Jarvik's characters to these groups, but it is obvious that many may without hesitation be placed under (1) and (2). Hence, the number of convergent characters required to sustain the present classification —these indubitably include the tooth plates—may not be unreasonably large. In any event, the conclusion at which Jarvik arrives: 'This raises the question if it is justified to distinguish between teleostomes and elasmobranchiomorphs' (1968a, p. 242) approaches, but does not coincide with, the opinion expressed below, according to which the classes Chondrichthyes and Osteichthyes may be united in the Linnean taxon Pisces.

I do not think it is common to find expressions of the view that Actinopterygii and Tetrapoda are closely related, but von Wahlert (1968, p. 12; author's translation) has

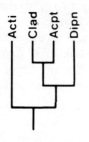

Figure 4.42. The relationship between Cladistia and the other 'osteichthyan' taxa. Acpt, Actinopterygii; Acti, Actinistia; Clad, Cladistia; Dipn, Dipnoi

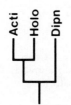

Figure 4.43. The relationship between Dipnoi and Holocephali. Acti, Actinistia; Dipn, Dipnoi; Holo, Holocephali

come to the conclusion: 'Thus we arrive at the conception that the Rhipidistia–Tetrapoda and the Actinopterygii form a natural group'. A similar view was advanced by Gregory (1951, p. 147): 'The palaeoniscoids share many basic skeletal features with their contemporaries the crossopterygians, or lobe fins . . . . This suggests that the two classes Actinopterygii and Crossopterygii were perhaps independently derived from different members of some unknown ancestral class'. Even if this statement is somewhat ambiguous because, undoubtedly, the implication of 'Crossopterygii' in this context is 'Actinistia + Rhipidistia', it nevertheless appears that the present classification endorses the quoted views, with the reservation, of course, that Dipnoi must be included in the mentioned non-actinopterygian taxa.

As explained above, the classification of Dipnoi was not made without hesitation, but it was finally decided to make this taxon the twin to Tetrapoda; as a consequence, many of the striking similarities between Dipnoi and 'Amphibia' become apotypic characters. This classification should be acceptable to those who endorse the affinity between Dipnoi and Caudata originally proposed by Holmgren (1933) and Säve–Söderbergh (1934). It may be objectionable to proponents of the taxon Sarcopterygii (='Crossopterygii' + Dipnoi) such as Romer (1955b) and Thomson (1969). However, it must be pointed out that the two viewpoints coincide if it is agreed that Actinistia are excluded from Crossopterygii. It is questionable, however, whether the classification is acceptable to those (e.g. Jarvik, 1968a) who claim that Dipnoi should assume a rather isolated position in the phylogenetic hierarchy.

When we proceed in the classification we may notice that while it is acknowledged by some, but not all, of the phylogeneticists concerned that Caudata and Salientia are more closely related to each other than to Gymniophiona, the present classification goes a step further by indicating that the latter are more closely related to Amniota than to the two 'amphibian' orders. It must be recalled, however, that another classification is possible, namely, the one which makes all three amphibian orders independent monophyletic taxa.

In the Amniota the classification has created havoc. Chelonia, which, together with Mammalia, were considered to be early side branches, have become one of the youngest taxa in the hierarchy, and Rhyncocephalia have been raised above Squamata. The classification of Aves as the twin taxon to Crocodylia is in agreement with current notions. Yet, considering that the old classification was based mainly on a character which, in the eyes of many taxonomists, does not give a satisfactory result, I wonder if the new proposal, founded as it is on many characters, may not cause less resentment than any of the other changes proposed.

As an overall conclusion to this comparison I think it is fair to state that, by and large, the new classification is a compromise between ideas put forward by various authorities, showing that all, or almost all, of them were correct in some respect, but that, lacking the cladistic procedure, they did not know how to reconcile their separate points of view.

Taxonomic consequences

It was proposed above that it may be convenient in phylogenetic classification to introduce a binary code to characterize the various taxa. This has been done in Figure 4.44 where, in order to avoid the repetition of long series of 1, the rule has been introduced that whenever two or more 1's succeed each other, uninterrupted by any 0, they are replaced by the corresponding decimal number. Lest misunderstandings arise, which may be the case if the latter is 10 or 11, points are placed between the digits according to a system exemplified in the figure.

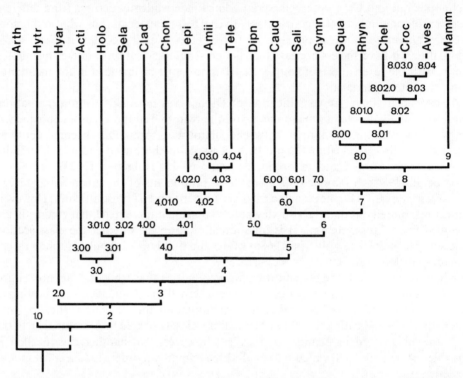

Figure 4.44. Digital classification of Vertebrata. Acti, Actinistia; Amii, Amiidae; Arth, 'Arthropoda'; Caud, Caudata; Chel, Chelonia; Chon, Chondrostei; Clad, Cladistia; Croc, Crocodylia; Dipn, Dipnoi; Gymn, Gymnophiona; Holo, Holocephali; Hyar, Hyperoartii; Hytr, Hyperotreta; Lepi, Lepisosteidae; Mamm, Mammalia; Sali, Salientia; Sela, Selachii; Squa, Squamata; Rhyn, Rhyncocephalia; Tele, Teleostei

It is seen that, for the degree of resolution adopted here, the taxon Vertebrata comprises 39 monophyletic taxa, one for each of the indicated numbers. As for the question of information retrieval, I think I should draw attention to the fact that cyclostomes are represented by taxa 1.0 and 2.0, fishes by 3.0, 4.0 and 5.0, amphibians by 6.0 and 7.0, reptiles by 8.00, 8.01.0, 8.02.0 and 8.03.0, Aves by 8.04 and Mammalia by 9.

Discussions about the 'primitiveness' of animals may be easily settled through this system. Dipnoi may be more primitive than cyclostomes in many ways, but the fact that their code number is 5.0, as compared to 1.0 and 2.0, shows that they are a younger taxon, and in this respect less primitive. Also, although Rhyncocephalia (8.01.0) are in many respects definitely more primitive than Squamata (8.00), their code number shows that they are younger. As to the other kind of 'primitiveness', the one due to lack of teleotypic advance, the present degree of resolution does not allow for a decision on this point. If the resolution is carried to lower levels the situation will be different, for then the taxa which have remained stationary will be distinguished by the fact that the numbers after the first zero in the code number will be very few; thus, the three genera of Dipnoi would be represented by 5.00 and 5.010 and 5.011, *Sphenodon* by 8.01.0. In the taxa where teleotypic advance has occurred, taxonomic divergence has resulted, with the consequence that several or many inferior taxa occur, requiring further binary digits in their code numbers.

Table 4.20
The most important non-terminal taxa in the present
classification of Vertebrata

1.	Vertebrata
2.	Hyperoartii + Gnathostomata
3.	Gnathostomata
3.0.	Actinistia + Chondrichthyes
3.01.	Chondrichthyes
4.	Cladistia + Actinopterygii + Dipnoi + Tetrapoda
4.0.	Cladistia + Actinopterygii
4.01.	Actinopterygii
4.02.	Neopterygii
4.03.	Halecostomi
5.	Dipnoi + Tetrapoda
6.	Tetrapoda
6.0.	'Amphibia'
7.	Gymnophiona + Amniota
8.	Amniota
8.0.	Reptiles + Aves
8.03.	Archosauromorpha

One objection against phylogenetic classification is the many taxonomic levels and hence taxa, which are required, a circumstance that may entail an unwieldy number of taxonomic names. Yet, as it turns out, most of the taxa in the suggested classification already have names; thus, if 'Amphibia' is accorded to Caudata + Salientia, then only nine out of the 39 taxa in Figure 4.44 are unnamed, viz., 2, 3.0, 4, 4.0, 5, 7, 8.0, 8.01 and 8.02. And of these it seems that only six really need to be named. These, listed in Table 4.20, which comprises the most important non-terminal taxa of the phylum, are 3.0, 4, 4.0, 5, 7 and 8.0. It may be noticed that some authors use Actinopterygii for taxon 4.0 (Cladistia + Actinopterygii); if this procedure is adopted, a separate name is required for taxon 4.01, Actinopterygii, according to the more conventional taxonomy. It must be admitted that as the resolution is continued a very large number of taxonomic levels will result, but it is difficult to see the need to name all of them. Some new names may be necessary, but by and large it may turn out that the old names, possibly with modifications as concerns their implication, may serve perfectly well even in the phylo-genetic classification.

Finally, if, for sentimental or didactic reasons, it should be deemed desirable to preserve the outlines of the present Linnean classification of Vertebrata, then I think this could be justified provided only one change is made, namely, that 'Chondrichthyes' and 'Osteichthyes' are suppressed and replaced by the old name 'Pisces'. It will be observed that the present class Amphibia will be formally equivalent to the class Pisces; if one can be accepted, the other can too, but in either case this classification gives a wrong impression of the phylogenetic kinship between the various taxa. An even better proposition may be to use the English names, 'cyclostomes', 'fishes', 'amphibians', 'reptiles', etc., for all those groups of animals which do not constitute monophyletic taxa and to use, as far as possible, the Latin names only for the latter. This expedient has been adopted in the present book, at least from the point where the non-monophyletic nature of a taxon has been established.

5

EVOLUTION OF THE VERTEBRATES

It was pointed out above that the phylogenetic classification of a given taxon constitutes a theory about the course by which the subordinate taxa came into existence.

Yet, however important may be the course of a particular instance of phylogenesis, our curiosity is not satisfied before we know also how the whole thing came about, i.e. the causes responsible for the occurrence of the evolutionary process, the end result of which we can observe. For this purpose we evidently need something more than a classification, namely, a theory of the mechanism of evolution. Such a theory may be anticipated to make predictions with respect to the course of evolution as revealed by phylogenetic classification. Moreover, since organic evolution is the most comprehensive biological process that has ever taken place, it may be expected that a good theory of evolution will make predictions of relevance to all biological disciplines.

As it happens, we do have two theories which claim to be concerned with the mechanisms of evolution, Darwinism and neo-Darwinism. It is often claimed that neither of these meets the requirement of a good theory of evolution, namely, to provide concise predictions which can be falsified by empirical observations concerning the process of phylogenetic evolution. The neo-Darwinian theory makes predictions about population–genetical events, but the bearing of these on evolution is, in my opinion at least, contestable.

In fact, I propose to demonstrate later in the present chapter that these two theories make a number of predictions which are falsified by empirical data, and that they are saved only by means of a number of *ad hoc* hypotheses. Before this can be done, it is necessary to put forward a theory which can manage without these auxiliary hypotheses. This will be the concern of the first of the sections to follow. Subsequently we shall subject the two theories developed here to further testing. The theory concerning the taxonomic divergence of Vertebrata, as embodied in their phylogenetic classification, makes various predictions about evolution in time and space, and some of these will be tested with reference to empirical data in two separate sections.

The theory about the mechanisms of evolution also makes predictions concerning the course of evolution, some of which will be confronted with known facts before, in the last section, the general implications of the present theory of evolution are outlined.

A THEORY OF EVOLUTION

On a previous occasion (1974) I tried to formulate a theory of evolution which I called the 'complete' theory. The present 'comprehensive' theory is a version, improved I hope, of the latter, from which it is distinguished in particular by the fact that it is stated in an axiomatized form.

In my earlier work I suggested that Darwinism is based upon six 'premises', viz., reproduction, inheritance, variation, mutation, competition and environment, and that the 'conclusions' might also be dealt with under six different headings, viz., isolation, selection, evolution, progress, divergence and direction. Here we shall also operate with these concepts, with the following modifications. One more premise, 'population', is introduced and 'variation' is replaced by 'tolerance'. Furthermore, the phenomenon of the direction of evolution was dealt with in the chapter on phylogenetic classification, and may be ignored here. After the theory proper has been developed, we shall discuss some associated problems in separate sections under the headings 'Dominance' and 'Fitness'.

The present theory is phrased with respect to animal evolution, but this does not mean that, if it is of any value at all, it cannot be applied to plants as well.

Population, reproduction and inheritance

(D5.1) A population P_i is a set of animals which, in a completely resolved phylogenetic classification, all belong to the same terminal taxon T_n.

Since T_n in this definition cannot be more comprehensive than a Linnean species it follows that the members of P_i all belong to the same species or subspecies, and possibly even to a phylogenetic subtaxon within the latter. If the members of P_i reproduce bisexually, then the implication of 'population' may coincide with that of 'Mendelian population' or 'deme'. Under very special conditions a population may consist of a pair of animals, of one gravid female or, if reproduction is not bisexual, of one individual. It is evident that there is a certain kinship between the concepts 'population' and 'taxon' when the latter is of the 'undefinable' type referred to in D5.1. There are, however, two differences, namely, that a taxon comprises successive generations, and that it may be represented by two or more geographically separated populations. As will be shown later, it will, under certain circumstances, be possible to equate these two entities.

The following two axioms will be stated about populations:

(A5.1) Successive generations of the population P_i arise through the process of reproduction, bisexual or otherwise, as the case may be. The number of descendants produced by a parent, or a pair of parents, exceeds one and two, respectively.

This axiom, which I shall call *Darwin's first axiom*, is formulated so as to avoid the exclusion *ab initio* of individuals with asexual reproduction from participation in phylogenetic evolution, since this would be in disagreement with empirical observations.

(A5.2) Except under the conditions specified in A5.18, the properties possessed by members of descendent generations of P_i, apart from those which may be specifically ascribed to the influence of environmental factors, were possessed, manifestly or latently, by their parent(s).

The notions of the Mendelian theory of inheritance, including the distinction between genotype and phenotype, are embodied in this axiom, which we may call *Mendel's axiom*. The word 'latently' may even be interpreted so as to allow for recurrent mutations of genes present in the ancestral genome. In spite of its name, the axiom does not exclude extranuclear inheritance.

On the basis of the relation between the numbers of individuals in successive generations, the following concepts may be defined:

(D5.2) Let $n_{i(n)}$ stand for the number of individuals in the nth generation of P_i, and $n_{i(m)}$ represent the corresponding number in the preceding generation. P_i is said to experience

$$
\begin{aligned}
&\text{Progression, if} && n_{i(n)} > n_{i(m)} \\
&\text{Survival, if} && n_{i(n)} = n_{i(m)} \\
&\text{Regression, if} && n_{i(n)} < n_{i(m)} \\
&\text{Extinction, if} && n_{i(n)} = 0
\end{aligned}
$$

These definitions do not exclude regression of a 'surviving' population in some generations, provided that this is compensated sooner or later by progression.

From D5.2 we may derive

(T5.1) Continuous regression leads to extinction.

Environment

The total space available to organic life, the biosphere, will first be subdivided through the following definition:

(D5.3) An environmental compartment (or compartment, for short) is a spatially continuous and circumscribed segment of the biosphere defined by certain physical parameters. The total of the defining parameters is called the physical extension of the compartment, and the variation obtaining with respect to each parameter its range. The physical variability of an environmental compartment is a summation of all the physical parameters in the extension with proper regard to their range.

It should first be emphasized that 'physical' is also meant to incorporate chemical and physico-chemical parameters, including the elements, air, water and land. It is evident that there must be a certain correlation between the spatial extension and the variability of the compartment. If, for instance, by the latter we imply the Pacific or Eurasia, then the temperature range must be much greater than if the compartment is a small tropical lake or island.

We shall now assert:

(A5.3) An environmental compartment is distinguished by the presence of many different populations, $P_1, P_2 \ldots P_n$, of animals (and plants), each of which constitutes a biological parameter of the compartment. The biological extension of an environmental compartment is a summation of all the biological parameters.

(D5.4) The inhabitable niche $N_{(i)}$ is a continuous and circumscribed space, constituting a part, or the whole, of a compartment, defined by the maximum values of the physical variability and of the biological extension which permit the survival of the population P_i. The biological extension of a niche is a summation of those populations in the biological extension of the compartment with which P_i is in direct biological contact, and upon which it is biologically dependent, or with which it is biologically affiliated.

The concepts 'biological contact', 'biological dependence' and 'biological affiliation' will be defined later.

Certain forms, for example migratory birds or amphibious mammals, may appear to inhabit more than one environmental compartment. The special difficulties presented by such populations have been disregarded in the following axiomatization.

We may now introduce some life into the niche by the following definition:

(D5.5) An inhabited niche N_i is a niche $N_{(i)}$ inhabited by the population P_i. There is only one N_i.

We can deduce the following theorem from D5.3, D5.4 and D5.5:

(T5.2) The physical variability and the biological extension of the niche $N_{(i)}$ are smaller than, or equal to, those of the environmental compartment to which the niche belongs. The physical variability and the biological extension of the niche N_i are smaller than, or equal to, those of $N_{(i)}$.

The following postulate will be made:

(A5.4) There is always at least one inhabitable, but uninhabited, niche in every environmental compartment.

D5.5, in combination with D5.1, implies that a population may in some cases comprise all living members of a low-ranking Linnean, or an undefinable phylogenetic, taxon but, as intimated above, nothing prevents subdivision of taxa of this rank into a number of separate populations each occupying their own niche. These niches must of course be spatially distinct, but they may also differ, within certain limits, with respect to their physical and biological parameters. It is seen that the reason why 'deme' or 'Mendelian population', in the sense of 'interbreeding community', cannot replace 'population' here is that niches may be so large that the members of a population cannot be regarded as interbreeding, but only as potentially interbreeding.

In introducing these environmental concepts, a few words on terminology may be justified. Bock and von Wahlert (1965) employ the concepts 'umgebung' and 'umwelt' with implications corresponding approximately to 'inhabitable niche' and 'inhabited niche', and two other pairs of related concepts are 'potential environment' and 'operational environment' (Bates, 1960) and 'potential niche' and 'operational niche' (Parker and Turner, 1961). As a reason for introducing the German words Bock and von Wahlert state that they want to preserve 'niche' for the implication of 'the total relationship between the whole organism and its complete umwelt' (Bock and von Wahlert, 1965, p. 282). The niche thus becomes a set of interactions, without any spatial circumscription, and furthermore, the 'the logical and quite useful extension of this argument is that the niche is the species' (p. 282). I am not convinced that this semantics will reduce confusion, but it appears that there is a tendency among ecologists to agree that 'niche' should mean 'something done' rather than 'somewhere to be', although it seems to contradict the original sense of the word.

Nevertheless, my terminology appears to be in accordance with that of Parker and Turner (1961) and the word 'niche' as defined seems to meet all the requirements of the physical and biological conditions which must be made of an 'ecological' as well as on a 'geographical' niche; the adoption of this name therefore seems justified. Should anyone, however, find the use of this word disturbing, it can be replaced by any word or symbol which is taken to stand for the given definition.

The problem of satiation, so important for evolution, may be approached in the following way:

(D5.6) The number of individuals of the population P_i in the niche N_i times the weight of an individual, i.e. the biomass, is called the capacity of the niche. The capacity of a given environmental compartment is the sum of the capacities of all the inhabited niches in the compartment.

(A5.5) Niches and environmental compartments are distinguished by having a maximum capacity which can only provisionally be exceeded.

The essence of this axiom was stated by Darwin (1859), and it may therefore rightly be called *Darwin's second axiom*.

(D5.7) When the capacity of a niche or an environmental compartment equals the maximum capacity, these entities are said to be sated; if it is below this value they are unsated.

From D5.6 and D5.7 we may derive:

(T5.3) In a sated niche N_i a constant and maximum number n_i of individuals of P_i prevails.

From A5.1, D5.2 and T5.3:

(T5.4) If the number of individuals of P_i in N_i is smaller than n_i, P_i will undergo progression; if it is equal to n_i, P_i will survive.

From T5.4:

(T5.5) Except for transitory periods, niches are always sated.

The implications of this statement were expressed by Darwin (1859), and we may therefore call T5.5 *Darwin's first theorem*.

From A5.4, A5.5 and D5.7 we may further conclude:

(T5.6) Environmental compartments are never sated.

Apparently, the mechanism of proliferation implied by A5.1 does not ensure restraint once satiation of the niche has been reached, and we must therefore conclude:

(T5.7) The number of offspring produced by the members of P_i inhabiting the sated niche N_i implies that the maximum capacity is exceeded.

Hence we may deduce:

(T5.8) In every niche a continuous elimination of individuals must occur so that the maximum capacity is not exceeded.

This theorem, embodying Darwin's notion about 'struggle for existence', I propose to call *Darwin's second theorem*. At this stage no specification is made about the individuals involved in the struggle leading to the elimination of some of them, but it may be recalled that Darwin presumed them to be members of the same or closely related species. The neo-Darwinian theory, of course, requires that the competing individuals must all belong to the same species.

We shall now introduce a constraint on the inhabitants of a niche:

(A5.6) Except under conditions specified in A5.7 and D5.19, the members of P_i cannot actively, i.e. by their own means, leave the niche N_i which they inhabit. Under conditions that are unpredictable they may, fortuitously and passively, be transferred to another niche.

From D5.4 and A5.6 it follows:

(T5.9) Except under conditions specified in D5.19, the members of populations inhabiting niches in a given environmental compartment cannot actively transgress the border of the latter. Under conditions that are unpredictable they may, fortuitously and passively, be transferred to another environmental compartment.

From D5.4 and D5.5 we may derive:

(T5.10) If, in the niche $N_{(i)}$ or N_i, the value for any of the physical and biological parameters exceeds the specified range, P_i can no longer inhabit the niche. The latter thus becomes uninhabitable to P_i.

(D5.8) When a niche $N_{(i)}$ or N_i becomes uninhabitable to P_i, it is said to be annihilated, and is represented by the symbol $\sim N_{(i)}$.

(T5.11) If the niche N_i is annihilated, the population P_i is eliminated.

This follows from T5.10 which, in combination with A5.6, allows them no opportunity to leave their former niche.

This theorem formulates a notion common among evolutionists, namely, that changes in the external conditions may lead to the extinction of living organisms. Needless to say, this event may allow the space concerned to become inhabited by another population that was barred from invading it as long as it was inhabited by the extinct population. As we shall presently see, in some cases it is the invasion of a new population which contributes to making the niche uninhabitable for the original population.

In order to account for spatial or geographical dispersal of populations we shall introduce the following axiom:

(A5.7) If N_i covers only a part of the expanse of $N_{(i)}$, then members of the population P_i will disperse until the two niches coincide spatially.

Since this dispersal implies an increase in the number of individuals of P_i, we may conclude from T5.5 and A5.7:

(T5.12) If the niche $N_{(i)}$ into which dispersal of the population P_i occurs is inhabited by a population P_j, members of the latter must be eliminated; if the niche is uninhabited, no elimination is involved.

The relation between the size of a niche and the size of an individual animal may be stated as follows:

(A5.8) The minimum size of a niche $N_{(i)}$ or N_i is directly correlated with the size of the individual members of the population P_i.

From this axiom we may draw the following conclusion:

(T5.13) Other things being equal, the number of potential niches in a given environmental compartment is smaller for populations of large than for populations of small animals.

The relation between the power of dispersal, or vagility, through locomotion and otherwise and the size of a niche is accounted for by this axiom:

(A5.9) The minimum size of a niche $N_{(i)}$ or N_i is inversely correlated with the vagility of the members of the population P_i.

Hence the conclusion:

(T5.14) Other things being equal, the number of potential niches in a given environmental compartment is smaller for populations of highly vagile than for populations of slightly vagile animals.

Furthermore:

(A5.10) A given niche $N_{(i)}$ can be $N_{(j)}$ only if the members of the populations P_i and P_j share a number of characters with respect to size, vagility, food preferences, habits of life, etc.

With reference to A5.10 we shall introduce a new concept, 'biological affiliation', defined as follows:

(D5.9) If $N_{(i)}$ is also $N_{(j)}$, then the populations P_i and P_j are said to be biologically affiliated.

P_i and P_j may be members of the same inferior taxon below the specific level, or they may be united only in a taxon of superior rank. We may also observe another kind of biological relation in the following axiom and definition:

(A5.11) All animals are members of the trophical hierarchy (links in a food chain).

(D5.10) If the members of the populations P_i and P_j are successive links in a food chain, then P_i and P_j are said to be biologically dependent. If P_i stands above P_j in the food chain, P_i is actively and P_j passively dependent.

Clearly, the lowest members in a food chain of animals are actively dependent upon populations of plants. Finally, we shall define the concept of 'biological contact':

(D5.11) The populations P_i and P_j are said to be in direct biological contact if the niches N_i and N_j overlap spatially, in part or wholly. If the niches are separated by but in direct contact with another niche, or by several which are mutually in direct contact, N_i and N_j are said to be in indirect biological contact.

The following axiom is submitted:

(A5.12) All populations which inhabit niches in the same environmental compartment are in biological contact, directly or indirectly.

Tolerance, selection, competition and progress

Living organisms vary to an amazing extent through the possession of a great array of characters, most of which, as we have seen, are actual or potential taxonomic characters. According to Darwin, none of these characters could have become established without contributing to the fitness of the individuals possessing them, and the fitness in turn would ensure their survival. However, it has never been possible to evaluate fitness independently of survival, and for this reason later evolutionists have frequently replaced 'fitness' by 'adaptation'. Yet this concept is as elusive as 'fitness', since it is an obvious truism to postulate that an organism is adapted to the environmental conditions under which it lives. Therefore it is also impossible to measure adaptation independently of survival, and thus one ends up in the same predicament as with fitness.

In my previous book (1974) I proposed that the theory of 'directive correlation' advanced by Sommerhoff (1950) might overcome this difficulty when it is realized that this concept may be equated with 'adaptability'. Since then I have come to the conclusion that even the word 'adaptability' is an unfortunate choice, because it may imply active participation of the animal in question. This situation undoubtedly occurs sometimes, but not necessarily always, and therefore it may be better to replace 'adaptability' with 'tolerance'.

We shall distinguish between 'physical tolerance' and 'biological tolerance'. The first concept is defined as follows:

(D5.12) The physical tolerance PT_i of a member of the population P_i equals the physical variability of the niche $N_{(i)}$.

We shall postulate:

(A5.13) Over long periods of time all physical parameters of any niche will be subject to fortuitous and extensive variations.

We have seen above (T5.10 and T5.11) that if any physical parameter exceeds the range specified in the physical variability of an inhabitable niche, the niche becomes annihilated, and if it happens to be inhabited by a population, the latter will become extinct. We may therefore conclude:

(T5.15) Over long periods of time the variations in the physical parameters will, through niche annihilation, lead to the elimination of populations. The chance of elimination is inversely proportional to the value of the PT.

It is presumed here that the tolerance characteristics of the eliminated animals distinguish whole populations; we shall discuss later the possibility that subpopulations within a population vary in this respect.

The kind of elimination dealt with here is of a particular type for which the following terminology may be proposed:

(D5.13) When the process of elimination in an animal population is not fortuitous, but may be referred to characters possessed by the individual animals, it is called selection.

The evolutionary process described by T5.15, which may be called 'physical selection', has the effect of eliminating populations with low values of PT. Provided that the continuous creation of new populations with higher PT is assured, an occurrence which will be stipulated in A5.18, then we can state the following theorem:

(T5.16) Niche annihilation through variation in the physical parameters of the biosphere leads to the progression and survival of populations whose members have ever-increasing PT.

This we shall call the *first theorem of progressive evolution*.

I am aware of the oversimplification implied by combining all the parameters of the physical extension into the PT. It is of course possible that a particular kind of animal has a large tolerance to all parameters but one, and thus a high PT, and yet variations in the last parameter may cause its extinction.

However, important as this kind of directed evolution may be, the one depending upon the 'biological tolerance' is unquestionably more important. We shall define this concept as follows:

(D5.14) The biological tolerance BT_i of a member of the population P_i is an algebraic summation of the populations in the biological extension of the niche $N_{(i)}$. Affiliated and passively dependent populations are added, actively dependent ones subtracted.

By defining the BT with reference to the populations which can potentially be tolerated, we face the difficulty that it is impossible to estimate this parameter by an enumeration of the populations which actually fulfil the stated requirements. From this it may appear that 'biological tolerance' is about as elusive as 'fitness', but, as we shall see later, it is at least possible to establish objectively the relative value of BT for various animal taxa.

We have defined PT and BT with respect to external parameters, but it should be evident that it is characters possessed by the individual animals that ensure the survival of the population to which they belong. This evidently implies that these parameters are related not only to individual survival, but also to fecundity and, in bisexually reproducing animals, to mating success (cf. Ayala, 1970).

The PT and BT possessed by individual animals must somehow be reflected in the properties of populations of animals. We may try to account for this in the following axiom:

(A5.14) The population potential PP_i of the population P_i in the niche N_i is a characteristic that is a function of those elements of the PT and the BT of the individual animals which respond to, or are directed towards, the parameters of the physical variability and the biological extension of the niche N_i, but is furthermore dependent upon the number of individuals n in the niche, increasing for small values of n, but passing through a maximum before the number n_i, corresponding to satiation of the niche, is reached.

In the definition of the potential of a population both PT and BT have been included, because it can be supposed that both parameters contribute to the PP. Under some circumstances, for instance under the conditions of competition to be discussed presently, it may, and will, be assumed that the BT is of decisive importance, while the PT can be disregarded. It should be pointed out that in fact it may be the density of the population rather than the number of individuals which is the contributing factor to the PP, but this distinction is, of course, immaterial as long as the size of the niche remains constant.

Various factors may be considered to account for the low value of PP when there are few individuals, but in bisexually reproducing animals mating success is one that is obvious. When larger numbers of animals are present, another set of factors may contribute to the reduction of the PP, for instance the scarcity of hiding places, difficulties involved in finding food, etc. Since the relation between population size and survival has been demonstrated by Gause (1934), A5.14 may be called *Gause's axiom*.

At this stage we need a further concept, 'biological equilibrium':
(D5.15) Two or more surviving populations are said to be in biological equilibrium if they inhabit niches in the same environmental compartment.

I shall further postulate:
(A5.15) When two populations, P_i and P_j, are in biological equilibrium, then $PP_i = PP_j$.

On this basis we can derive the following theorem:
(T5.17) All surviving populations in a given environmental compartment have the same PP.

T5.17 may be called the *theorem of coexistence*. It is necessary to stress that 'surviving' here does not have the implication of 'existing' or 'prevailing', but the exact meaning defined in D5.2.

Since the PP is a function of the number of individuals, it appears to follow that the reciprocal oscillations which have been observed in the number of individuals in various predator and prey populations may be interpreted as the adjustments in PP required to maintain the biological equilibrium. Apart from phenomena of this kind, it may be anticipated that when biological equilibrium obtains in an environmental compartment, only minor changes in population number should be possible. However, in T5.12 we have allowed for the phenomenon of dispersal into niches that are either uninhabited or inhabited by another population. The former alternative is not very interesting because it may be presumed that dispersal and progression occur until the original population density, and hence PP, are restored.

If, on the other hand, the dispersal of one population occurs at the expense of another, then evidently the two populations cannot be in equilibrium, and consequently their PP's cannot be identical. Provided the regression of one population is not due to physical selection, we may conclude that the variation of the physical parameters lies within the variability of the niche concerned, and hence that the imbalance is solely due to differences in the biological tolerance; in terms of this parameter there must be a distinct inequality between the two populations. If we call the latter P_i and P_j, then $BT_i = P_1 + P_2 + \cdots P_n + P_j - P'_1 - P'_2 \ldots$ and $BT_j = P_1 + P_2 + \cdots P_n + P_i - P'_1 - P'_2 \cdots$. If P_j is the population undergoing regression, then it follows that P_i can support $P_1 + P_2 + \cdots P_n + P_j - P'_1 - P'_2 \ldots$ without undergoing regression, while P_j cannot attain this in the presence of P_i. Thus we have $BT_i > BT_j$.

The situation discussed here may be represented by the following axiom:
(A5.16) If the populations P_i and P_j are biologically affiliated and in direct biological contact, and $BT_i > BT_j$, then N_j is also $N_{(i)}$.

From T5.12 and A5.14 we may now conclude:

(T5.18) Under the conditions specified in A5.16, P_i will progress through invasion of N_j, causing regression of P_j. Through this event PP_j will increase, and if biological equilibrium is attained before, or when, the maximum value of PP_j is reached, both populations will survive; if not, P_j will become extinct.

The situation envisaged by T5.18 calls for a new concept, 'competition', which may be defined in the following way:

(D5.16) Competition always obtains between populations that are biologically affiliated and in direct biological contact. The selection which ensues if the populations are not in biological equilibrium is called competitive selection.

This definition allows T5.18 to be named the *theorem of competitive selection*.

We have in T5.17 concluded that the PP is the same for all populations in an environmental compartment. Nothing can be stated about the value of this parameter in different compartments; equality may or may not obtain. We can, however, state the conditions for success when, as submitted in T5.9, members of a population pass from one compartment to another:

(T5.19) If members of a population P_i pass from one environmental compartment to another, then P_i has a chance of becoming established in the new compartment, provided a niche $N_{(i)}$ obtains there. If $N_{(i)}$ is uninhabited, P_i may progress and survive if the value of PP is of the same order in the two compartments; if $N_{(i)}$ is inhabited, the same outcome requires that the PP is higher in the original than in the new compartment.

Whether or not a population will be extinguished as the outcome of the competitive selection envisaged by T5.18, the PP in the compound niche will be raised to a new and higher value. Hence we can derive a further theorem from T5.17 and T5.18:

(T5.20) Competitive selection between any pair of populations will lead to an increase of the PP's of all the surviving populations in the same environmental compartment.

This theorem may be called the *theorem of adjustive evolution*. Evidently, this adjustment must in most cases consist in a change in population density rather than in a change in the BT.

If competitive selection leads to the extinction of a population, then the maximum value of its PP is below that of the competing population. We may assume that this situation is related to the properties of the individual animals, and we shall therefore conclude with reference to A5.16:

(T5.21) Competitive selection leading to the extinction of a population entails occupation of the niche in question by a population with a higher BT.

Provided that the continuous creation of new populations with higher BT is assured, an occurrence which will be stipulated in A5.18, we can now state the *second theorem of progressive evolution*:

(T5.22) Niche annihilation through competitive selection leads to the progression and survival of populations whose members have ever-increasing BT.

Without knowledge about the relative contributions of the BT and the population number to the PP of competing populations, it is impossible to state the exact requirements for the establishment of equilibrium, but the following formulation may be acceptable:

(A5.17) When the difference in BT between two competing populations does not exceed a certain value, then biological equilibrium can be established between the populations.

On the basis of this axiom a *theorem of exclusion* may be phrased as follows:

(T5.23) When the difference in BT between two competing populations exceeds a certain value, then the one with the lower BT will be extinguished.

This theorem was first stated by Gause (1934) and may be called *Gause's theorem*. T5.23 is stated with reference to populations, without regard to taxonomic levels. This evidently implies that whenever competition between two populations, irrespective of their taxonomic affiliation, leads to the extinction of one of them, then this event must be classed as progressive evolution, since it implies replacement of a certain population by another whose BT is higher.

However, it seems to follow that as we approach the inferior levels in the phylogenetic hierarchy the differences in BT must become smaller and smaller, and consequently the various populations should easily reach biological equilibrium, entailing that no competitive selection occurs. This is evidently in disagreement with Darwinism, according to which competition is strongest between members of closely related species or of the same species, and with neo-Darwinism, according to which only the last alternative is of importance for evolutionary progress.

How do these alternative predictions fare when confronted with empirical observations? The fossil record shows that phylogenetic evolution has been associated with large-scale extinction. Very often it is possible to demonstrate a temporal and spatial connection between the appearance of new, and the disappearance of old, forms. It is usually surmised that this connection is also causal, but it is generally not possible to establish this point. However, in some cases the causality seems certain, thus, for instance, the extinction of various marsupials in South America seems to be a direct consequence of the invasion of eutherians from the north. And even from historical times we know of cases where transfer, usually by man, of animals and plants between the continents have caused regression and, occasionally, extinction of indigenous forms.

Phenomena of this kind are predictions of the comprehensive theory, and are not incompatible with Darwinism. But they contradict the neo-Darwinian theory, which accounts for coexistence as the outcome of 'adaptation', each population being uniquely specialized to the particular niche it occupies. This explanation may be valid in some instances, but in general it appears that this hypothetical environmental differential and the corresponding adjustments on behalf of the living organisms belong to the class of neo-Darwinian *ad hoc* hypotheses which cannot be falsified, because they do not predict anything but those phenomena they are supposed to account for. In fact, the neo-Darwinian predictions seem to be falsified by the empirical observations discussed above, for example the fatal outcome of the competition between marsupials and true mammals. Surely, if the members of subspecific taxa are 'adapted' to separate niches, then the same should be true of animals as different as the two mammalian subclasses.

This shortcoming of the theory has of course been recognized by its adherents. Fisher (1958) has tried to remedy it by invoking something called 'deterioration of the environment'. But since it is absolutely impossible to demonstrate this phenomenon in any other way than by regression or extinction of a population or taxon, it seems that Fisher's explanation is still another neo-Darwinian *ad hoc* hypothesis.

Turning to the other end of the scale we face the question: Are the members of the same species engaged amongst themselves in a 'struggle for existence'? I believe that numerous observations suggest a negative answer, but I am sure that it will be possible always to find an 'explanation' in neo-Darwinian terms. I shall therefore mention only one example which seems to corroborate the predictions of the comprehensive theory.

Since the constraint exerted by the capacity of the niche relates primarily to the bio-mass, one consequence of the postulated lack of eliminative power between the members of a population might be that the number of individuals inhabiting a niche is relatively high, the size of each animal being correspondingly smaller. This solution, which prob-ably cannot be realized in all animal taxa, seems to be corroborated by various empirical observations, thus, for instance, by the changes in size occurring in populations of Scottish Red Deer after transfer to New Zealand (cf. J. S. Huxley, 1932). If my interpre-tation is true, the lesson to be learnt is that competitive selection between members of the same population cannot lead to what might otherwise be considered the optimal 'phenotype', and hence that the mechanisms of selection envisaged by the Darwinian theories in many cases are not very efficient. For a further discussion of the shortcomings of natural selection as an evolutionary agent, the reader may refer to Grassé (1973).

It must be emphasized, though, that T5.18 does not exclude competitive selection within a species, or a subgroup of species. Populations have been treated here as being uniform with respect to BT and, incidentally, also to PT. It is, of course, a general observation that almost all properties are distributed normally among the members of a population, and it may therefore be anticipated that a distinct difference obtains between the most and the least successful (tolerant) members of any population. It is consequently possible to divide a population into subpopulations with respect to their BT, and as long as the differential exceeds that envisaged by T5.23, members of the least successful subpopulations will be eliminated by competitive selection.

After this has been allowed, it must nevertheless be admitted that there is a strong difference of emphasis with regard to the involvement of competitive selection between the Darwinian theories and the present one. According to the latter, the circumstance that populations of two inferior taxa, for example subspecies, can live peacefully together in biological contact is based on the fact that the PPs of the populations are identical, and hence the BTs of the individual animals closely similar.

The present discussion seems to place us in an embarrassing predicament, for although according to T5.8 a continuous elimination of animals must occur, apparently members of the same population are not very cooperative in this enterprise. There are, however, various expedients which may be invoked to circumvent this difficulty. Thus, it may be envisaged that the BT increases in the course of the life cycle. In some animals this has the result that smaller or younger individuals are devoured by larger or older ones. The interference of natural selection under these conditions may be anticipated to lead to an accelerated rate of development and growth, but not necessarily to an improvement of any other properties.

Predators may assist in the process of elimination. If this occurs during embryonic and larval stages, there is reason to believe that the selection is fortuitous, especially in those cases where the number of progeny is very large. But even if predators may, under certain circumstances, ensure an increase in the BT of larvae or adults, it is probably the reduction in numbers caused by them that is the most important factor in ensuring the survival of the remaining individuals.

Another mechanism which may serve to eliminate surplus members of a population has been studied by Christian (1971). This author observed that, under conditions of crowding in various small mammals, some individuals may react by undergoing a hormonal derangement leading to their death. From a neo-Darwinian point of view it is difficult to accept that this mechanism can be efficient for any protracted stretch of time, for through natural selection the members of the population should gradually become more and more endocrinally stable. And in any case it is hard to see that any other characters should be affected.

In the neo-Darwinian literature it is quite commonly argued that differential reproduction may replace Darwinian differential elimination as an agent in natural selection (cf. Simpson, 1953; Dobzhansky, 1970). How can this completely un-Darwinian notion be acceptable in a Darwinian theory?

In order to understand this we must analyse the phenomenon of elimination. It follows from Darwin's second theorem (T5.8) that in a sated niche the number of off-spring must equal the number of parents, the excess progeny being eliminated. This *total elimination* (TE) comprises *fortuitous elimination* (FE) and non-fortuitous elimination, or *natural selection* (NS; cf. D5.13). Thus, $TE = FE + NS$. This point was discussed by Darwin (1885), who stated the opinion that in Nature FE is very often quantitatively more important than NS.

Darwin's theory aims to account for the fact that most often evolution has been progressive, leading for instance to animals with better vision, better locomotion and all those other properties which are important in the 'struggle for existence'. This clearly requires that these various faculties face challenge, and this happens only when the animals are living in Nature. The non-fortuitous elimination occurring under these conditions, may be called *ecological selection* (ES), and I believe that Darwin would unhesitatingly endorse the equation $NS = ES$.

Under the conditions prevailing, say, in a *Drosophila* cage, it is possible to abolish virtually completely those inorganic and organic agents which are or may be responsible for the process of elimination. When this is done it is found that parents (or mothers) vary with respect to the number of their progeny. This differential reproduction may be referred to the presence of debilitating genes in the parental genomes. If the most fertile female is used as standard, it follows that an elimination has occurred among the offspring of the less fecund ones. Being non-fortuitous, this elimination is selection, and may be called *genetic selection* (GS). Since this phenomenon must occur even under natural conditions, we have $NS = ES + GS$. Thus, when properly corrected, the elimination equation in classical Darwinism states: $TE = FE + ES + GS$, where the three components are arranged in accordance with their likely numerical importance.

The situation represented by the population-genetical experiments has a measure of other-worldliness; if they were allowed to go on, all the resources of our planet would soon be required for feeding fruit flies. Having no practical experience, I can only guess how this potentially threatening situation is coped with in the laboratory, but I believe that all but a few particularly interesting flies are thrown out once they are counted and the experiment finished. Thus, since elimination must occur, all the animals which were not genetically selected, are fortuitously eliminated, and the neo-Darwinian elimination equation is thus: $TE = FE + GS$.

If the population-genetical experiments are to be of any value for our understanding the mechanism of evolution, they must correspond to a natural situation. And this is possible, for 'natural selection may occasionally occur even if all the progeny survive in some generations. Consider a species that gives several generations per year; if food and other resources are not limiting during the warm season, the population may increase geometrically with no mortality among the progeny. Selection will nevertheless take place if the carriers of different genotypes have produced different numbers of offspring or have developed to maturity at varying rates. A winter season may reduce the numbers of individuals drastically but unselectively' (Dobzhansky, 1970, p. 97). This quotation is a close rendering of a laboratory experiment in terms of 'natural conditions'. As is seen, the elimination equation in this example is also $TE = FE + GS$; it is thus clearly possible to fit a hypothetical species into the neo-Darwinian equations and experiments. But how many real organisms would do so? And how is it possible to select for anything but fecundity and rate of development? What would the kingdom Animalia look like, if these were the only targets of selection?

From the preceding discussion it appears that ecological selection, the true driving force in Darwin's theory, has been abolished in the neo-Darwinian theory. From this I shall conclude that the two theories are quite disparate, as was observed also by Macbeth (1971). Is this the clue to why adherents of neo-Darwinism prefer the name 'synthetic theory', as it has next to nothing to do with Darwin's theory?

To round off this discussion let me mention that there are natural situations where differential reproduction is of importance, namely, whenever unsatiated niches occur. We shall discuss this point later, but it may be mentioned here that it is typical of human populations, mainly in historical times, where agricultural technology has ensured a substantial increase in the maximum capacity of the niche of *Homo sapiens*.

Isolation

From what we have seen above, the requirement for the survival of an animal population is that it is able to adjust its PP to the highest value attained by any other population in the environmental compartment where it lives. One way to achieve this is to reduce the number of animals in the niche inhabited by the population, but if the BT is too small, then this expedient will not work. In that case there is still one possibility open to ensure survival, namely, isolation. This phenomenon may be defined as follows:

(D5.18) If, through the occupation of a particular niche N_i, the members of the population P_i avoid competition with other populations that would otherwise cause regression and extinction of P_i, the latter is said to be isolated. If the isolation may be referred to properties possessed by the animals themselves, it is called non-random isolation; if it occurs fortuitously, it is called random isolation.

The observant reader will possibly point out that isolation is no problem: all that is required is that the population terminates the direct biological contact with the inimical populations. However, this may not be as easy as it sounds if, as we have stated axiomatically, a population cannot by its own means leave the niche it inhabits. And should an empty niche be available, then it is also available for competing populations (cf. A5.16).

It is possible to distinguish two kinds of non-random isolation, which may be characterized as 'specialization' and 'protection', respectively. Both of these phenomena presuppose the occurrence of mutations, as postulated in A5.18.

(D5.19) Specialization is said to occur when a population P_i raises its PP, either through including one or more populations from the environmental compartment as passively dependent members of the extension of the niche $N_{(i)}$, thereby raising the BT, or through an increase in the physical variability of the niche $N_{(i)}$, thereby raising the PT.

The populations referred to in the first clause may obviously comprise plants as well as animals. It will be envisaged that, under some circumstances, raising the PT may allow invasion of a new niche or environmental compartment.

Examples of specialization include transfer from fresh to salt water and *vice versa*, from aqueous to terrestrial life, etc. These cases represent colonization of new environmental compartments and usually concern rather superior taxa, but ecologists have recorded numerous instances of specialization in inferior taxa which may reasonably be interpreted as invasion of new niches.

Specializations of an intermediate kind are adoption of a parasitic or sessile existence, while the selection of a particular material as the sole, or the main, source of food illustrates the situation where the BT is raised through including new dependent populations. As an example of the latter kind, ants, apparently, are regarded by most animals as rather unappetizing. Yet the selection of ants as the staple food has permitted the survival of five widely different families of mammalian anteaters, all representing superior, i.e. old, taxa, suggesting that these animals owe their survival to their choice of food. This kind of specialization will clearly reduce the competition from other populations. That the animals are not on a par with non-isolated ones is evident from the fact that the survival of such populations is utterly dependent upon the availability of one particular stock of food.

The phenomenon of protection may be introduced in the following axiom:

(D5.20) Protection is said to occur when a population P_i raises its PP through excluding one or more actively dependent populations from the biological extension of $N_{(i)}$, thereby raising the BT.

Among the devices that may provide protection are large, well-protected eggs, shells and carapaces of various kinds and any other property which may aid escape from a potential predator, in short, various kinds of 'passive' defence. The importance of this kind of protection for evolutionary stabilization, i.e. survival, has been discussed by Schmalhausen (1949). Body size may also be important in the present context. Thus, while the increase in body size which has been observed in many phylogenetic lines suggests that larger bulk entails an increase in BT, the survival of many small animals in such lines indicates that their body size involves a measure of specialization or protection which enables them to establish biological equilibrium with their larger counterparts and with the other populations present in the environmental compartment.

From the theorem of coexistence we may conclude:

(T5.24) Non-random isolation involving passage of a population P_i between niches within the same environmental compartment is successful only if it ensures that the PP is raised to the value distinguishing the biological equilibrium of the compartment.

The conditions for progression and survival in passage between different compartments are stated in T5.19.

We shall now deal with the notion that specialization leads to 'degeneration', although the discussion will be generalized by also including protection. First, it follows from D5.12, D5.14 and A5.14:

(T5.25) Elements of the PT and the BT which respond to, or are directed towards, parameters absent in the niche N_i do not contribute to the PP of the population P_i.

From this follows:

(T5.26) Elements of the PT and the BT which do not contribute to the PP may be lost without entailing a decrease in the latter.

This theorem is generally valid, but it may be of particular importance in association with non-random isolation because, as we have seen, this phenomenon usually involves elimination of environmental parameters from the niche.

However, from the theorem of competitive selection it follows that the progression and survival of a population requires that its PP is greater than the one with which it competes. Hence, the fact that certain characters may be lost without a decrease in the PP does not imply that the 'degeneration' entailed by their loss actually occurs. In order to account for this event we must first postulate that mutations involving the losses of characters may arise; this will be done in A5.18. Secondly, it cannot be excluded that such losses may entail gains in the PP, and this circumstance must increase the likelihood that they occur. This is stated in the following axiom:

(A5.17) If the loss of characters specified in T5.26 involves an increase in the PP, the chances are great that they will be lost in the course of time.

Examples of this kind of loss are numerous, for example eyes in cavernicolous animals (Vandel, 1965), appendages in sessile animals and, perhaps, teeth in anteaters. We do not know how blindness is an advantage to cave-dwelling animals, but it seems reasonable to presume that if it were not, loss of eyes would not be so common as it is. A consequence of such losses is, naturally, that although the specialized animal can establish biological equilibrium with organisms in surrounding niches by having the same PP as the latter, it can never leave its chosen niche.

It thus seems possible to conclude:

(T5.27) Under certain circumstances, in particular in association with nonrandom isolation, the possibility exists that losses in PT and BT may occur.

If this theorem is correct, it follows that, from the point of view of progressive evolution, non-random isolation may, but must not, be a side-issue. Since T5.27 to some extent corresponds to Cope's 'law of the unspecialized' (1896), I propose to call it *Cope's theorem*. It is important to stress that the theorem is a probabilistic prediction, not a law.

Since the phenomenon of random isolation ensures the fortuitous survival of a population, we may infer from T5.19:

(T5.28) Random isolation must always involve the occupation of a niche in an environmental compartment distinguished by a PP which is lower than or equal to that of the original compartment.

Random isolation may arise in a number of different ways. Geographical events like inundations and continental movements may subdivide an environmental compartment into two or more parts. But actual transfer of animals may also occur, the vector being inorganic or organic factors. From the following discussion it will appear that random isolation is a very important factor in the survival of relict animal forms. However, occasionally it may be difficult to decide whether random or non-random isolation prevails. Thus, while it seems reasonably certain that *Sphenodon* and *Latimeria* represent the first alternative, it is less certain whether the survival of the lungfishes, at least the African and South American ones, is due to luck or to some kind of specialization or protection.

Mutation and divergence

In order to account for the continuous creation of new forms of life, it is necessary to introduce the concept of 'mutation', defined as follows:

(D5.21) A mutation is a modification in one of the vectors of biological inheritance.

As I have discussed previously (1974) and in Chapter 1, the vectors in question are not only nuclear factors like genome, chromosomes and genes, but also cortical and cytoplasmic factors, and the inheritance of mutations cannot therefore always be described

by Mendelian genetics. In Chapter 1 I have also put forward my view on the relation between the effect and the frequency of mutations (cf. Figure 1.1). Here I shall state it in a
more formal way. First of all we must ensure that mutations do occur:

(A5.18) Mutations arise fortuitously from time to time in any population of animals.

Concerning the quality of mutations, the following axiom may be adequate:

(A5.19) It is possible to distinguish between advantageous—or positive—mutations,
which increase the PT or the BT, and disadvantageous—or negative—ones, which
decrease the value of either of these parameters. The frequency of occurrence of the
former is extremely low compared to that of the latter.

This axiom amounts to stating that destruction is more likely than construction and,
because the former represents disorder, the latter order, A5.19 may to some extent
represent a special case of the second law of energetics.

With respect to the size of mutations, the following axiom may be stated:

(A5.20) On the molecular and subcellular level, all mutations are likely to be small
events. Regarding the size of their effects on the PT and the BT, it is possible to distinguish
a sequence of mutations, ranging from infinitesimal to very large. The frequency of
occurrence of the mutations decreases in the direction of that sequence.

The last clause in this axiom was demonstrated experimentally by Timofeeff-Ressovsky
(1935), and it may therefore be appropriate to call A5.20 *Timofeeff-Ressovsky's axiom.*

It follows that certain mutations, even clearly demonstrable ones, may have such
slight effects on the PT and BT that it is impossible to decide whether they are positive
or negative; in short, neutral mutations may occur. The occurrence of neutral mutations
on the molecular level has recently been the subject of much discussion (cf. Kimura and
Ohta, 1971; 1972).

Certain mutations will exert their effect through epigenetic modification of the course
of ontogenesis. About these we shall postulate:

(A5.21) When a mutation acts through interference with epigenetic mechanisms, the
size of its effect is correlated with the stage of development at which the action takes
place. The earlier this occurs, the larger the effect is likely to be.

The essence of this axiom may be traced back to notions advanced by von Baer.
Since in recent times it has been advocated most ardently by Goldschmidt (1940), I
propose to call it *Goldschmidt's axiom.*

The correlation between the mutation characteristics dealt with in A5.19 and A5.20
is stated in the following axiom:

(A5.22) There is no correlation between the quality and the size of a mutation, except
the one related to their frequencies.

From A5.19, A5.20 and A5.22 we may conclude:

(T5.29) Large, positive mutations are possible, but extremely unlikely, events.

This theorem, the implication of which is *Natura facit saltum*, has been accepted by
quite a few biologists, even in Darwin's closest entourage (cf. Provine, 1971), as a
necessity, or at least as a possibility, that cannot be rejected *a priori.* As T. H. Huxley
claimed, in a letter to Darwin on 23rd November, 1859: 'You have loaded yourself with
an unnecessary difficulty in adopting *Natura non facit saltum* so unreservedly' (L.
Huxley, 1900, Vol. 1, p. 176). Yet, none of Darwin's early critics argued the necessity of
saltations, i.e. large-scale mutations, more cogently than Mivart (1871), and I therefore
propose to call T5.29 *Mivart's theorem.* The stand that large-scale mutations are
necessary to explain phylogenetic evolution was later taken up by Goldschmidt (1940),

but his views have been consistently rejected by geneticists and other adherents of neo-Darwinism. One reason for this is undoubtedly Goldschmidt's attempts to account for the mechanism of large mutations on the subcellular level. As mentioned in Chapter 1, recent discoveries seem to vindicate Goldschmidt's views by showing that taxonomic divergence may be correlated with karyotypic changes (Wilson, Maxon and Sarich, 1974; Wilson, Sarich and Maxon, 1974; Wilson *et al.*, 1975). The opposition is also partly directed, however, against the notion of large beneficial mutations. The basis of this dissent is to be found in what may be called 'Fisher's axiom', according to which, so it is said, large mutations are always deleterious. We shall take up this question in a later section.

Before we proceed to discuss the influence of mutations on the course of evolution we may first make the restriction that we shall be concerned here only with mutations increasing the PT or the BT. Negative mutations will be weeded out, and neutral ones cannot persist due to their own effects. They may, however, be preserved through 'genetic drift' (cf. Kimura and Ohta, 1972; Ohta, 1974).

We may distinguish a series among the positive mutations with respect to the size of the change they cause in the PT and the BT. According to A5.20 their frequency will be inversely related to this change. There can be no sharp distinction between common and rare mutations, but if we nevertheless attempt to make one I propose to start with a reference to the following quotation: 'Events, such as chromosome rearrangements, duplication or deficiency of a nucleotide sequence, or polyploidy . . . may occur only once in the history of a [taxon]. These are hardly amenable to any effective quantitative treatment [as concerns changes in gene frequencies]' (Crow and Kimura, 1970, p. 368). Since changes of this kind are found in many animal taxa, it may be presumed that they have become established through inbreeding. In any case, it may be that the frequencies of the rare, positive mutations implied by T5.29 are of the same order as those mentioned in the above quotation; indeed, they may be the outcome of events of this kind. Hence, as a conceptual distinction between the several kinds of mutation the following definition may be proposed:

(D5.22) An existing mutation or gene is one which is amenable to population–genetical treatment. When the frequency of the mutation is too low to permit this, the mutation is said to be a unique event or an innovation. When the effect of the mutation on the selection coefficient is too slight to permit population–genetical treatment, the mutation is said to be neutral or quasi-neutral.

From A5.19, A5.20 and D5.22 we may derive:

(T5.30) Quasi-neutral mutations produce insignificant, existing mutations small, and innovations large, changes in the PT and the BT.

We shall now try first to outline the influence of existing genes, those studied by geneticists, on the course of evolution, beginning with the following axiom:

(A5.23) In the population P_i of bisexually reproducing animals, recombination of the existing genes, possibly modified by environmental factors, will entail that the members of P_i become distributed normally with respect to their PT and BT.

In this axiom it is admitted that the PT and the BT are properties which vary not only between populations but also within populations. It should be noted that evolution through recombination of existing genes is a phenomenon in which only bisexually reproducing animals can participate. There is no such constraint with respect to innovations.

From D5.12 and A5.23 we may infer:

(T5.31) If the physical variability of the niche N_i exceeds the PT possessed by the members of P_i in the lower range of the PT distribution, then the latter will be eliminated through selection.

Likewise, from D5.14 and A5.23:

(T5.32) If the biological extension of the niche N_i is such that the BT possessed by the members of P_i in the lower range of the BT distribution is exceeded, then the latter will be eliminated through selection.

Since the PT and BT are determined by heritable factors, the process of selection described in T5.31 and T5.32 must have the following consequence:

(T5.33) In the course of time the range of distribution of PT and BT will be narrowed so that no animals have values of these properties lower than those determined by the physical variability and the biological extension of the niche. The resulting distribution of PT and BT represents the maximum obtainable under the given external conditions and with the given repertoire of existing genes, and cannot be further improved through selection.

This theorem may approximately express the implications of the neo-Darwinian notion of 'normalizing selection'. If we want further evolution to occur, as implied by changes in PT and BT, it is necessary that the environment is changed.
Thus:

(T5.34) If the physical variability and the biological extension of the niche are increased, the process of selection will be resumed until new maximum values are reached.

This theorem corresponds, I believe, to the neo-Darwinian concept of 'directional selection'. We may thus infer:

(T5.35) As long as changes occur in the environment, selection and, hence, evolution will take place within the range specified by the existing genes.

Although, I hope, essentially correct, the foregoing presentation is, of course, a very cursory outline of the modern population–genetical theory of evolution. But there is no reason to go into more detail. First of all, the theory has been expounded on so many occasions, recently by authorities like Crow and Kimura (1970) and Dobzhansky (1970). Secondly, in spite of many attempts to demonstrate its applicability to actual evolutionary events, the collected evidence has not been very convincing (cf. Lewontin, 1974). This fact may indicate that although the neo-Darwinian theory is unquestionably an evolutionary theory, it does not hold the central position that has hitherto been postulated.

We shall now turn to a discussion of the implications of innovations for the process of evolution. From T5.29 we know that mutations involving substantial increase in PT and BT may occur, but that they are extremely rare. Hence we may infer:

(T5.36) The population distinguished by the possession of an innovation must be represented by a single individual or, in the case of bisexually reproducing animals, possibly by the offspring of a single individual. In the latter case inbreeding must therefore be the means by which evolutionary innovations become established.

If the innovation is concerned with the PT, then it cannot become established through selection under the given physical conditions. Two alternatives seem available to account for its preservation, either that the physical variability of the environment changes by chance simultaneously with the appearance of the innovation, a possibility which may be ignored, or else that the population becomes isolated in a new niche. Hence:

(T5.37) The population P_i, distinguished by the possession of an innovation involving the PT, can become established only through random or non-random isolation. The latter alternative presumes that the innovation itself enables the members of P_i to invade a new niche or environmental compartment.

The situation is different with respect to innovations affecting the BT, for here the possibility exists that the members of the new population may compete with their ancestors in the old niche. Thus:

(T5.38) If the increase in BT incurred by an innovation is so large that, though few in number, the new population P_i has a PP_i larger than the PP_j of the population P_j from which it sprang, then P_i will progress, and P_j regress and be extinguished unless it becomes randomly isolated.

Obviously, if the differential in BT is so large that the population can progress in spite of its initial small size, then it is also so large that biological equilibrium cannot be attained. Under the given premises it is thus seen that adjustment of population size serves no useful purpose when the establishment of innovations is involved. Apparently, it is of use only when a niche is invaded by another population, and in 'direct' struggle, for instance between predator and prey.

If the situation described by T5.38, when a new population becomes established through extinction of its ancestors, takes place repeatedly in the same niche, then we will have a succession of populations conforming to the second theorem of progressive evolution.

If the innovation involves an increase in BT, but not large enough to ensure the progression of the population possessing it, the latter can become established only through isolation, random or non-random as the case may be.
Thus:

(T5.39) The population P_i, possessing an innovation which raises the BT but does not ensure progression in the niche in which it arose, can only become established through random or non-random isolation. The latter alternative presumes that the innovation itself enables the members of P_i to invade a new niche.

The phenomenon dealt with in T5.37 and T5.39 may be called 'taxonomic divergence', defined as follows:

(D5.23) Taxonomic divergence is said to occur when a population P_i, distinguished from the ancestral population P_j by the possession of an innovation (taxonomic character), survives in a niche N_i different from N_j.

From T5.37, T5.38 and T5.39 follows:

(T5.40) The survival of a population distinguished by the possession of an innovation involves either progressive evolution or taxonomic divergence.

As we have seen, features which are due to existing mutations cannot usually be used as taxonomic characters. The properties which arise through innovations may in most cases be presumed to be inherited according to the rules of Mendelian genetics. However, due to the substantial advantage they confer, it may be anticipated that, through in-breeding, they rapidly become established in the population to the extent that they comply with the definition of taxonomic characters given above.

If this is true, it follows that the populations arising through taxonomic divergence belong to different taxa, and the latter are, of course, twin taxa. We may thus conclude:

(T5.41) The outcome of taxonomic divergence is the dichotomous splitting of a taxon T_j into two twin taxa T_{j+1}.

From T5.37, T5.39 and T5.41 it follows:

(T5.42) In any set of twin taxa one is always non-randomly or randomly isolated. Both taxa may, however, be randomly isolated.

The last provision is required because no predictions can be made about fortuitous events. The situation envisaged may be exemplified by the two families of Monotremata, and possibly by the lungfishes. This alternative is important for 'divergence through dispersal', as discussed in the next section.

We shall now introduce the concept of 'dominance', defined in the following way:

(D5.24) Of a pair of twin taxa the one which is not isolated is called the dominant taxon.

There are certain difficulties involved in the phenomenon of dominance which require a special section devoted to this problem. However, this does not prevent us operating with the concept in the present context. Thus, we may further conclude:

(T5.43) If a monophyletic taxon is resolved in such a way that at each bifurcation except the last there is one terminal and one non-terminal taxon, then one of the most inferior twins is a dominant taxon; all the other taxa are either randomly or non-randomly isolated.

A new practical concept may now be defined, viz., 'dominant phylogenetic line':

(D5.25) A dominant phylogenetic line is a phylogenetic line composed of taxa which have each been a terminal dominant taxon at one stage during the course of evolution. From T5.43 and D5.25 we may infer:

(T5.44) In any monophyletic taxon there is one, and only one, dominant phylogenetic line.

(T5.45) Each dominant taxon has arisen from a dominant taxon.

The latter theorem conforms quite closely to the following statement: 'I have argued throughout that the law of evolution consists in the origination of successive forms from the dominant group then alive' (Gaskell, 1908, p. 395), and I therefore propose to call T5.45 *Gaskell's theorem*.

It is important to observe that although, according to Cope's theorem, non-random isolation (specialization) may entail losses in BT, these need not always occur. Since, furthermore, gains in this parameter are subject to chance, the possibility exists that under adverse physical conditions a dominant taxon may be extinguished and a new dominant taxon arise from a non-randomly isolated one. But this is no exception to Gaskell's theorem since the phenomenon of dominance refers to competition between populations, and thus primarily to the BT.

It is very important to realize that if it is possible to resolve a non-dominant taxon, then this will in turn comprise a dominant and an isolated taxon, etc. We can therefore deduce:

(T5.46) With the degree of resolution envisaged in T5.43, all the non-dominant phylogenetic lines coincide with the dominant one except for the last taxon. With further resolution, each of the non-dominant phylogenetic lines will contain a new dominant phylogenetic line.

As discussed in Chapter 2, it may be presumed that many innovations have been acquired before a taxon, at least a superior one, came into existence. According to T5.38, some of these steps may have occurred within a particular niche, a population possessing a particular innovation causing the extinction of the original one. By definition, this phenomenon constitutes evolutionary progress. Hence we may conclude:

(T5.47) Innovation and extinction are necessary and sufficient conditions for non-divergent progressive evolution.

This theorem should not be taken to imply that progressive evolution usually occurs without taxonomic divergence. The normal course of events may rather be that extensive divergence takes place, and that in one of these taxa a particular innovation is acquired which permits dispersal over large spaces, thereby causing the extinction of many more or less related taxa. It appears that this interpretation conforms with the fossil record (cf. Eldredge and Gould, 1972).

To the believer in the availability of an inexhaustible source of mutations which may ensure the complete adaptation of the members of an animal population to its environment, the large measure of extinction revealed by the fossil record is an embarrassment. This is demonstrated by the following quotations: 'The frequency of extinction is a great puzzle to me. Far too little attention has been paid to the factors responsible for this failure and breakdown of natural selection' (Mayr, 1960, p. 141); and 'Death and extinction are antitheses of biological progress' (Dobzhansky, 1974, p. 311).

In contrast to this neo-Darwinian perplexity, it will be noticed that the present theory actually predicts the observed extinction. As we shall presently see, progressive evolution may occur without extinction, when divergence occurs through non-random isolation. But this always involves non-dominance; in the dominant phylogenetic line progress requires extinction. In this context it should be recalled that each superior taxon contains many dominant phylogenetic lines.

Since non-random isolation, initially at least, involves an increase in either PT or BT, we may further conclude:

(T5.48) Innovation and non-random isolation are necessary and sufficient conditions for divergent, progressive evolution.

According to Darwin (1885), Wagner (cf. 1868) first emphasized the importance of isolation for taxonomic divergence. I therefore propose to call T5.48 *Wagner's theorem.*

Since random isolation does not require an increase in the PT or BT, and thus does not entail progressive evolution, we may state:

(T5.49) Innovation and random isolation are necessary and sufficient conditions for divergent, non-progressive evolution.

It should be noted that the innovation may occur either in the randomly isolated taxon or in its twin.

In connection with the problem of taxonomic divergence, we shall return to the problem of genetic incompatibility. As appears from T5.48 and T5.49, isolation is an essential aspect of taxonomic divergence. This will ensure that little, if any, gene exchange occurs between the two taxa arising through the divergence. The innovation may occasionally involve genetic incompatibility, but if not, this condition may be expected as the outcome of innovations occurring later.

We have seen above that an innovation is at first possessed by a very small number of animals. This must mean that, whether these individuals remain in their ancestral niche or become isolated, they must act as founder populations which, through inbreeding, ensure the propagation of the innovation. Since this state of affairs does not necessarily imply genetic incompatibility with the ancestral population, we shall call it 'sexual isolation', defined as follows:

(D5.26) Two populations are sexually isolated if, irrespective of their possible genetic compatibility, their members do not interbreed.

Hence we may infer with reference to T5.47, T5.48 and T5.49:

(T5.50) Biological evolution requires sexual isolation for each step of progress or divergence associated with the acquisition of an innovation.

As this statement in part reflects the ideas on the evolutionary importance of small isolated populations propounded by Wright (1931), I propose to call T5.50 *Wright's theorem*.

Populations of placental mammals are often small, comprising closely related individuals. Bush (1975) has proposed that this social structure ensures the isolation required for the establishment of innovations through inbreeding. This circumstance may be responsible for the rapid evolution in this taxon (Wilson *et al.*, 1975).

Because the view of mutations and their effects propounded above deviates from current notions in some respects, I shall briefly summarize the correlation between size of mutational effect and evolutionary impact implied by the preceding discussion. Although undoubtedly mutations cover a continuous range with respect to their effects, I shall here distinguish three classes, the quasi-neutral, the intermediate and the large-scale.

In principle the first kind ought to be imperceptible and, in fact, it is only through studies of the amino acid substitutions which have occurred during the evolution of various proteins that the existence of purportedly neutral mutations has been recognized (Kimura, 1968; 1969; King and Jukes, 1969; Kimura and Ohta, 1972; Ohta, 1974). Involving only infinitesimal changes in the phenotype, quasi-neutral mutations cannot become fixed in the genome through Darwinian selection. Consequently, they can only spread through random genetic drift in small populations. On this presumption the limiting factor in the fixation of quasi-neutral mutations is the rate at which they occur. The actual rate will be somewhat lower, varying from one protein to another, because a number of amino acid substitutions are disadvantageous (cf. Ohta, 1974). Obviously, quasi-neutral mutations cannot be of evolutionary significance except when they fortuitously pave the way for later changes with greater effect. A particular feature of protein evolution is that, in agreement with the postulate made above, the rate of mutation is higher than that observed in genetic studies. The fraction of these mutations which may be classed as neutral is disputable (Kimura, 1968; King and Jukes, 1969).

Except for those of microorganisms and moulds, the mutations studied by geneticists have traditionally to a large extent had clearly visible effects. In accordance with the present terminology the mutations themselves, which in many instances are undoubtedly amino acid substitutions, have been subjected to epigenetic amplification. Under these circumstances they may significantly affect the selection coefficients, so that they are subject to natural selection.

This does not imply, of course, that mutations which can be demonstrated only by molecular–biological means are always quasi-neutral, and thus exempt from selection. Rather, it has been observed that even on the molecular level Darwinian evolution may take place (Ayala, 1971; Goodman, Moore and Matsuda, 1975).

It may be anticipated that some amino acid substitutions, affecting for instance regulatory proteins, may have large-scale effects. And the same is true if mutations of this kind concern regulatory genes. In addition, large effects may also arise from chromosome and genome mutations. The typical feature of large-scale mutations is that the advantages they confer on their bearers are so great that they can either eliminate their rivals through direct competition or avoid selection through non-random isolation. Thus evolution is non-Darwinian at both ends of the mutational spectrum, and only in the intermediate range may neo-Darwinian principles be expected to be of phylogenetic importance.

Evolution

Two essential tasks of a theory of evolution are to account for the origination and for the survival of the populations (taxa) which we can observe in Nature. In an attempt to elucidate these questions, we shall begin by concluding from T5.43:

(T5.51) The survival of a taxon T_j implies that it is either a dominant, a non-randomly or a randomly isolated taxon.

As concerns the dominant taxa, it may be observed that the progressive, non-divergent evolution described by T5.38 is associated with a steady increase in the BT. Clearly, any step of reversion would entail a reduction in BT which could not become established. Hence we may deduce:

(T5.52) Progressive, non-divergent evolution is irreversible.

The implication of this theorem is, of course, that advantageous characters, once gained, cannot be lost again. As was discussed above, such losses may occur, but only under conditions of isolation, random or non-random as the case may be. In its wording and intent, T5.52 bears some resemblance to Dollo's law (1893), and may therefore be called *Dollo's theorem*. It should be emphasized that Dollo's law has a wider scope (cf. Gould, 1970).

There is one phenomenon which is of special interest in association with dominant taxa, namely, 'divergence through dispersal'. If a new dominant taxon arises at some location, then the spatial confines of the whole environmental compartment in which it arose, perhaps the ocean, perhaps a continent, may coincide with its inhabitable niche, because none of the existing affiliated populations can ever mobilize sufficiently high PP to attain biological equilibrium, and hence cannot prevent the dispersal of the members of the new taxon.

In the course of a dispersal of this kind, new innovations, and hence new taxa, may arise. If these represent increases in the PP, then the new taxa may spread into the niches of the related taxa, but if they involve neither specializations nor significant changes in the PP, then the several taxa may disperse side by side, because none of them has a PP that allows the invasion of the niche occupied by another one. Under these circumstances we seem to have the simultaneous existence of two dominant twin taxa, in apparent contradiction with T5.42. It must be stressed first, however, that the dominance of these taxa is primarily measured *vis-à-vis* those they replace. Nonetheless, it is evident that situations like those described here require either that T5.42 is rejected or else that we rate one or both of the taxa as randomly isolated. I believe that the latter alternative is fully justified for, on the given premises, it must be a matter of luck whether or not the innovation becomes established as a taxonomic character, is lost through extinction of the incipient taxon, or is dissipated as a Mendelian character in the ancestral population (taxon). Since the last alternative is excluded if the innovation involves genetic incompatibility, the chances of establishment are evidently increased if this condition prevails. The situation envisaged here may account for the occurrence of sympatric species in many instances.

The process of divergence through dispersal described here is evidently related to so-called 'adaptive radiation'. The particular feature of the latter phenomenon is that, during the early history of a new taxon, representing, we may presume, a dominant one, taxonomic divergence proceeds unusually rapidly. We may account for this phenomenon by the following axiom:

(A5.24) In the archetype of a new taxon a number of potential modifications exist which, if realized, will constitute inferior taxa within the new one. The probability of the origination of any of these is inversely proportional to the number which have already come into existence.

The concept of 'archetype' was defined in D2.13. With reference to this axiom and the preceding discussion we may derive:

(T5.53) The dispersal of a dominant taxon is likely to be associated with taxonomic divergence.

It will be shown subsequently that this statement represents a view put forward by Willis (1922), and I therefore propose to call T5.53 *Willis's theorem*. It is unfortunate, however, that the phenomenon discussed here is called 'adaptive radiation', because the adjective is teleological and the noun an ambiguous synonym for 'taxonomic divergence'. I believe it is preferable to talk of 'divergence through dispersal', as I have done above. It seems that T5.53 coincides also with one of the main theses of Croizat, namely that 'dispersal [equals] form-making + translation in space' (1962, p. 17). However, I find this terminology confusing, for is not 'dispersal' = 'translation in space'? This, at least, is the sense in which I have used the concept here.

We may furthermore conclude from A5.24:

(T5.54) Within any taxon, the rate of taxonomic divergence is likely to decrease gradually with time.

The two stages of taxonomic divergence implied by T5.53 and T5.54 correspond to the first two phases of Schindewolf's theory of typostrophism (1950). The third phase, typolysis, comprises two aspects according to Schindewolf, degeneration and extinction. The present theory also predicts extinction as the last phase in an evolutionary cycle, but this event occurs primarily because new forms with higher BT arise, and not as a consequence of degeneration. I propose to call T5.54 *Schindewolf's second theorem*. It should be mentioned that Vandel's views on evolution (1958) in many respects coincide with those of Schindewolf, and hence also with those put forward by me.

As concerns non-dominant taxa, we may first focus attention on the process of isolation, which, from T5.48 and T5.49, we know to be of great importance for taxonomic divergence. We shall make the following postulate:

(A5.25) The probability of isolation, non-random as well as random, increases with the number of available niches.

Hence from T5.13 and T5.14:

(T5.55) Other things being equal, the probability of taxonomic divergence within a given taxon is inversely proportional to the size and the vagility of the members of the taxon.

The implications of this statement have been set out independently by Bush (1975), and I therefore propose to call T5.55 *Bush's theorem*.

The prediction of this theorem may be tested by reference to many sets of taxa. Thus, the effect of size can be evaluated by comparing Insecta and Crustacea, Gastropoda and Cephalopoda, small and large birds, small and large mammals, etc. It is important to note that randomly and non-randomly isolated groups should not be compared; it would make little sense to oppose Cephalopoda to Monoplacephora. With respect to vagility, it is enough to mention the fact that certain highly vagile animals, for example whales and many birds, occur in a limited number of species throughout the biosphere, while sessile animals are frequently subdivided into a number of subspecies and local races. T5.55 was corroborated by Wilson *et al.* (1975), who showed that in mammals there is an inverse correlation between body size and rate of evolution, as estimated by karyotypic variability.

From T5.55 we may also derive:

(T5.56) The body size–frequency curve for the terminal taxa within a superior taxon will be right-skewed.

The testing of this theorem of course demands a reasonable variation of size within the superior taxon in question. It has been corroborated by Hutchinson and MacArthur (1959), and it may therefore be appropriate to call it *Hutchinson and MacArthur's theorem*. It was pointed out by Stanley (1973) that the steep fall of the left flank of the curve may be predicted on the assumption that the influence of the factors determining the minimum size is abrupt (Figure 5.1).

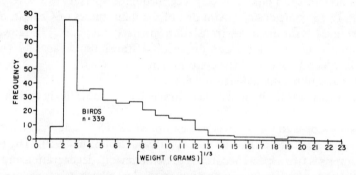

Figure 5.1. Histogram of cube roots of body weights for recent flying birds. (Reproduced from S. M. Stanley, after Magnan and Poole, *Evolution*, **27** (1973), by permission of The Society for the Study of Evolution)

It is possible to make some further inferences with respect to the body size of animals. For one thing, we may derive:

(T5.57) Innovations affecting body size will become established if they increase the PT or the BT.

I shall submit the following axiom:

(A5.26) An increment or decrement in body size allows for non-random isolation in a new niche.

This axiom carries little conviction if regarded from a Darwinian viewpoint, according to which all changes, including those of size, are infinitesimal. It will later be postulated that various lines of evidence tend to support the notion that size variations occur in large steps, and if this is true, then A5.26 gains in credibility.

From this axiom we may infer:

(T5.58) As long as the physiological limits to body size have not been reached, there are always potential niches corresponding to sizes outside (and sometimes also within) the prevailing range of body sizes.

If we can rely upon chance to ensure that the proper mutations arise, we may also conclude:

(T5.59) In the absence of factors which limit body size to some particular range, evolution will tend to establish body sizes covering the total range between the physiological limits.

I shall now submit one further axiom which is substantiated empirically, at least for mammals.

(A5.27) The members of new taxa of superior rank are usually of small body size.

From T5.59 and A5.27 we may now conclude:

(T5.60) In a new taxon of superior rank the prominent direction of evolution will give rise to animals with larger body size as long as this is associated with an increase in the PP.

This theorem states the principle of orthogenesis, first proposed by Haacke (1893). I propose to call T5.60 *Haacke's theorem*. If this theorem is accepted, and also that size modifications are generally associated with various kinds of allometric growth, then it is possible to account for several evolutionary sequences (cf. Simpson, 1944; Gould, 1974). Besides morphological changes, allometry may lead to advantageous modifications of various kinds, for example ecological, physiological and even energetic (cf. Gould, 1966; Van Valen, 1976). Therefore, if increase in size does not lead to non-random isolation, it may be anticipated that the larger form is dominant *vis-à-vis* the smaller one and causes its extinction.

The following postulate will be made:

(A5.28) Extreme body sizes entail a reduction in the BT, except when they allow for non-random isolation (protection). Under certain circumstances they may entail an increase in the PT.

The protection procured through extreme body size involves, of course, hiding for small animals and defence for large ones. We may conclude:

(T5.61) Under conditions where protection through non-random isolation is of advantage, i.e. when actively dependent populations exist in the biological extension of the niche, innovations involving modifications towards extreme sizes will become established until the contribution of the non-random isolation to the BT is offset by the simultaneous reduction in this parameter due to the size.

As is well known, the permissible size range of an animal taxon depends on life habits, especially upon food. Thus, the mean size of insectivores and rodents is smaller than that of carnivores, and the latter in turn smaller than the mean size of perissodactyls and artiodactyls. This circumstance implies that for relatively small animals only the lower range of extreme body size is attainable, while for relatively large ones it is the higher range. This point is well established by the size distribution of rodents and artiodactyls (Valverde, 1964).

We may further deduce:

(T5.62) If the non-random isolation referred to in T5.61 is changed into a random isolation, i.e. if actively dependent populations are excluded from the extension of the niche, innovations involving modifications towards intermediate sizes will become established.

This theorem, which has been corroborated through observations of dwarf and giant forms on isolated islands (cf. Thaler, 1973), I propose to call *Valverde's theorem*.

Extreme body sizes may also increase the PT. Here it may suffice to mention that in endothermic animals heat preservation is improved with the size of the body. It may therefore in general be presumed that large animals can survive in colder environments than smaller animals belonging to related species.

It is of particular interest to test this proposition on that evolutionary riddle *Megaloceros giganteus*, the 'Irish elk'. It is generally accepted that the enormous antlers of this animal were a terrible embarrassment, and that consequently *Megaloceros* was an evolutionary failure, whose extinction was predictable. But this view entails a very tricky problem, viz., how could selection allow such misfits to appear in the first instance?

I shall submit that the size of *Megaloceros* was not reached through an accumulation of small increments, but in a one-step event, from an animal half as large. This postulate, which will be further substantiated in a later section, is corroborated by Figure 5.2. It is also seen that the awkward size of the antlers was a simple consequence of allometry.

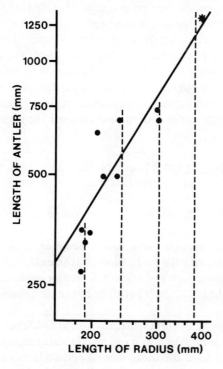

Figure 5.2. The allometric relationship between length of antler and length of radius in 10 species of Pleistocene cervids and *Megaloceros*. The latter is distinguished by an asterisk. The first dashed line is drawn through the mean value of the four smallest species. The other dashed lines represent the values arrived at by successive multiplication with $\sqrt[3]{2} = 1.26$. It is seen that apart from two instances the values for the length of the radius perfectly fit the prediction that the body volumes stand in the ratios $1:2:4:8$. Based on data published by Gould (1974)

In this way we are relieved of the burden of explaining how *Megaloceros* came into existence, but we still face the problem of accounting for the fact that it progressed and survived. There were no large contemporaneous predators, so neither antler nor body size were necessary for the purpose of defence (Gould, 1974). But *Megaloceros* lived in a rather inclement climate, and it is possible that its bulk permitted it to invade territories which were inaccessible to smaller animals because the temperature was too low. If this is true then we should expect from A5.28 that *Megaloceros*, together with many other large animals, for instance the woolly mammoth, became extinct once the climate improved sufficiently to permit invasion by smaller animals.

If we were to submit the axiom that the number of available niches is much smaller in an aquatic than in a terrestrial environmental compartment, a reasonable proposition it seems, then we might also derive the theorem that taxonomic divergence is much less extensive in the former than in the latter case, a prediction which is borne out by empirical observations (cf. Hutchinson, 1959).

Evolution must of necessity have been directed towards higher and higher levels in the food chain. However, the possibilities for interaction are so manifold that it seems nearly impossible to state this correlation unambiguously with respect to phylogenetic classification. Yet one point seems certain, that a dominant taxon is never below its twin in the food chain. It should also be noted that there are always empty niches above the prevailing range of the food chain, for the number of organisms which can live in a certain niche is not affected by the fact that some individuals are eaten.

It has been pointed out that one aspect of non-random isolation involves specialization. It is sometimes argued against Cope's rule that in certain cases specialization does not involve an evolutionary *cul-de-sac*, but may give rise to successful, and highly divergent, taxa. In point of fact this outcome is predictable from A5.25:

(T5.63) Other things being equal, the probability of taxonomic divergence following non-random isolation (specialization) depends upon the availability of niches implied by the isolation.

The notion inherent in this theorem is possibly of long standing, but since it was expressly stated by Simpson (1953), I propose to call it *Simpson's theorem*. The prediction implied by T5.63 may be supported by reference to two extreme cases of specialization, namely, ants as the main source of nourishment, as found in various ant-eating mammals, and flight, as found in most birds. It is evident that, in spite of the success of the ants themselves, their use as a diet must imply a severe constraint on the number of available niches, whereas the property of flying, besides involving protection, has opened up a number of new niches to the birds.

Isolation and innovation are the essential factors involved in the origination of new taxa. The first of these has just been dealt with. Concerning the latter it must be observed that it is a chance event, which makes it impossible to predict where and when it will occur. Nevertheless, if it is true, as postulated in Goldschmidt's axiom, that the mutations responsible for innovations may operate through interference with epigenetic processes, then it may obviously be concluded that the more intricate the ontogenesis, the greater the likelihood of innovations. Thus:

(T5.64) Other things being equal, the rate of taxonomic divergence is likely to be greater in taxa with a highly complex ontogenesis than in taxa with a simple one.

The complexity referred to here may in particular arise through interference by hormones and other physiological regulatory mechanisms. Without stating it formally, I submit that the physiological sophistication thus implied is correlated with the BT, and perhaps with the PT. Recalling that these parameters increase in the course of evolution, we may conclude:

(T5.65) The rate of taxonomic divergence is likely to increase with the numerical rank of the taxon.

The implication of this statement was formulated by Darwin (1885, p. 291) as follows: 'There is some reason to believe that organisms, high in the scale, change more quickly than those that are low', and I therefore suggest calling it *Darwin's third theorem*. The theorem seems to predict relatively high rates of evolution in Arthropoda, Teleostei, Aves, Mammalia and possibly Mollusca, and when we also consider the influence of size, we should expect that prominent among these taxa are Insecta and, to some extent, Mollusca. These predictions are all corroborated.

It seems likely that the following statement is correct:

(A5.29) The rate of mutation is proportional to the number of zygotes produced.

Consequently, the origination of innovations, and hence of taxonomic divergence, must be influenced by the fecundity, as measured by the mean number of zygotes produced per adult in a given time. Thus:

(T5.66) Other things being equal, the rate of taxonomic divergence is likely to be higher in taxa with a high fecundity than in taxa with a low one.

This theorem may possibly be corroborated by the great difference in taxonomic divergence observed in Selachii and Teleostei.

In the present context it should be recalled that according to the Darwinian theories changes in the environment constitute the driving force in evolution; when these occur, selection will act to drive the animals towards a new optimum of adaptation. Since the various elements in the inorganic and organic environment interact in multifarious ways, one would, *a priori*, expect that evolution had proceeded at comparable rates in the various phylogenetic lines. As we have just discussed, some taxa have evolved very fast, while others have remained almost unchanged for hundreds of millions of years, for example *Lingula*, *Latimeria* and *Sphenodon*.

Simpson (1953) has tried to account for this state of affairs, but I doubt if these 'explanations' have any predictive value. Many evolutionists accept these differences as the outcome of different 'selection pressures', the rate of evolution being proportional to the latter. Thus, for instance, it is intimated that *Latimeria* has been exposed to little or no selection pressure, implying that its environment has remained essentially constant for millions of years while that of its close neighbours has changed over and over again during the same period. This appears a most unreasonable idea but, in addition, the notion of 'selection pressure' is an *ad hoc* hypothesis. It serves to explain differences in rate of evolution, but it can only be recorded by differences in rate of evolution.

According to the comprehensive theory, the important mutations are so rare that they cannot be expected to occur in all phylogenetic lines. Hence, evolutionary rates may differ substantially from one phylogenetic line to another. The peculiar feature about the various animals mentioned above is not that they did not evolve, but that they survived. In the normal course of events they should have become extinct. The stochastic occurrence of mutations, together with some of the preceding theorems, suffice to account for the observed differences in rate of evolution without recourse to any form of 'selection pressure'.

At this stage I shall terminate the axiomatic presentation. This decision does not imply that further theorems cannot be derived, with or without the introduction of accessory axioms, but the continuation of this work must depend upon whether or not it is found to be useful for evolutionary studies.

Dominance

It was concluded in a preceding section that one of each set of twin taxa is a dominant while the other is an isolated taxon.

If we draw the phylogenetic dendrogram of Vertebrata as required by T5.43, it will look as shown in Figure 5.3. Here the dominant phylogenetic line is seen to be marked out by the numbers 1 to 9. The latter number, which stands for 8.1, shows that Mammalia, the dominant taxon in Vertebrata, is the only one that ends in 1; all the other taxa end in 0.

It should be possible to decide which of the latter are randomly and which non-randomly isolated. It may be that Dipnoi belong to the former, while Actinistia + Chondrichthyes, Cladistia + Actinopterygii, 'Amphibia' and reptiles + Aves may be allocated to the latter category with some certainty. The remaining taxa, Hyperotreta, Hyperoartii and Gymnophiona, are more difficult to classify; it may be that elements of both randomness and non-randomness are involved in their survival.

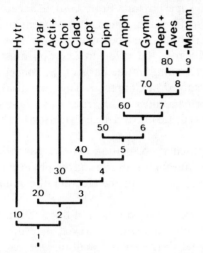

Figure 5.3. Resolution of phylum Vertebrata to establish dominance and non-dominance in the various taxa. The line 1–9 is the dominant phylogenetic line in the phylum. Acpt, Actinopterygii; Acti, Actinistia; Amph, 'Amphibia'; Choi, Chondrichthyes; Clad, Cladistia; Dipn, Dipnoi; Gymn, Gymnophiona; Hyar, Hyperoartii; Hytr, Hyperotreta; Mamm, Mammalia; Rept, reptiles

At this stage it may be appropriate to discuss the element of subjectivity involved in the concepts of 'isolation' and 'dominance'. Thus, if Vertebrata originated in the sea but Hyperoartii in fresh water, then clearly the taxon Hyperoartii + Gnathostomata originally became isolated from the ancestors of Hyperotreta; yet there is no doubt that the former taxon is the dominant one, so if we classify the latter as 'isolated', the implication is rather 'non-dominant'. Of course, the fact that Hyperotreta have survived, although members of various taxa included in the twin taxon have invaded the sea, must mean that the former are isolated in one way or another. It is easy to see that exactly the same arguments can be advanced to support the classification of Dipnoi + Tetrapoda as the dominant taxon *vis-à-vis* Cladistia + Actinopterygii.

Proceeding to resolve some of the isolated taxa in Figure 5.3, we may begin with the taxon 3.0, Actinistia + Chondrichthyes (Figure 4.44). In this case we can without hesitation classify the former as randomly isolated and Selachii as the dominant taxon. It may be a little more difficult to decide whether Holocephali are non-randomly or randomly isolated.

In the taxon 4.0, Cladistia + Actinopterygii (Figure 4.44), Teleostei are the dominant taxon, Amiidae and Lepisosteidae and presumably Cladistia are randomly isolated, while Chondrostei appear to be non-randomly isolated.

Of the taxa Caudata and Salientia, the younger age, the geographical distribution and the taxonomic divergence of the latter leave no doubt that it is the dominant taxon within 'Amphibia'.

In the taxon 8.0, reptiles + Aves (Figure 4.44), either Crocodylia or Aves must be the dominant taxon. It appears that the latter should be chosen, even though, through the power of flying, most birds are isolated relative to the terrestrial reptiles. Rhyncocephalia are evidently randomly isolated, the other three taxa non-randomly. It is interesting to note that the oldest taxon, Squamata, is today the most successful.

It seems to me that the agreement between the phylogenetic classification and the predictions made by the theory of evolution is quite satisfactory.

The present discussion shows that the dominant phylogenetic line of the taxon Vertebrata leads straight from their invertebrate ancestors to Mammalia. Even without any cladistic analysis it seems reasonably certain that the dominant phylogenetic line for the invertebrates must end within the annelid superphylum. From Gaskell's theorem it follows that when these two lines are joined to form the dominant phylogenetic line for Animalia, then the non-dominant twin to Vertebrata must be found among or within the major taxa in the annelid superphylum. Clearly, the taxon in question must be either Arthropoda or Mollusca. Thus, simple reasoning on the basis of the concept of 'dominance' leads to the same result as that obtained in Chapter 3. This conclusion is further corroborated by the fact that Echinodermata must be regarded as a side-issue, a well-protected, and thus non-randomly isolated, taxon of a rank superior to those mentioned above.

The preceding discussion shows that the postulated correlation between taxonomic rank and dominance and non-dominance may be useful for testing a given phylogenetic dendrogram. In many instances it may even be employed for the erection of classifications.

From Cope's theorem it follows that once a taxon has become non-dominant it has little chance of becoming dominant again in its former realm; however, it seems that there is one possibility open of achieving this goal. If the PT of the dominant taxon is inferior to one or more non-dominant ones, then an untoward change in the physical variability of the environmental compartment may result in the dominant taxon being extinguished; should this occur, a non-dominant taxon which can invade the niches left behind may become dominant. As we shall discuss in a following section, this explanation seems to account for the rise of Mammalia at the end of the Mesozoic.

A special case is represented by *Homo sapiens*. His closest relatives, the primates, are clearly not the dominant taxon in Mammalia. It is not known which taxon assumes this position; offhand, one thinks of the large carnivores. Whether this idea is inspired by folklore I do not know, but at least they are high up in the food chain. If, nevertheless, we admit that *Homo* is the dominant taxon in Animalia then, I think, the reason is that man has gone beyond the scale by which all other animals measure the BT. Man, un-aided by the cultural tradition which helps to clothe, feed and arm him, is not the head of the kingdom. But with this assistance he has reached a value of BT—and PT—which no animal or plant can ever aspire to. Hence, since no other species can establish bio-logical equilibrium with this particular one, the future of evolution, for better or worse, lies in the hands of His Majesty *Homo sapiens*.

J. L. Travis (1971) has discussed the criteria used by J. S. Huxley (1959) and by Simpson (1967) to define the phenomenon of dominance, showing that the views of these authors imply various logical inconsistencies. Travis (1971, p. 374) states: 'Under [Simpson's] distinction of sequences of dominance, it seems to me that I can choose any group fairly numerous at a given time and find some criterion by which to establish it as dominant in a dominance sequence defined by the chosen criterion. Such a procedure would render the concept of dominance virtually meaningless since almost any life form could be shown to be dominant on some criterion'.

It will be realized that the concept of 'dominance' put forward here corresponds very well to the first sentence in the quotation. The essential point is that there are not one, but many, dominant taxa; that, indeed, almost every taxonomic bifurcation implies the separation of a non-dominant from a dominant taxon. Therefore, the conclusion in the second sentence above is correct, but that does not render the present concept of 'dominance' meaningless.

Fitness

To the general public the essence of Darwinism is embodied in Spencer's maxim:
(1) The survival of the fittest.
Much discussion about the validity of Darwinism has been centred around the implications of this statement (e.g. Grene, 1969). However, it is apparent that both of the significant words in the latter are so ambiguous that it seems impossible to use it for the intended purpose before these concepts are clearly defined.

The first point to decide is the nature of the entity which survives, and to which thus the epithet 'fittest' applies. The most obvious interpretations, and maybe the only possible ones, are that reference is made to either 'taxon' or 'individual'. Thus, (1) may be stated either as:
(2) The survival of the fittest taxon
   or
(3) The survival of the fittest individual.
The meaning of 'survival' is clearly different in these two formulations. A taxon can survive for very long periods of time because, through the process of reproduction, it is ensured that new individuals come into being possessing the characters of the taxon. Even without defining 'fittest' we must admit that (2) makes sense in an evolutionary context, and that it may well represent the intended implication of (1).

Individuals may reproduce themselves, but the fact that they are bound to die means that 'survival' cannot be applied to individuals in the same sense as to taxa. 'Survival' may, however, refer to the conditions of competition envisaged by Darwin's 'struggle for existence', so that (3) should read:
(4) The survival of the fittest individuals under competitive conditions.
It is possible that (4) represents an interpretation of (1) which might have been accepted by Darwin, but the statement, although perhaps of great ecological interest, is of little importance from an evolutionary point of view. Yet if we introduce the premise that the fittest individuals are members of the fittest taxon and that (4) $\supset$ (2),* then we arrive at an implication which appears to be a meaningful element of a theory of evolution:
(5) The survival of members of the fittest taxon under competitive conditions $\supset$ the survival of the fittest taxon.
Clearly, 'survival' occurs in two distinct senses in (5), and this may possibly explain some of the confusion which has arisen from the several attempts to interpret the meaning of (1).

Because in the present context the 'survival of a taxon' is the most important concept, it seems necessary to define it as precisely as possible. As a first approximation one might propose:

The taxon T survives $\supset$ the taxon T is not extinct

---

*The sign $\supset$ means 'implies'.

In conjunction with (1) this implies that, without further specification, all taxa of existing organisms, from bacteria to the highest plants and animals, are 'fittest', i.e. of equal fitness. If this view is adopted, then it will appear that, with reference to A5.15, 'fitness' must stand for PP. However, since, according to T5.17, this parameter is the same for all surviving populations in a given environmental compartment, the use of the superlative is not indicated; rather, one should phrase (1) in the following way:

(6) The survival of the (taxa or populations) fit (to adjust their PP to the level prevailing in a particular environmental compartment at the present time).

Although this statement, at least on the basis of the present theory, is meaningful from an evolutionary point of view, it is evident that the suggested interpretation of (1) can never have the meaning intended by Darwin, for even though the adjustments required for the establishment of biological equilibrium involve increases in PP, they do not entail the origination of organisms with higher and higher 'fitness', which surely is implied by his 'struggle for existence' and by the use of the superlative in (1).

In order to approach an interpretation of (1) which is in better harmony with the original intent it is necessary to refer to the concepts of 'progression', 'survival', 'regression' and 'extinction' in D5.2. Here 'survival' is defined as a *status quo* with respect to the number of individuals in successive generations of a given taxon or population, corresponding possibly to an intuitive idea of the concept, although not necessarily the one implied by Spencer and Darwin.

If we imagine that in the course of evolution members of two different taxa have had the opportunity to compete with each other, and that those of one taxon were 'fitter' than those of the other, then, presumably, (5) should be phrased:

(7) The survival of members of the fittest taxon under competitive conditions $\supset$ the progression of the fittest taxon.

With reference to T5.18, we may infer that in this statement the implication of 'fitness' is the BT. If, for the sake of brevity, we write only the implication of (7) and introduce the present terminology, we have:

(8) The progression of the (taxon with the) highest BT.

This statement, which coincides with the theorem of competitive selection, corresponds, I believe, quite closely to the idea implicit in (1). However, competition usually occurs between members of two populations or taxa, and accordingly it would be more appropriate to introduce the comparative rather than the superlative in (8), thus:

(9) The progression of the taxon with the higher BT.

The taxon referred to in (9) is, of course, a dominant taxon. As a consequence of Darwin's second axiom (A5.5) it follows that the non-dominant, non-isolated, taxon must regress and eventually become extinct, the progression of the dominant taxon becoming at this stage replaced by survival. With the introduction of (9) we thus impose upon evolution a direction, a progress towards fitter and fitter organisms (cf. Williams, 1970).

As we have seen above, the competition which leads to extinction cannot lead to taxonomic divergence, so, if the interpretation of (1) represented by (9) is presumed to be exhaustive, then the former can only represent a part of the truth. Rather, taxonomic divergence can occur only if the members of various taxa, through isolation, become exempt from competition with those of dominant taxa. Since isolation may allow that, in spite of a relatively low BT, the PP is raised to the level of the dominant taxon in the environmental compartment, the survival of non-dominant taxa which is essential in taxonomic divergence is stated by (6).

We have distinguished above between non-random and random isolation. In the former instance the survival involves an active contribution on the part of the members of the taxon, this situation may therefore be represented by

(10) The survival of the cunning (enough to avoid competition).

In the latter case the survival is due to the fact that those dominant taxa which could replace the isolated one happen to be absent. Although in a certain sense (6) may also be true in this case, it would seem more appropriate to replace it with:

(11) The survival of the lucky.

Thus, if we postulate that (1) should be a complete rendering of the mechanisms of evolution, then the formulations in (9), (10) and (11) should be implied. And thus it appears that survival (and evolution) is not a matter of 'luck or cunning', as suggested by Butler (1887), but rather of *dominance* or *cunning* or *luck*.

It is interesting to reflect that, from a special point of view, a new and meaningful interpretation of Spencer's axiom becomes possible. If the survival of a taxon depends upon its exemption from competition with a fitter taxon, then obviously no members of a fitter taxon have ever had the opportunity to invade the niche occupied by the taxon in question, either because the latter is dominant or because it is isolated. Hence, we may submit that the following statement:

(12) The survival of the fittest (taxon that ever occupied a given niche) is a formulation of a situation which corresponds to a biological reality.

I hope I have been able in the present section to bring out the point that, although Spencer's maxim has been of great publicity value to Darwinism, the process of evolution is too complex to be covered by such a simple statement. And it is therefore rather pointless to try to evaluate Darwin's theory on this basis.

In population genetics fitness is ascribed to genotypes rather than to populations. The concept of 'genotype fitness' (relative Darwinian fitness) has in turn led to the concept of 'genetic load', referring to the number of deaths a population must sustain to eliminate deleterious genes (Haldane, 1957). Wallace (1968; 1970) has shown that the population geneticists' calculations on the genetic load lead to incongruities, primarily because in their mathematical fervour they forget to think in biological terms. Thus, 'the need to remove excess progeny in order to stabilize population size is generally overlooked in discussions of genetic loads' (Wallace, 1970, p. 78). Hence, the genotype fitness values employed by the population geneticists represent the mean of the survivors of a much larger number of individuals.

Table 5.1
Relative and absolute fitness of polymorphic and monomorphic populations,
based on data in B. Wallace (1968)

	Relative fitness	Equilibrium population (per cent)	Absolute fitness	Fitness of monomorphic population
AA	0·80	56	0·94	1·00
Aa	1·00	38	1·18	1·00
aa	0·40	6	0·47	1·00
Fitness of equilibrium population	0·85 $(0{\cdot}8 \times 0{\cdot}56 + 1 \times 0{\cdot}38 + 0{\cdot}4 \times 0{\cdot}06)$		1·00 $(0{\cdot}94 \times 0{\cdot}56 + 1{\cdot}18 \times 0{\cdot}38 + 0{\cdot}47 \times 0{\cdot}06)$	

In agreement with the definition of survival employed here, Wallace suggests that the fitness of a surviving polymorphic population is 1.00, implying that the average number of progeny equals the average number of parents. The genotype fitness is an index of the relative number of offspring left by the mothers of each of the various genotypes when they occur together in a polymorphic population. Therefore, if one of the genotypes progresses under these conditions, then the corresponding subpopulation must have a fitness higher than, and not equal to, 1.00, as assumed in population–genetical calculations. As shown by Wallace, the correct value is obtained by dividing the relative fitnesses by the relative fitness of the population (Table 5.1).

To this I shall only add that, except for lethal genes, which of course may give population fitnesses of 0, all other monomorphic populations should have fitness 1.00. Relative fitness values are valid only when two or more genotypes occur together, and in this situation the 'fittest' is not distinguished by survival, but by progression, in complete agreement with the conclusion reached above. One difference between Wallace's absolute fitness and the BT, which is here suggested as a measure of fitness, is that the former only deviates from 1.00 during phases of competition, while the latter has been subject to a steady increase during the course of evolution.

EVOLUTION IN TIME

The information embodied in the fossil record relates to the evolution of form in time and in space. As regards the temporal aspect of evolution, the subject of the present section, the recorded data bear on the times at which the several new forms appeared. Since the phylogenetic classification of Vertebrata implies predictions about the temporal succession of the origin of the various taxa, it follows that the classification can be tested through confrontation with palaeozoological observations.

The spatial distribution of living and extinct animals is a function partly of the time and the location at which they arose, and partly of various geographical factors. If the biosphere had been freely accessible to all forms at all times, then the age and the availability of niches would be the only factors preventing the ubiquitous presence of all dominant taxa. However, the existence of various kinds of geographical barriers, which have undergone momentous modifications in the course of the history of our planet, implies that the observable distribution of animals and fossils embodies a temporal aspect besides taxonomic age, namely, the age of the various changes of the terrestrial surface. Hence, even zoogeographical observations can be used to test the phylogenetic classification.

Using data from these two disciplines, such a confrontation will be made in the following sections.

The palaeozoological record

In the present section we shall test separately the classification of cyclostomes, fishes, 'Amphibia' and Amniota. Before we can begin this task it is necessary to establish some rules to be followed in the interpretation of the fossil data.

RULES OF INTERPRETATION

It will be recalled that in the dendrograms depicting the phylogenetic kinship between the various taxa the vertical lines represent those periods of time during which a certain taxon has come into existence. So far, only aesthetic considerations have dictated the length of these lines, and the time-scale, i.e. the ordinate, is consequently completely arbitrary. Therefore, if it is necessary to lengthen or shorten some or all of these vertical lines in order to establish the correct temporal relations between the various taxa, then this expedient is completely justified on the basis of the given premises.

On the other hand, the horizontal lines represent the event of isolation which gave rise to two existing twin taxa and this, being a unique event, must have occurred at a definite point of time; therefore, these lines must always remain horizontal. In other words, the testing of the classification must imply that the dendrogram can be modified so as to conform with the fossil record without slanting the horizontal lines or, of course, changing their relative position.

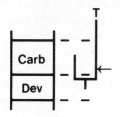

Figure 5.4. Diagram illustrating the principle adopted for the utilization of actual palaeozoological observations in the drawing of dendrograms in which the ordinate represents an absolute time-scale. The arrow indicates oldest recorded find of a member of the taxon T

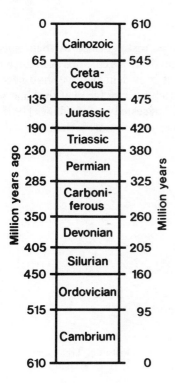

Figure 5.5. The absolute time-scale used in the drawing of the following dendrograms

Another point to note concerns the application of chronological records to the phylogenetic dendrograms. Here it may first be observed that we cannot ascertain all the characters of fossil specimens, only those pertaining to the skeleton. But every taxon is distinguished by a number of characters besides these, and we shall never know whether the animals preserved in the fossil record possessed any or all of these. Hence, even if we were sure that some fossils really represent the first animals in possession of the skeleton of a particular taxon T_j, we cannot be sure that they were actually members of T_j. However, this particular point must be neglected in the present context.

Furthermore, the earliest recorded finds of T_j are not necessarily the fossils of the first animals having the typical skeleton. All we can infer is that animals with the T_j skeleton did not originate later than the fossils. Since, apparently, the precision with which palaeozoology can work does not allow for a discrimination in time exceeding about one-third of the various geological periods, it follows that if the taxon T_j has been recorded first from the Lower Carboniferous then we must allow its possible origination in the Upper Devonian, and thus there may be some justification for drawing the horizontal line representing the separation of T_j from its twin taxon below the line of transition between Devonian and Carboniferous (Figure 5.4). The usefulness of these proposed 'corrections' may be questioned, of course, since they displace the lines of branching without abolishing the sources of error alluded to above.

The time-scale used in the present section is shown in Figure 5.5. It represents a compromise between those published by Colbert (1969) and by Olson (1971). The means of the duration of the various periods obtained from these sources have consistently been raised so that the last digit is 0 or 5.

CYCLOSTOMES

The oldest known vertebrates are the ostracoderms, distinguished by their calcified integumental skeleton (Stensiö, 1958). Since only animals with a calcified skeleton have a chance of becoming preserved in the fossil record, it is evident that we shall never be able to know whether these animals were the first vertebrates, or whether they had some uncalcified ancestors, perhaps similar to *Myxine*, as suggested by Spjeldnaes (1965).

However, if the Arachnid theory should be further corroborated, then this will un-

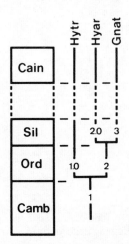

Figure 5.6. Phylogeny of the cyclostomes. Gnat, Gnathostomata; Hyar, Hyperoartii; Hytr, Hyperotreta

questionably turn the scales in favour of the first alternative, implying that we possess, or may be able to acquire, a reasonably good insight into the earliest history of Vertebrata.

If this is true, then it also follows that the two taxa of living cyclostomes are descendants of calcified ancestors and maybe, as claimed by Stensiö (1958) and Jarvik (1968b), are derived from two separate extinct groups. Acceptance of this tenet implies that the power of calcification was lost independently in the two cyclostome taxa. This contention is by no means improbable, considering that loss or reduction of calcification has occurred on several occasions in the course of the evolution of Vertebrata. We may consequently infer that the extinct cyclostomes must be represented by side-branches to the taxa 1, 1·0 and possibly 2 in Figure 4.44.

On the assumption that Hyperotreta are derived from Heterostraci, known from the Middle Ordovician (Stensiö, 1958), we must place the branching of the taxon 1 into 1·0 and 2 in the Lower Ordovician. If, furthermore, we accept that Hyperoartii are derived from Cephalaspidomorphi, as submitted by Stensiö·(1958), then the fact that the latter appeared first in the Middle Silurian implies a splitting of taxon 2 in the Lower Silurian (Figure 5.6).

The reasoning adopted here leads to the inference that the creation of taxon 2 covered a span of 60 million years, a very long time compared to that required for the formation of many other vertebrate taxa. This fact may lead us to suspect some of the premises employed, but in support of the conclusion reached it may be mentioned that, as we have seen above, Hyperoartii are distinguished by the possession of a variety of non-morphological characters in common with Gnathostomata, characters which, we may presume, it took a considerable time to acquire.

FISHES

The earliest known members of Actinistia, Diplocercidae, have been found in the Upper Devonian (Lehman, 1966b) so, according to the proposed rules, the earliest possible origin of this taxon is Middle Devonian. On the other hand, if we presume that Cladoselachii, which are found in Middle Devonian beds, belong to Chondrichthyes, then we must place the branching of taxon 3·0 in the Lower Devonian (Figure 5.7). It appears that there is a discrepancy of one-third of a geological period here between the predictions based on the classification and the fossil record.

The first well-known members of Palaeonisciformes, representatives of taxon 4·0, are found in deposits from the Middle Devonian (Lehman, 1966a), indicating, according to the rules adopted here, an origin in the Lower Devonian. Yet, even this date is too late to be compatible with the proposed classification since, as we shall presently discuss, Dipnoi, representing taxon 5·0, already existed in the Lower Devonian. The branching of taxon 4 into 4·0 and 5 must therefore have occurred no later than the Upper Silurian, as indicated in Figure 5.7. In support of this contention it may be mentioned that W. Gross (1968) has described some scales from the Upper Silurian which, although the ganoine layer is missing, resemble the scales of Devonian Palaeonisciformes. Gross therefore suggests that these scales belonged to very early members of this taxon. It should be noted that the scales are reminiscent in some features of those of some acanthodians and rhipidistians. The oldest finds of Cladistia are, possibly, Upper Cretaceous (Lehman, 1966c), too late to be of any interest in the present context. Whatever views are held concerning the affinities of these fishes, it is generally agreed that they must have originated before the Devonian (Daget *et al.*, 1964; Jarvik, 1968b), and I have therefore placed the branching between taxa 4·00 and 4·01 just below the transition between Silurian and Devonian (Figure 5.7).

The classification of the fossil Chondrostei, 'Subholostei' and 'Holostei' is so ambiguous that it is impossible for a non-specialist to estimate the origin of these taxa. It seems certain that Teleostei arose in the Upper Triassic or Lower Jurassic (Romer, 1966), yet this information cannot be recorded in Figure 5.7 in a formally correct way when the other actinopterygian taxa are ignored.

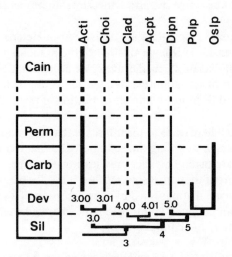

Figure 5.7. Phylogeny of the fishes. Acpt, Actinopterygii; Acti, Actinistia; Choi, Chondrichthyes; Clad, Cladistia; Dipn, Dipnoi; Oslp, Osteolepiformes; Polp, Porolepiformes. The number of taxon 4.0, Cladistia + Actinopterygii, is omitted. The heavy vertical lines represent 'crossopterygian' fishes

The earliest recorded members of Dipnoi are Dipnorhynchidae from the Lower Devonian (Lehman, 1966b). From this it follows that taxon 5 may have already split in the Upper Silurian, but since taxon 4 purportedly branched at this time, I have placed the branching of taxon 5 at the transition between Silurian and Devonian (Figure 5.7).

The split in question must concern Dipnoi and the aquatic ancestors of Tetrapoda. As regards the latter, all specialists agree that these are represented by Rhipidistia, a group of extinct 'Crossopterygii'. The Rhipidistia are subdivided into two groups, Porolepiformes and Osteolepiformes. Since the oldest known form of the latter is *Osteolepis* from the Middle Devonian (Lehman, 1966b), it may be proposed that the split between the groups took place in the Lower Devonian (Figure 5.7).

In choosing Osteolepiformes as a reference for the time of the split between the two rhipidistian groups we neglected *Porolepis*, known from the Lower Devonian (Lehman, 1966b), and it may therefore be presumed that Porolepiformes are represented in the short branch representing the incipient taxon 6 and possibly also in the branch uniting the latter with taxon 5·0 (Figure 5.7). If this is true then, of course, Rhipidistia is not a monophyletic taxon (cf. Jarvik, 1972; Andrews, 1973).

On the basis of the established dendrogram we can now determine the dimensions of the 'crossopterygian' realm. Thus, since these fishes are represented in taxon 3·0, we may infer that they were already existing in taxon 3, the ancestors of all living gnathostomes, at least towards the end of the period of creation of this taxon. They also occur in the presumptive taxon 6, as represented by Rhipidistia, and we cannot exclude that they were also represented in the early stages of the lines leading to Chondrichthyes, Dipnoi and even to Cladistia + Actinopterygii. According to Figure 5.7 it thus follows that 'Crossopterygii' is a synonym for 'extant Gnathostomata'.

From the distribution of 'Crossopterygii' indicated in the figure it is seen that the upper Silurian and the lower Devonian was the time of 'adaptive radiation' of this stock. It also appears that several important events in the evolutionary history of Gnathostomata must have taken place in the Silurian, and since no fossils of members of this taxon have been found before the lower Devonian it must be acknowledged that in this respect the fossil record is incomplete. Yet the phylogenetic classification allows us to give quite a detailed description of what we should expect from the Silurian fossils of Gnathostomata. Thus, these fishes should, of course, have an intracranial joint, they should be achoanate, the maxillaries should be missing, and the teeth should have no plicidentine. Furthermore, they should have archipterygial or archipterygium-like lobed fins, an unconstricted notochord and a heterocercal tail. The only taxa in which animals of this type might occur beyond the Silurian appear to be 3·0, leading to Actinistia and Chondrichthyes and, possibly, Porolepiformes.

If the present classification is correct, it should be possible to allocate the various groups of extinct fishes to their proper places in the diagram. This may appear to be a very ambitious undertaking, but an attempt will nonetheless be made. We may confidently assert that there is no room for the origin of gnathostomous animals, not affiliated to 'Crossopterygii', at any place in our dendrogram except during the early stages of taxon 3. Since in Figure 5.7 the latter covers the time from early to middle Silurian, it follows that such animals should have originated in the early Silurian and, according to the convention adopted here, have been recorded not later than the middle Silurian. If the earliest finds are younger than that it implies either that, as for 'Crossopterygii', the fossil record is incomplete in the Silurian, or else that they have originated from some extinct taxon. Since we cannot distinguish between these alternatives it follows that if and when this situation arises, we must refrain from classifying the animals in question with respect to the extant forms.

Placodermi first appeared at about the Silurian–Devonian boundary (Romer, 1966), and according to the present rules they should consequently have originated at the transition between Middle and Upper Silurian. This is clearly too late to warrant their classification as pre-'crossopterygian' Gnathostomata. And therefore the description of Placodermi as 'a series of wildly impossible types which do not fit into any proper pattern; which do not, at first sight, seem to come from any possible source, or to be appropriate ancestors to any later or more advanced types' (Romer, 1966, p. 33) appears to be quite a close approximation to the truth.

The case is better for Acanthodii, for finds of spines and scales appear to make it possible to trace this group back to the beginning of the Upper Silurian (Romer, 1966). Hence, it may be proposed that Acanthodii are a twin group to Gnathostomata, which originated at the beginning of the Middle Silurian. This classification has been implied or proposed by Watson (1937) and von Wahlert (1968), and rejected by Miles (1973).

The fact that in various features they exhibit similarities with 'Crossopterygii' (Romer, 1966) evidently supports this classification (Figure 5.8). Romer envisages that the prime of these fishes occurred in Ordovician–Silurian fresh waters, but this early origin is not compatible with the time suggested for the origin of Gnathostomata. It is possible that this estimate is wrong, but it should be pointed out that it is supported by data published by Romero-Herrera *et al.* (1973). On the basis of the amino acid sequence in myoglobin, these authors calculated the separation between Hyperoartii and Gnathostomata to have occurred about 420 million years ago, and they proposed that this dichotomy could be equated with the earliest evidence of Acanthodii. According to Figure 5.8 these animals must have originated between 430 and 440 million years ago, an exquisite agreement in the present context.

It is also possible to trace some of the history of taxon 3·01, Chondrichthyes, through the fossil record. I have already submitted that, on the basis of finds of Cladoselachii, it must be presumed that this taxon branched off from Actinistia in the Lower Devonian. The earliest representative of this line is *Cladoselache*, found in Middle Devonian deposits (de Saint-Seine, Devillers and Blot, 1969). The very short snout in this animal testifies to its non-selachian origin. The absence of claspers shows that if indeed *Cladoselache* belongs to Chondrichthyes, then it must be situated in a very early side branch, for all other animals discussed here possess this kind of intromittent organ. One difficulty in the present classification is to account for the broad-based cladoselachian fins. From a formanalytical point of view these represent a 'primitive' form to which the various other fin types can be traced back. Yet 'possibly (although not too probably) these fins are derived . . . from smaller, narrow-based types' (Romer, 1966, p. 39), and this possibility becomes a necessity if the present classification is correct.

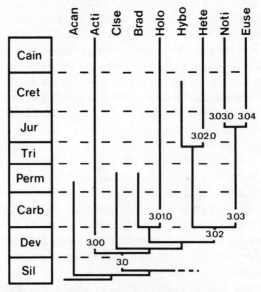

Figure 5.8. Phylogeny of Acanthodii, Actinistia and Chondrichthyes. Acan, Acanthodii; Acti, Actinistia; Brad, Bradyodonti; Clse, Cladoselachii; Euse, Euselachii; Hete, Heterodonti; Hybo, Hybodonti; Noti, Notidanoidei

According to the rules, we have placed the bifurcation between Actinistia and Chondrichthyes in the Lower Devonian. From the Upper Devonian we know of the existence of a group of fishes, Bradyodonti, which were in possession of claspers but, in contrast to all other sharks, earlier as well as later ones, had tooth plates rather than teeth. This implies that in the remaining part of the Devonian we must have two bifurcations, one representing Cladoselachii and clasper-possessing chondrichthyans, and one subdividing the latter into Bradyodonti and toothed sharks. Since the rules dictate that the latter branching must be located in the Middle Devonian, I submit that the first branching occurred around the transition between Lower and Middle Devonian (Figure 5.8). The Ctenacanthidae, closely related to Cladoselachii, and also lacking claspers, display some affinity to the later members of Chondrichthyes (Romer, 1966), and it is therefore possible that they should be represented as a side-branch to the taxon comprising the extant forms.

The Bradyodonti themselves became extinct in the Permian, but before this occurred they had, according to current notions, given rise to the Holocephali. Most of the known fossil members of this taxon are from the Jurassic, but the family Deltoptychiidae, of the suborder Menaspoidei, is recorded from the Lower Carboniferous (de Saint-Seine *et al.*, 1969). If this classification is correct we may, deviating slightly from the rules, place the branching at the transition between Devonian and Carboniferous (Figure 5.8).

In the other branch Romer distinguishes two lines, one comprising the extinct Hybodonti + the extant *Heterodontus*, the other all the remaining living forms. It should be noted that this distinction is not generally accepted (cf. Compagno, 1973). Heterodontidae first appeared in the Lower Jurassic (de Saint-Seine *et al.*, 1969); the splitting of this line from the remaining Hybodonti may therefore be placed at the Upper Triassic. Similarly, the modern sharks, Euselachii, are first found in the Upper Jurassic (de Saint-Seine *et al.*, 1969), and may therefore be separated from the remaining 'archaic' sharks, Notidanoidei, in the Middle Jurassic. Whether or not the latter is a monophyletic taxon is a problem which will not be touched upon here. Since Hybodonti appeared in the lower Carboniferous, we shall also locate this bifurcation at the transition between Devonian and Carboniferous.

As appears from Figure 5.8, the Devonian represents the period of 'adaptive radiation' of the marine chondrichthyans, as was the Silurian for the fresh-water 'crossopterygians'. This circumstance supports the present classification rather than one which alleges an early and independent origin of Chondrichthyes. For the sake of completeness it should be mentioned that no attempts will be made to establish the relationship of the pleuracanths.

'AMPHIBIA'

As indicated by their name, Amphibia represent the transitional link between aquatic and terrestrial vertebrates, the latter represented by Amniota. By adopting this definition we may expect to find amphibian animals not only in taxa 6·0 and 7·0, representing the living forms, but also as extinct side-branches to taxa 6, 7 and 8 in the phylogenetic dendrogram (Figure 4.44). Obviously, if the same name is used for all these groups confusion is sure to arise. It is not uncommon among palaeozoologists to attempt to remedy this situation by introducing the name 'Lissamphibia' to designate the three orders of living amphibians. According to the present classification this is not sufficient since Amphibia is not a monophyletic taxon. We shall, however, ignore taxon 7·0

(Gymnophiona) here and reserve the discussion of taxon 8·0 for the following section; consequently we can focus attention upon taxa 6, 6·0 and 7 in the present context.

Owing to the fact that the aquatic ancestors of Tetrapoda, Osteolepiformes, died out in the Lower Permian, we must expect to find terrestrial forms at least as far back as the Carboniferous. The animals which most convincingly represent this transition are Ichthyostegalia, found in Upper Devonian deposits in Greenland (Jarvik, 1952). We may therefore submit that a bifurcation occurred in the Middle Devonian, one branch representing the continuation of Osteolepiformes, the other leading to Ichthyostegalia (Figure 5.9). Before the latter became extinct, they must have branched off the line leading to the extant Tetrapoda. The animals which according to the present classification gave rise to Amniota + Gymnophiona are known with reasonable certainty to be represented by Labyrinthodontia, a group which thus must constitute a side-branch to taxon 7. The Labyrinthodontia are divided into two orders, Temnospondyli and Anthracosauria. Presuming these two groups to be twins, we must locate their origin in the Lower Carboniferous, since Anthracosauria first appeared in the middle of this period (Figure 5.9).

What is now missing in the dendrogram is the twin of Labyrinthodontia, the line which is assumed to lead to 'Amphibia'. According to more or less general consensus the search for the ancestors of the latter should be directed towards Lepospondyli. If we assume that this subclass is the twin to Labyrinthodontia, we must introduce a further branching of the twin to Ichthyostegalia, and since Lepospondyli can be traced back to the Lower Carboniferous, we shall submit that this dichotomy occurred at the transition between Devonian and Carboniferous (Figure 5.9).

One further bifurcation must have occurred, one line leading to the extinct Lepospondyli, the other to 'Amphibia'. The first certain record of the latter taxon is *Triadobatrachus* from the Lower Triassic, which permits us to locate the separation between

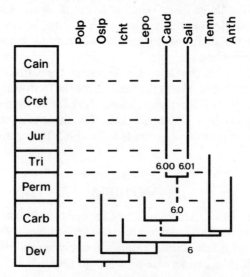

Figure 5.9. Phylogeny of 'Amphibia'. Anth, Anthracosauria; Caud, Caudata; Icht, Ichthyostegalia; Lepo, Lepospondyli; Oslp, Osteolepiformes; Polp, Porolepiformes; Sali, Salientia; Temn, Temnospondyli

Caudata and Salientia in the Upper Permian. Yet the Caudata must have originated much earlier, in point of fact at some time during the Carboniferous, since Lepospondyli disappeared in the Lower Permian. As we have no record which permits us to establish at what time during this span of 60 million years the bifurcation took place, I have arbitrarily located it in the Middle Carboniferous (Figure 5.9).

The palaeozoological data discussed in the present section, as well as the postulated phylogenetic relationships between the various groups, are all based on Romer (1966).

AMNIOTA

The Anthracosauria are supposed to have given rise to the order Cotylosauria, the 'stem reptiles', in which Captorhinomorpha is the most important suborder. These have been found in sediments from the Lower Pennsylvanian, i.e. Middle Carboniferous, thus almost as far back as Temnospondyli and Anthracosauria. We may, however, propose the time of separation between the latter and Cotylosauria to be at the transition between Lower and Middle Carboniferous.

At a very early stage the stem reptiles divided again, giving rise to Synapsida, the mammal-like reptiles, and Diapsida. The primitive synapsids are represented by Pelycosauria, which were present in the Upper Carboniferous, suggesting a time of branching in the middle of this period. The Pelycosauria in turn gave rise to Therapsida, known in abundance from the Middle Permian, indicating that they must have arisen at the latest in the Lower Permian. The Therapsida survived until the Lower Jurassic, but before then they had given off the line leading to Mammalia. The time at which this occurred is not known, but according to Romer (1966) mammals may have come into existence towards the end of the Triassic. In Figure 5.10 the branching has been placed in the

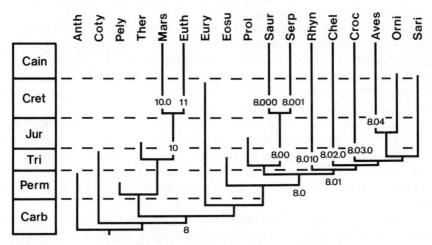

Figure 5.10. Phylogeny of Amniota. Anth, Anthracosauria; Chel, Chelonia; Coty, Cotylosauria; Croc, Crocodilia; Eosu, Eosuchia; Eury, Euryapsida; Euth, Eutheria; Mars, Marsupialia; Orni, Ornithischia; Pely, Pelycosauria; Prol, *Prolacerta*; Rhyn, Rhyncocephalia; Sari, Saurischia; Saur, Sauria; Serp, Serpentes; Ther, Therapsida. This classification is exclusive because taxon 9.0, Monotremata, is missing. It has not been neglected, however, in the numbering of the other mammalian taxa. The numbers of taxa 8.02 and 8.03 have been omitted

Middle Triassic. Since, as will be discussed later, observations suggest that the branching between Marsupialia and Eutheria occurred in the lower Cretaceous, this dichotomy has been introduced in the figure, so consequently Prototheria are missing in the dendrogram.

The line giving rise to reptiles and Aves is represented by the subclass Diapsida, the earliest members of which were Eosuchia. The oldest remains of the latter are found in the Upper Permian, and they may thus have originated in the middle of this period. However, following Romer (1966), I shall submit that, before Eosuchia had arisen from Captorhinomorpha, a branching occurred in the late Carboniferous giving rise to the aquatic and amphibious Euryapsida (Figure 5.10).

According to Romer (1966, p. 127) 'it may be that primitive eosuchians gave rise to the advanced archosaurs; it seems, however, definite that the eosuchians were the direct ancestors of the lizards and the snakes of the order Squamata'. A forerunner to the latter, *Prolacerta*, is found in the Lower Triassic, so we need a subdivision of Eosuchia in the Upper Permian, of which one line leads to *Prolacerta* + Squamata, the other to the remaining reptiles + Aves. The first true lizards appear in the Middle Triassic, and the dichotomy between *Prolacerta* and Squamata may consequently be placed in the Lower Triassic. The first fossil snakes are from the Cretaceous, and accordingly a division between Sauria and Serpentes in the Lower Cretaceous is proposed in Figure 5.10.

The other eosuchian line should, following the present classification, divide into two lines, one leading to Rhyncocephalia, the other to Chelonia + Crocodylia + Aves. Since representatives of the former taxon existed during most of the Triassic, we can place the time of bifurcation no later than at the transition between Permian and Triassic.

With this splitting we should, in principle, leave the Lepidosauria and deal with the 'ruling reptiles', Archosauria. The first forms belonging to this subclass were Thecodontia, which made their appearance in the Triassic, suggesting that they originated at almost the same time as Rhyncocephalia.

If the currently accepted classification is correct, Chelonia should, presumably, be the twin taxon to the remaining reptilian orders + Aves, implying that they are represented by a branch originating in Upper Carboniferous or Lower Permian times. In point of fact, the earliest fossil turtles date back to about Middle Triassic, suggesting an origin in the lower part of this period, a situation which is seen to conform very well with the present classification.

Crocodylia, the youngest of the surviving reptilian taxa, are distinguished by the fact that its members are very little advanced above the level of the original archosaurian stock. For this reason it is difficult to establish the exact time at which the history of the taxon was initiated, but fossils from the Middle Triassic may belong to this taxon, so we shall place the bifurcation at the transition between Lower and Middle Triassic (Figure 5.10).

Before we discuss the origin of the birds we shall try to allocate the dinosaurs to their proper place in the dendrogram. These animals belong to two orders, Saurischia and Ornithischia, both of which date back to the Upper Triassic (Romer, 1966; Thulborn, 1971). It is not quite certain that these orders are twins, but since alternative classifications in the present dendrogram appear extremely implausible, it is supposed here that they arose through a bifurcation from their thecodontian ancestors in the Middle Triassic, whence they survived to the end of the Cretaceous.

The earliest known bird, *Archaeopteryx*, is from the Upper Jurassic, suggesting an origin in the Middle Jurassic. Concerning the ancestors of Aves, Romer (1966, p. 166) states the following: '*Archaeopteryx* was already definitely a bird, but was still very close to the archosaurian reptiles in most structures and was obviously descended from that group. The pelvis is suggestive of relationship to ornithischians, but on the other hand, the limb structure parallels that of the carnivorous saurischians in many regards. Surely the birds arose, independently of either dinosaur stock, from Triassic thecodonts, and there is little in the structure of some of these small bipedal thecodonts to debar them from being avian ancestors.'

Romer's suggestion that the birds originated before the Middle Triassic leaves an extensive lacuna in the fossil record. This is avoided if it is submitted that Aves arose through a bifurcation from dinosaurian ancestors in the Middle Jurassic. As appears from the above quotation, the birds exhibit affinities with both orders of dinosaurs; in Figure 5.10 it has been proposed, in accordance with Galton (1970), that they branched off from Ornithischia. It should be mentioned, however, that a detailed investigation of *Archaeopteryx* has led Ostrom (quoted by Bakker, 1971) to suggest that the birds arose from animals belonging to Saurischia.

VERTEBRATA

A survey of the classification of all the groups that have been dealt with in the present section is shown in Figure 5.11. It is seen that, within the limits of accuracy dictated by

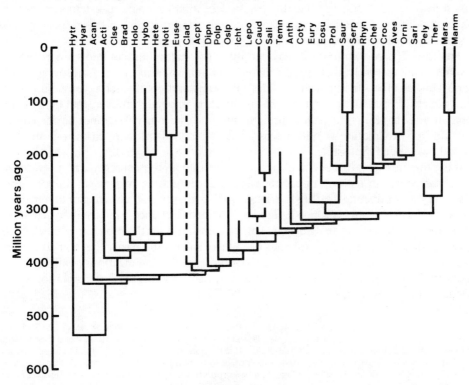

Figure 5.11. Phylogeny of Vertebrata. For the significance of the various abbreviations the reader is referred to the captions to Figures 5.6, 5.7, 5.8, 5.9 and 5.10

the fossil record, it is possible to account for a large number of fossil taxa without having to accept gaps in our knowledge except in two cases, namely, Cladistia and 'Amphibia', besides Gymnophiona and Prototheria, which are not represented in the figure. This appears to constitute excellent support for the present classification and also suggests that the fossil record is much more complete than is usually believed.

The zoogeographical record*

The spatial distribution of living and extinct animals is partly a function of animal evolution and partly of the geographical changes to which the surface of the planet has been subjected in the course of time. There are a number of other factors that may influence the geographical distribution of animals, and since the interaction of these several agents is complex, it will be necessary in this case also to establish certain rules of interpretation for an appropriate evaluation of the data. However, these are not so important in the present context, and this point will therefore be postponed to a following section.

Confining ourselves to the two factors mentioned above, it follows that if the course of evolution is known, zoogeographical observations may be used to test theories concerning the history of tellurian geography, or, conversely, if the latter is known, the data may be used to test theories concerning the course of evolution, i.e. phylogenetic classifications. The reason why zoogeography can thus supply information about evolution in time is, of course, that the dispersal of animals contains a temporal aspect. The main difference between palaeozoology and zoogeography in this respect may be that the fossil record dates much further back in time.

Traditionally, zoogeographers have adopted the standpoint common among geologists until quite recently, that apart from minor, but in some instances zoogeographically important, changes, the present outlines of oceans and continents have remained unchanged as far back as we can tell. This assumption implied, however, that many living animals must have arrived at their present station fortuitously. It cannot be denied that this kind of explanation is both necessary and justified in many cases, but surely it should be avoided if possible.

It has therefore greatly benefited zoogeography that, in the last few years, a revolution has occurred within geology, due to the rapidly accumulating evidence in support of the old theory of continental drift (J. T. Wilson, 1973); this reversion has meant that many of those instances of location which had to be explained as chance events may be accounted for as the outcome of active dispersal. This circumstance partly implies that much of the existing zoogeographical literature has become obsolete, and partly that a great new era is about to be inaugurated in this branch of biology. The publications available at the moment show that many authors are busily revising the present interpretation of the zoogeographical record.

*Croizat has published some very comprehensive studies on biogeography, summarized in (1962). In this work he assumes a critical attitude towards neo-Darwinian interpretations of evolution, based especially on the geographical distribution of animals and plants. It thus appears that Croizat and I are fighting the same combat and one might therefore expect that Croizat would be frequently quoted here and elsewhere in my book. That this is not the case can be explained on two grounds. First, much of the work of Croizat concerns details outside the scope of the present work. Second, with respect to generalizations I do not concur with Croizat. The reason possibly is that I do not understand Croizat (this, at least, is what I have been told by some of his adherents). Unfortunately, that does not change the situation, nobody can expect me to quote work beyond my understanding.

Although the various continental separations involved in the break-up of Gondwanaland and Laurasia are known in outline, the times at which these events occurred, and hence their sequence, are still under discussion. Recent studies (Larson and Pitman, 1972; Hays and Pitman, 1973) suggest that the most important plate motions may have taken place much later than has hitherto been presumed (Figure 5.12). However, there seems to be agreement that the splitting up of Gondwanaland, geographically by far the most extensive aspect of continental drift, began at the end of the Triassic, about 200 million years (my) ago. Unfortunately, none of the branchings in the principal classification (Figure 4.44) occurred as late as that, and in the more complete dendrogram (Figure 5.12) there are only three dichotomies, Marsupialia–Eutheria, Sauria–Serpentes and Aves–Ornitischia, which can be subjected to test on this score. The discussion of the first of these pairs of taxa will be reserved for a later section.

If we make the assumption that New Zealand was separated from Gondwanaland at a rather early stage—the geological sources are not explicit on this point (cf. Dietz and Holden, 1973), but much biogeographical evidence supports this notion—then, barring chance dispersal, we should not expect any snakes on New Zealand. This is borne out and, in addition, the presence among saurian taxa of only the 'primitive' families Gekkonidae and Scincidae (Darlington, 1957) reveals the state of evolution in the taxon Sauria when New Zealand became separated. From the dendrogram (Figure 5.11) we may conclude that this event occurred before the Lower Cretaceous, i.e. more than 125 my ago.

Since snakes are found in all other parts of Gondwanaland, as well as in Laurasia, it must be concluded that, if we can exclude fortuitous dispersal, the break-up of the former occurred after the dichotomy of Squamata, thus not earlier than about 125 my ago. This inference is supported by the views quoted above on the age of the continental drift.

Owing to their great vagility, which strongly increases the possibility of both active and passive dispersal, it is difficult to use the geographical distribution of birds for predictions in the present context. However, the Southern continents, including New Zealand, are distinguished by the presence of the flightless birds known as 'ratites' and, if we can assume that their shortcoming has been a distinctive feature during the whole of their evolutionary history, a point that is subject to dispute (Mayr and Amadon, 1951) but which may be supported by chemical investigations revealing close affinities between these birds (Osuga and Feeney, 1968), then we must infer that they arrived in New Zealand before its separation from Gondwanaland. This will imply, with reference to Figure 5.11, that the latter event occurred after the Middle Jurassic, i.e. less than 165 my ago. This estimate may be biased because the timing of the origin of Aves refers to the first appearance of *Archaeopteryx*.

Apart from some bats, no native mammals occur in New Zealand and, following the same line of reasoning, we may thus infer that Monotremata arose after the separation between this country and Australia.

It must be pointed out that, according to some geologists, New Zealand split off from Gondwanaland some time between the Upper Cretaceous and Early Tertiary (cf. Keast, 1971). If this view is accepted, the little sense which the present attempt to account for the composition of the New Zealand fauna has contributed must be abandoned in favour of the traditional explanations based on the occurrence of chance events. Yet, whichever choice is made, there are many facts, for instance the past presence of palms and absence of dinosaurs, which make this country a 'biogeographical riddle' (Keast, 1971).

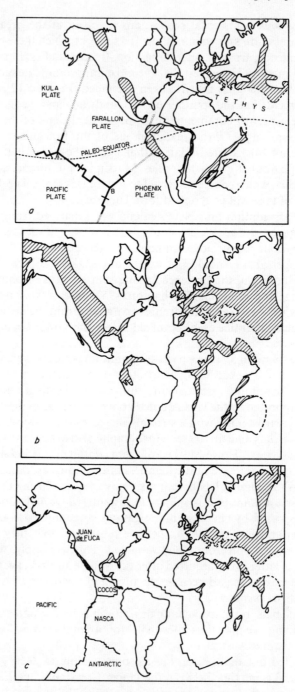

Figure 5.12. Continental drift during the Upper Cretaceous and the Early Tertiary. a, 94–100 my ago; b, 70–85 my ago; c, 40–50 my ago. (Redrawn from J. D. Hays and W. C. Pitman, *Nature,* **246** (1973) by permission of the authors and *Nature*)

It appears that the information about the age of taxa which can be deduced from the geographical distribution of animals primarily concerns ones younger than those dealt with here. Indeed, a cursory inspection of any zoogeographical text reveals that there are numerous classificatory problems awaiting their rational solution on the basis of the theory of continental drift (cf. Cracraft, 1973).

Among the projects which might be rewarding, I would like to mention that the proper phylogenetic classification of various animal taxa present in New Zealand, Australia, India, Madagascar, Africa and South America might contribute to the solution of the puzzle concerning the temporal succession of the separations between these various land masses. This research has been initiated by Brundin in his work (1966) on the trans-antarctic relationships between several taxa of Insecta.

EVOLUTION IN SPACE

While undoubtedly the most important contribution of palaeozoology pertains to evolution in time, zoogeography is, for obvious reasons, of particular value for the study of evolution in space. However, just as there is a temporal aspect involved in the dispersal of animals, so there is a spatial element inherent in the fossil record, and both sources may thus supply significant information about evolution in space. It is, however, rather difficult to keep the two sets of data separate; in most cases observations on recent distribution must be supported by fossil data in order to derive firm conclusions. Hence the discussion in the present section will not be subdivided with respect to these disciplines.

The dispersal of animals is an extremely complex phenomenon, and the evaluation of the available data requires great discrimination. Therefore I propose to discuss briefly some rules of interpretation in the first subsection below.

Since the continents in the Palaeozoic purportedly formed one large land mass, Pangaea, then data on geographical distribution cannot contribute very important information about this phase of evolution. The available information therefore refers particularly to the effects of the enormous changes in planetary geography which occurred as a consequence of the continental drift which began in the Late Mesozoic. The taxa concerned therefore are mostly inferior to those dealt with in the present book. This fact, coupled with the uncertainty that still exists about the timing of the individual stages in the break-up of Gondwanaland, means that I can contribute very little to the discussion of evolution in space, but some few points related to the dispersal of 'Amphibia' and Mammalia will nevertheless be outlined in the following pages.

Rules of interpretation

Darlington (1957) has set down seven zoogeographical working principles. To do so is, I believe, a very sound approach, which Darlington has stated as his first principle. In the present section I propose to state some rules which to some extent are reformulations of those of Darlington.

(1) Zoogeographical work must be based on theories from which falsifiable predictions can be made, and attention should be focused on those data which can be used for testing the theories.

(2) The dispersal of animals occurs through random and non-random events (passive

and active dispersal). The outcome of the former cannot be predicted and is therefore of little interest from a zoogeographical point of view.

(3) If the geographical distribution of some animal taxon is to be used for testing a phylogenetic classification, the history of the surface of the earth must be correctly known.

(4) Zoogeographical work must deal with monophyletic taxa.

(5) The various clues employed by zoogeographers, the number clue, the differentiation clue, the area clue and the continuation of area clue are valid only for dominant taxa which have subsisted for a rather extensive period of time.

As an illustration I may mention the use of the number clue to ascertain the time of isolation between Africa, Australia and South America on the basis of the similarities in their floras and faunas (Keast, 1971). If this is done, one observes that the last two continents have more taxa in common than either of them share with Africa. From this it is tempting to conclude that the latter was separated from both the other continents before they in turn parted from each other. This is probably true, but when it is recalled that African fauna and flora have undergone much more extensive changes than those of the other continents, the possibility exists that the outcome of the numerical analysis is strongly biased because of extinction, and in that case the conclusion arrived at may be wrong.

(6) Like all other scientists, zoogeographers should formulate bold theories which can be falsified through confrontation with empirical facts.

As an example, we may mention the filter bridge theory of Simpson (1940). An instance of a bad theory is one implying that dispersal has involved chance events (rafting). This mechanism, which may be traced back to Darwin (1859), is an absolute necessity for explaining certain zoogeographical facts. But it should always be regarded with distrust and avoided as far as possible because it is a *cul-de-sac* from a scientific point of view.

My aversion to explanations involving random dispersal is shared by Croizat (1962), and for the same reasons. But if I understand him correctly, he wants to explain geographical distribution by a 'law of biogeography' (1962, p. 629). As discussed above, it cannot be denied that a number of biogeographical laws may be formulated, but it seems equally certain that random dispersal has occurred in many cases. It is therefore wrong in this context to adopt the Aristotelian antithesis between chance *or* law; dispersal, as well as many other aspects of phylogenetic evolution, involves both chance *and* law.

Because they ignored the continental drift, the early biogeographers had to rely on random dispersal to an improbable extent. But opting for the opposite view, as Croizat seems to do, implies either an enormous amount of convergent evolution or the existence of geographic conditions in the past for which there is no support at present. Neither of these alternatives seems to lead to more acceptable explanations of phylogenetic evolution.

'Amphibia'

It was observed above that the separation between Caudata and Salientia occurred 'before' the Upper Permian, and this must, presumably, imply that the former arose in the Lower Permian or even earlier. These animals should therefore have had plenty of opportunity to spread over Pangaea. And yet Caudata are today largely confined to the northern temperate zones; the only deviation from this rule is that some members of the taxon occur in South America. There seem to be two possibilities to explain the observed distribution: either Caudata have during most of their history been confined to Laurasia, their presence in South America being the outcome of a relatively recent invasion; or else they were widely distributed over the continents in the past, but have disappeared from the tropics everywhere except in South America. These two alternatives were posed by Darlington (1957), who himself prefers the second. It seems difficult, however, to accept this proposition considering that Caudata are conspicuously absent from Australia and New Zealand. Particularly in the latter location, where so many relicts have survived, it seems unlikely that competition has been so severe that members of this otherwise reasonably successful taxon could not meet the challenge.

The distribution of Salientia is worldwide, implying that they are found on most of the various territories that once made up Gondwanaland. Although random dispersal must be accepted as accounting for some of the observed localizations, there is no doubt that, once the temporal sequence of the several geographical separations is precisely known, we shall possess a powerful instrument for testing the phylogenetic classification of this taxon. In the present context it may suffice to note that the most primitive among living Salientia, represented by the family Leiopelmidae, are found as relicts in North America and in New Zealand. The latter location supports the purported superior rank of the taxon and may also help to establish the age of the extant forms of Salientia.

Mammalia

The history of the geographical distribution of mammals before the era of the continental drift theory was written primarily by Matthew (1915) and by Simpson (1940). In recent years several revisions of this work have been published (e.g. Cox, 1970; Keast, 1971; Fooden, 1972; Jardine and McKenzie, 1972). On the basis of these and various other sources I shall attempt to survey this very interesting problem in the present section.

(1) We may begin by noting that fossils which may be classed with certainty as Prototheria are lacking. Nevertheless, the survival of this taxon in the Australian region in, obviously, random isolation intimates that it was reasonably well established in Gondwanaland before the rifting of this continent began. This presumption is supported by the fact that the earliest known mammals have been found in the Upper Triassic of Southern Africa (Crompton and Jenkins, 1968). Since there is good evidence that Marsupialia, with which they exist in the Australian region, entered there through a different route, it may be inferred that Marsupialia were absent from, or at least not widespread in, Gondwanaland.

(2) Both marsupial and placental mammals were present in North America in the Middle Cretaceous (Slaughter, 1968). Lillegraven (1969) has advanced the reasonable proposition that the common ancestors of the two therian taxa were marsupials, and this leads to the inference that the latter must have arisen not later than the Lower Cretaceous (Cox, 1970); this point is confirmed by amino acid sequence analyses on myoglobin and haemoglobin, suggesting that the marsupial–eutherian divergence occurred about 130 my ago, thus in the early Lower Cretaceous (Air *et al.*, 1971). Most likely this event occurred either in Europe or in North America.

(3) The absence of Theria from Gondwanaland can, under these conditions, be explained only on the assumption that the break-up of Pangaea was well under way when this taxon arose. According to Dietz and Holden (1973), the separation between Laurasia and Gondwanaland was complete by the end of the Triassic. By comparing their timing for the subsequent stages of continental rifting with that claimed in later work (Larson and Pitman, 1972; Hays and Pitman, 1973), it appears that these authors may have located the separation in too early a period, but it should be safe to postulate that it had taken place by the end of the Jurassic.

(4) If the statements in (2) and (3) are true, then the only passage from Western Laurasia to South America available for the marsupials and placental mammals must have gone through Northern Africa. If this is correct, the conspicuous absence of Marsupialia in Africa noted in (1) must be accounted for. However, this may not be too difficult for, according to Cooke (1972), the northeastern corner of Africa was separated from the rest of the continent by a strait—or bay—of the Tethys sea. Under these circumstances there might have been a land connection between Western Europe and South America in the Upper Cretaceous. Since the connection between Europe and North Africa was not permanent, passage by this route must have involved a certain measure of chance. This may suffice to explain why some mammals from the Holarctic zone, for example Multituberculata, did not reach South America. The mammals which in one way or another reached South America in the Cretaceous or in the Tertiary, before it was united with North America through an isthmus, are representatives of Marsupialia (Didelphidae), Condylarthra, Notungulata, Edentata, Caviomorpha and Ceboidae (Patterson and Pascual, 1972). Of these the first two taxa are known to have been present in South America in the Upper Cretaceous, and the next two in the Early Tertiary. This suggests that they may have migrated through North Africa.

It is noteworthy that these four taxa all belong to the primitive mammalian taxon Palaeotherida, while various members of the more advanced taxon Neotherida were present in North America in the Upper Cretaceous, i.e. at the time when or before these animals are supposed to have invaded South America (cf. Hoffstetter, 1970). The migrations of the Palaeotherida proposed here are shown in Figure 5.13. It is possible to explain the absence of Eutheria in Australia on the assumption that the antarctic connection with South America functioned as a filter, allowing the passage of Marsupialia only, but it is equally possible that the latter were the only mammals to arrive early enough to utilize the land bridge between South America and Australia.

(5) The caviomorph rodents and the ceboid monkeys do not appear in the South American fossil record before the Lower Oligocene, and this is so late that great difficulties are involved in accounting for their presence there.

Before we discuss this problem we may let the traditional views be heard. The statements quoted here concern the New World monkeys, but similar opinions are held with respect to the rodents. First the palaeozoologist: 'More important is the fact that by including the monkeys of both South America and the Old World in a common unit, we imply that we are dealing with a natural group and that they had a common monkey ancestor. This is highly improbable. It seems certain that the ancestors of the South American monkeys reached that continent from the north. But no North American primate ever evolved beyond the tarsioid stage, and hence New World and Old World higher primates represent two independent lines of advance and should be classed as distinct groups' and 'presumably the platyrhines gained entry from the north by "island hopping" down islands which may have existed in the Central American and Panamanian region' (Romer, 1966, pp. 220–1).

And the zoogeographer: 'Nor is the problem insoluble. There are too many solutions.... Monkeys may have reached South America through North America in spite of the absence of fossils there. This, I think, is the most probable solution. If the group was then forest living and arboreal, it would have been the less likely to leave fossils in North America and the more likely to reach South America across narrow water gaps. Or South American ... monkeys may ... be products of parallel evolution, and not directly related to their Old World counterparts' (Darlington, 1957, p. 361).

It appears that these authorities offer us two solutions, namely, that the various similarities between Old World and New World monkeys are all due to convergence, or that the monkeys may have migrated from the Old World to South America through North America, without leaving any fossils there.

(6) As I have stressed on several occasions, it is unwise to use convergence as an explanatory expedient, unless absolutely unavoidable, because it rules out the use of deductive reasoning in the solution of our problems. Nevertheless, since convergence is a factual possibility, it is necessary to investigate the extent to which it is supported by age determinations through chemical methods.

On the basis of immunological data Sarich (1970) has come to the conclusion that Primates arose 65 my ago, at the transition between the Cretaceous and the Palaeocene. This value must be an underestimate, for Primates are found in the Upper Cretaceous in North America (Hoffstetter, 1970b), but the value can hardly exceed 75–80 my. Judging from the palaeogeographic data (Figure 5.12), the possibility exists that South America and Africa were not widely separated from each other at that time, and therefore it cannot be excluded that some prosimian primates could have reached the former

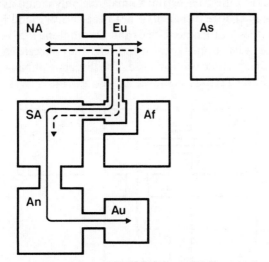

Figure 5.13. Dispersal of Marsupialia (——) and Palaeotherida (– – –) from the Middle Cretaceous to the Early Tertiary. Af, Africa; An, Antarctic; As, Asia; Au, Australian region less New Zealand; Eu, Europe; NA, North America; SA, South America

continent without too much 'chance dispersal' being involved. However, in that case these animals should have accomplished two equally improbable feats, to reach the simian stage independently of the identical event occurring simultaneously in Africa, and to do that without leaving a trace for about 35–40 my. But this solution seems to be excluded by the fact that immunological data suggest a time of separation between Old and New World monkeys of about 35–40 my (Sarich, 1970). The latter value corresponds to the Lower Oligocene, the time at which the primates first appear in the South American fossil record (Hoffstetter, 1969). This also indicates that the ages calculated by Sarich are somewhat of an underestimate, but in any case it seems that these results, together with other data discussed above, lead to the conclusion that Primates most probably reached South America from Africa, and that this occurred not earlier than the late Eocene.

It must be pointed out, however, that since fossils of Old World monkeys date also from the Lower Oligocene (Hoffstetter, 1969), the passage might just as well have occurred in the other direction, except that the complete absence of prosimian Primates from South America appears to rule out this alternative.

(7) The data pertaining to the caviomorph rodents are not so comprehensive as those dealing with the Ceboidae. They appear in South America at the same time as the monkeys (Hoffstetter and Lavocat, 1970). The similarities between the African and South American Hystricomorpha are, however, so close in various anatomical details that a phylogenetic affinity is indicated (Lavocat, 1969). And once more it appears that the common ancestors to these animals cannot have originated before the Early Tertiary, a date which thus fixes the lower time limit for the passage between the continents.

(8) At the time these representatives of Neotherida arrived, South America was definitely separated from Africa by an Atlantic Ocean which, although narrower than today, must have been quite wide. For these terrestrial animals the only means available to cross this barrier must have been chance dispersal, if not necessarily 'island hopping'. So far we must accept the hypotheses of Romer (1966) and Darlington (1957) quoted above. That many other mammals existed in Africa at the time they succeeded in their adventure supports the notion that fortune was involved. The dispersal of Neotherida in the Upper Cretaceous and the Early Tertiary is shown in Figure 5.14.

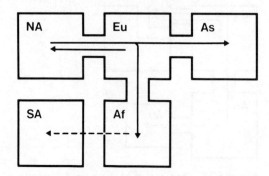

Figure 5.14. Dispersal of Neotherida from the Upper Cretaceous to the Early Tertiary. The broken line indicates the fortuitous crossing of the South Atlantic by members of Rodentia and Primates

MECHANISM OF EVOLUTION

In the present section we shall discuss three points concerned with the mechanism of evolution, viz., the extinction, the dispersal and the origin of animal taxa.

Extinction of taxa

It follows from Darwin's second axiom that inhabited niches are always satiated and hence that the progression of one form must entail the regression and extinction of another one. Yet if evolution is essentially an intraspecific event, then this successive replacement of form must be envisaged as a very slow process which in principle may be confined to the same niche, provided, of course, that the niche is gradually undergoing the changes which are required for natural selection to carry out the work ascribed to it. Except for the difference in opinion as regards the origination of innovations, this is essentially the mechanism of evolution which is assumed to occur in a niche occupied by a dominant taxon according to the present theory.

The latter does, however, allow the possibility that a taxon may invade niches occupied by completely different species or higher Linnean taxa and cause their extinction. However, extinction may also ensue as the consequence of a failure to adjust the physical tolerance to the variability of the niche. An example of this seems to be provided by the extinction of the dinosaurs. This problem has been the subject of several recent papers, but it will be interpreted here with particular reference to the present theory of evolution.

The reptiles that were ultimately to lead to the Mammalia, the pelycosaurs, branched off from the ancestors of the extant reptiles and birds in the Carboniferous. As far as can be told from the fossil record, they must have been the dominant terrestrial animals throughout the Permian and Triassic (Figure 5.15), suggesting that the ancestors of Squamata, Rhyncocephalia and Chelonia belonged to isolated taxa. However, with the rise of Archosauria, particularly the dinosaurs, they clearly lost this position. This raises a very interesting problem, namely, the origination of a dominant taxon from an isolated one. On the basis of the present theory, the available data must be interpreted by postulating that the ancestors of the dinosaurs acquired an increase in BT leading to their invasion of niches occupied by the mammal-like reptiles. Under these circumstances the latter had only two options, to become non-randomly isolated or extinct. As evidenced by the fossil record, both expedients were adopted. Thus, all the large forms of these animals are found to disappear (Figure 5.15), evidently because they occupied niches which the dinosaurs could invade. However, a number of small animals survived, and on various lines of evidence Bakker (1971) suggests that they were nocturnal, a habit which was made possible by thermal insulation provided by an outer covering of hair. Hence, 'the survival and success of mammals during the Mesozoic was probably due to their invading an adaptive zone—small-body-size nocturnality—the dinosaurs could not enter and that lizards could exploit only in part' (Bakker, 1971, p. 656).

Evidently, the disparity in BT between the dinosaurs and the mammal-like reptiles was so large that the latter had no possibility of restoring biological equilibrium except through isolation. As far as this parameter (BT) is concerned, evolution had clearly with the dinosaurs reached a level that was unattainable for any animals living in the Mesozoic. It may therefore be inferred that a difference in PT to the disadvantage of the

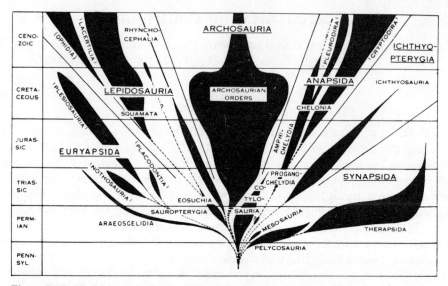

Figure 5.15. Phylogeny of reptiles. The width of the various lines indicates the relative abundance of the taxon in question. It is easily seen that the progression of one taxon is associated with the regression of one or more of the others. (Reproduced from A. S. Romer (1966), *Vertebrate Palaeontology* (3rd ed), by permission of The University of Chicago Press)

dinosaurs, associated with extensive changes in one or more parameters in the physical variability of the environment, caused the decline of the ruling reptiles. Since we know with fair certainty that the dinosaurs were endothermic (Bakker, 1971), there is reason to accept the proposition that a change towards a colder climate was their downfall (Russell, 1965). In accordance with this notion, apart from Crocodylia, the only Archosauria which have survived to the present day are Aves, which, like the ancestors of the mammals, possessed means to prevent excessive heat loss. Thus, the subsequent divergence of Mammalia may, it seems, be exclusively the outcome of climatic changes; had the latter not taken place, our ancestors might have continued their existence as a small and insignificant taxon, which survived thanks to non-random isolation through specialization.

The phenomenon of extinction, disclosed by the fossil record and also, incidentally, by observations made on recent animals, shows that very often one animal taxon may, through competitive selection, be replaced completely by one with which it is more or less distantly related. Even though drawings of this kind may not be quantitatively correct, this type of event is readily borne out by Figure 5.15.

Such phenomena may not be specifically ruled out by neo-Darwinism. It is, nevertheless, somewhat of an embarrassment to this theory that animals which are 'adapted' to a particular niche cannot survive in competition with animals which have been 'adapted' to a completely different environment.

Dispersal of taxa

Darlington (1957) has summarized the process of 'evolution of pattern' which can be derived from zoogeographical data. This question will be discussed in the first subsection to follow. Subsequently, an evaluation of Willis' theory (1922) on 'Age and Area' will be presented.

PATTERNS OF DISTRIBUTION

According to Darlington (1957), the zoogeographical record testifies to repeated dispersal cycles, involving the origination of new taxa which undergo divergence as they disperse. In association with the progression of such taxa others undergo regression and extinction, unless they happen to become randomly isolated. This phenomenon is explained as the outcome of competition between a dominant and a non-dominant, non-isolated taxon, and in terms of the present theory it implies that innovations have originated which increase the BT of the members of the new taxon over and above that distinguishing the old one.

But there is also evidence that dispersal may occur without extinction, namely, when it involves the invasion of new, hitherto uninhabited areas (environmental compartments), for instance inclement climatic zones. This exemplifies non-random isolation, following specialization involving an increase of the PT. Spreading into new areas is, of course, only a particular case of the more general notion of 'invasion of new niches' which is presented as a consequence of specialization.

Darlington (1957) suggests that the main directions of evolution originate in the Old World tropics, spreading from there towards all parts of the world. This observation is interpreted on the basis of the axiom that the origin of mutational innovation is proportional to the number and size of the taxa present in a given region. This, I think, is a very reasonable proposition which might be incorporated in the present theory.

From the preceding discussion it appears that, in principle at least, I accept the concept of 'centre of origin'. In fact, it is a consequence of the theory of evolution which I support that *every phylogenetic taxon, whatever its rank, has originated in a singular geographic location, in a numerically very small population.*

Croizat (1962) and Croizat, Nelson and Rosen (1974) have argued against the notion of 'centre of origin'. If the latter implies that one may look for the origin of a taxon somewhere near the midpoint of the present area of distribution, then, of course, this criticism is justified. In fact, many things may happen in the course of the dispersal of a taxon, both in the physical and the biological extension of the environmental compartment. It is therefore very optimistic to expect that present distribution can give any clear ideas about the geographical origin of a given taxon. But this does not imply that there has not, at some time in the past, been a site of origin for every taxon.

Also, the various instances of random transfer of animals between different environmental compartments corroborate the notion that in isolation biological equilibrium may be established at different levels of PP, so that when the transfer involves a taxon which is dominant in the new location it may successfully invade niches occupied by endemic taxa or, if it is specialized, occupy unexploited niches.

At this stage it may be appropriate to discuss one of Darlington's (1957) clues, that pertaining to continuity of area. Darlington infers that a taxon whose included taxa are discontinuously dispersed cannot be dominant in its environmental compartment. It is probably correct to state that if the discontinuity has arisen through the fortuitous creation of physical barriers in the course of time, then the taxon may be non-randomly isolated, and hence may, but need not, be dominant within its realm. If this explanation does not apply, then the taxon is probably randomly isolated. The discontinuous distribution of Gymnophiona and Dipnoi can be accounted for by continental drift, yet the distinct differences in divergence and dispersal between the two taxa suggest that the former are non-randomly, the latter probably randomly, isolated.

Indeed, rather than discontinuous distribution, it seems that confinement to a narrowly circumscribed area is the best indicator of a randomly isolated taxon, as exemplified by *Sphenodon* and *Latimeria*.

AGE AND AREA

Willis' theories (1922) on 'Age and Area' and 'Size and Space' have been stated by him as follows: 'The area occupied . . . by any group of allied species . . . depends chiefly . . . upon the ages of the species of that group' (p. 63) and 'On the whole . . . the larger families and genera will be the older, and will therefore occupy the most space' (p. 113).

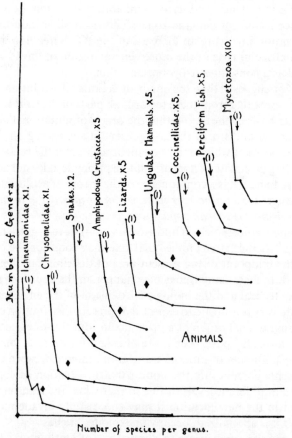

Figure 5.16. Hyperbolic relation between size and number of genera in various animal families. Each curve has been displayed from the origin by a distance indicated by the diamond nearest to the left of the curve. (Reproduced from J. C. Willis (1922), *Age and Area*, by permission of Cambridge University Press)

Willis based his case on the demonstration of certain remarkably regular numerical relations obtaining between distribution and taxonomic size, relations which clearly must have some significance. In spite of this fact, his work has been largely ignored by biogeographers, except for a very aggressive criticism of Willis published by Fernald (1926) which, although undoubtedly correct in certain aspects, fails to distinguish between Willis' observational facts and his interpretation of them.

It appears that the 'Age and Area' theory may be dismissed rather easily, for Willis introduced the premise that the rate of dispersal is the only factor limiting the area occupied by a species, and the quoted statement is a direct conclusion of this. It may be observed that Willis (1922) presumes that a taxon can spread freely, and this is in agreement with the predictions of the present theory, provided the taxon is dominant, and in that case the dispersal will be compensated by the suppression of some other taxa. The latter clause is not accepted by Willis, because it must mean that some taxa occupy a small area because they are old ones on the way to extinction, a contradiction of his theory. Willis does not exclude this possibility, but considers it of no significance. The problem arising through the rejection of Darwin's second axiom about the satiation of the biosphere was 'solved' by Yule (1924) through the proposition that, as the number of species increases, the number of individuals in each is correspondingly reduced. In fact, the number of taxa has increased in the course of evolution (Valentine, 1969). Since taxonomic divergence presupposes isolation, it may be inferred that in the majority of cases the origination of new taxa has been based upon a specialization which makes some hitherto unexploited niche inhabitable. Furthermore, it may be anticipated that some of the evolutionary progress which has occurred in plants has involved better utilization of the solar energy reaching the surface of the earth. These factors taken together show that it is unnecessary to presume that the observed taxonomic divergence has involved a reduction in the size of the individual taxa. In any case, Yule's (1924) proposition is contradicted by the observed scope of extinction.

In fact, one basic flaw in Willis' reasoning is that he completely neglects the phenomenon of dominance. From the preceding discussion it appears that in any phylogenetic hierarchy the dominant taxon is one of the twin taxa of highest numerical rank, and thus one of the youngest. Owing to its dominance, we may also expect it to be the largest, or one of the largest, taxa. This is a direct falsification of Willis' theory.

Yet in a certain sense there is some truth in Willis' theory, for it is in fact valid for a phylogenetic line. This is seen from the following reasoning.

The areas occupied by each of a set of twin taxa T_{j+1} never coincide completely.

Since the taxon T_j comprises both twin taxa, the area occupied by it must be the sum of those occupied by only one T_{j+1}, and the area which may be common to both. Hence:

The area occupied by the taxon T_j is larger than the one occupied by the taxon T_{j+1} in the same phylogenetic line.

Since we have found that in a phylogenetic line the taxon T_j is older than the taxon T_{j+1} (T2.29), it follows:

In a phylogenetic line the area occupied by a taxon increases with its age.

It will never be known if Willis would appreciate this vindication of his theory.

By and large, the theory on 'Size and Space' fares a little better, and it is also much better substantiated. As shown in Figure 5.16, Willis has demonstrated that within a certain family there is an inverse relation between the number and the size of the genera, i.e. there are many genera with one or a few species, while there are few genera with many species. This is not what is stated in the above quotation but, since Willis has in fact shown that the largest genera occupy the largest areas, his statement is evidently correct, apart from the reference to age.

The correlation between size and area implies that all species must occupy commensurable areas. Since this would require an enormous regularity with respect to the factors which, according to Darwinian theories, are responsible for the appropriate environmental adaptation of each individual species, Willis claims, rightly in my opinion, that this kind of correlation may constitute a falsification of the theory of natural selection. Instead, Willis (1922) suggests that the process of dispersal is itself associated with taxonomic divergence, and this contention is identical with the one which I have called Willis' theorem.

Willis also postulated that the characters acquired during this process of divergence through dispersal may involve single large-step mutations, and also that evolution proceeds from the top to the base of the phylogenetic hierarchy (Willis, 1940), together with various other points which conform with the present theory.

Origination of taxa

It might be possible to discuss many different problems in the present context, but I shall take up only one question, evolutionary innovations which, according to the present theory, may involve far-reaching modifications in form and function, as well as other properties.

This stand is in direct contrast to Darwin's theory, in which one of the fundamental premises is that evolutionary modification results from the directional accumulation of the small variations existing between individual members of a species. Since genetics is concerned with the inheritance of these variations, adoption of this premise in neo-Darwinian theory met with no difficulties.

Yet, as already discussed, many biologists in the course of time have contended that some of the decisive evolutionary alterations must have occurred in large steps. Among the first to criticize Darwin on this point was Mivart (1871), who quoted numerous examples where the incipient stages of an innovation, for instance the bat's wing, could not possibly be of selective advantage, but rather the opposite. Consequently, Mivart argued, such features must have arisen as one-step modifications. This postulate in no way falsifies a theory of evolution involving selection, but unfortunately Darwin could not see this, and therefore felt compelled to reject it. He was, however, in an awkward situation, knowing very well from his experience with cultivated plants and animals that large-scale modifications (sports) occur from time to time. This difficulty he circumvented with the following postulate: 'But as species are more variable when domesticated or cultivated than under their natural conditions, it is not probable that such great and abrupt variations have often occurred under nature, as are known occasionally to arise under domestication' (Darwin, 1885, p. 201).

Whether true or not, this is hardly a counterargument, for given sufficient time even highly improbable events may take place. Darwin therefore continued his plea in the following way: 'My reasons for doubting whether natural species have changed as abruptly as have occasionally domestic races . . . are [that] according to our experience, abrupt and strongly marked variations occur in our domesticated productions, singly and at rather long intervals of time. If such occurred under nature, they would be liable . . .

to be lost by accidental causes of destruction and by subsequent inter-crossing; and so it is known to be under domestication, unless abrupt variations of this kind are specially preserved and separated by the care of man' (pp. 201–2).

Thus, Darwin does not exclude large-scale mutations in Nature, but he does not believe they can be preserved through inbreeding. But how can we be sure of that? In my opinion it would be a most likely event if, say, an innovation entails that a new environmental compartment may be invaded. At any rate, it would seem that one single instance of inbreeding occurring in Nature would suffice to falsify this hypothesis, unless it is required that the inbreeding should concern organisms possessing innovations. In that case it might be difficult to falsify the hypothesis because innovations are so rare, but then, on the other hand, Darwin's postulate approaches an *ad hoc* hypothesis.

The neo-Darwinians have a similar attitude towards large-scale mutations. This was stated by Fisher (1958, p. 44) in what may be called *Fisher's axiom*:

'A considerable number of such [large-scale] mutations have now been observed, and these are, I believe, without exception, either definitely pathological (most often lethal) in their effects, or with high probability to be regarded as deleterious in the wild state.'

There are three points to be noted here. First of all the positivist position. Referring to Figure 1.1, it may be acknowledged that the statement is empirically correct, but theoretically irrelevant as regards the extremely rare beneficial mutations which purportedly have revolutionized phylogenetic evolution. Secondly, *prima facie*, Fisher's axiom and T5.29, which asserts that large-scale, positive mutations are possible but extremely unlikely events, are not incompatible. It would, nonetheless, be incorrect to postulate that this formal agreement is of any practical significance, for while T5.29 implies that these rare events are of great evolutionary significance, the basic intent of Fisher's statement is that large-step mutations cannot be ascribed a role of this kind. Thirdly, the axiom is a typical *ad hoc* hypothesis. Its main purpose is to preserve the neo-Darwinian theory, based on Mendelian population genetics. For if the highlights of evolutionary change are mutations occurring with extremely low frequency but having very high selection coefficients, then the neo-Darwinian equations become of minor importance for the explanation of phylogenetic evolution.

We thus face two opposing theories, one contending that evolution has involved a relatively limited number of large mutational steps plus, indisputably, a large number of small ones, the other that only the last possibility obtains. The usual way to settle the issues involved in scientific disputes of this kind is to confront the predictions of the respective theories with empirical observations. Unfortunately, since we are dealing with theories of evolution, a process which, as we know, has lasted for millions of years, both parties may with some justification claim that the time during which relevant observations have been made is too short to allow demonstration of the evolutionary changes anticipated by their theory.

It therefore appears that if we are ever to succeed in testing these theories it must be on the basis of the already existing products of evolution, i.e. extinct and extant animals and plants. We have already, in the second chapter, discussed various chemical innovations, the origination of which must be regarded as large-step innovations. Yet since most debate on evolution has been concerned with morphological characters it would seem natural to confine the present discussion to the latter. Among morphological characters we may distinguish between quantitative and qualitative properties. As regards the latter, it has been argued above on several occasions that relatively trivial changes in the interplay between the epigenetic mechanisms which are responsible for the execution of ontogenesis may entail very large modifications, but as long as we know so little about these mechanisms such arguments may not carry much weight.

The situation is simpler with respect to the quantitative aspect, particularly the body size of animals, because this is a property which can easily be ascertained through measurement. We may thus propose two alternatives to account for the observed variation in animal body size, either that the changes occur in many small increments or decrements, as envisaged in the Darwinian theories, or that few large steps have been involved.

As they stand, neither theory makes any specific predictions about the distribution of body sizes that might be expected in the various taxa of the animal kingdom, but in regards to the first theory it appears that if the evolution of the species included in a particular taxon is still progressing towards the respective optimal body sizes, then the latter parameter might be anticipated to be more or less evenly distributed between the observed maximum and minimum values. If the body size is of little adaptive value, the same situation might be expected, even if temporary equilibria had been reached. The neo-Darwinian theory does not predict, but it does not exclude either, that certain body sizes are optimal and that, consequently, when equilibrium is reached, the various species within a higher taxon should be grouped with respect to the body size of their members. If the change in body size occurs in large but irregular steps, then a more or less even distribution of body size might be anticipated while, if the modifications obey some rule, a grouping of species might be anticipated.

So far the predictions of the two theories appear to be quite similar, but there is one substantial difference. For if this rule predicts that the ratios between the means of the various groups exhibit simple mathematical relations, this prediction can be tested empirically. Certainly, it cannot be excluded that, should the optimal body sizes, which according to Darwinian notions must be reached through natural selection, fall in groups, the mean values of the latter might stand in a simple mathematical relation to each other. This, however, cannot be predicted by a theory of population genetics and would require the introduction of further axioms in neo-Darwinism.

As regards the large-step theory, a grouping may be the outcome of the mechanism of size variation itself if we introduce the postulate that, at least in certain animal taxa, the phylogenetic variation in body volume always occurs in steps which are powers of two, upwards or downwards, as the case may be. If this is correct we should expect that, when the species within a given taxon are arranged with respect to their average body size, a grouping should appear such that the ratio of the mean values of two successive groups is a power of two if reference is made to volume or weight or of $\sqrt[3]{2} = 1 \cdot 26$ if length is used.

It will readily be envisaged that attempts to test this proposition face a number of difficulties, particularly with respect to the availability of appropriate data. Nevertheless, it has been found that a remarkable agreement obtains between prediction and observation (Løvtrup, Rahemtulla and Höglund, 1974) when the member species of various taxa of Mammalia and Aves are subjected to a cluster analysis designed to establish the significance with which a set of data can be arranged in groups (Engelman and Hartigan, 1969).

It is not possible here to present all the results obtained, but those representing one of the mammalian and one of the avian taxa are shown in Tables 5.2 and 5.3. In addition,

Table 5.2
Cluster analysis on species of Carnivora

Species group	Length (mm)				Significance of group	Relative body size
	Observed mean value	Standard deviation	Expected value	Ratio E/O		
1–15	311	55	[311]	[1·00]	0·95–0·99	1
—	—	—	392	—	—	2
16–30	482	41	494	0·97	0·95–0·99	4
31–53	643	58	622	1·03	0·95–0·99	8
—	—	—	784	—	—	16
54–61	889	69	988	0·90	0·95–0·99	32
—	—	—	1244	—	—	64
62–73	1410	160	1568	0·89	> 0·99	128
—	—	—	1976	—	—	256
74–79	2317	242	2489	0·93	> 0·99	512

Source: E. P. Walker (1968), Vol. 2.

Table 5.3
Cluster analysis on species of Piciformes

Species group	Length (in)				Significance of group	Relative body size
	Observed mean value	Standard deviation	Expected value	Ratio E/O		
1–20	8·1	1·3	[8·1]	1·00	0·95–0·99	1
—	—	—	10·2	—	—	2
—	—	—	12·9	—	—	4
21–22	16·5	2·1	16·2	0·98	0·95–0·99	8

Source: C. S. Robbins, B. Bruun and H. S. Zim (1968).

it may be mentioned that only in one taxon, Pinnipedia, were the body sizes found to vary in the ratios 1:2:4:8; in all other taxa gaps of one or two steps were found—thus in Primates the ratios were found to be 1:8:64:512.

It thus appears that the postulate made above is corroborated. This finding does not, of course, falsify Fisher's axiom, for neither this nor any of the predictions of the Darwinian theories are concerned with the body size of animals except for the implication that, whatever distribution of this parameter we happen to observe, it must represent the optimal 'adaptation' under the given external conditions, i.e. that the recorded values must represent the 'best of all possible body sizes in the best of all possible words'. And then, by introducing the hypotheses that in Primates the optimal body sizes are represented by 1:8:64:512, in Pinnipedia by 1:2:4:8, etc., the prevailing state of affairs is 'explained' on the assumption that, according to Fisher's axiom, these body sizes have been reached through the laborious accumulation of minuscule size variations in processes which have all but reached completion.

Unfortunately, this expedient involves the serious error of using *ad hoc* explanations. Thus it will be realized that the amendments to the neo-Darwinian theory proposed to account for the observed body sizes do not make any predictions except that the optimal body sizes stand in the observed quantitative relations to each other. But they do not explain why just these sizes happen to be 'optimal'. If the present case was an isolated phenomenon it might be dismissed but, as discussed above and in what follows, the same approach is frequently used when evolutionary facts have to be explained with reference to the neo-Darwinian theory.

The consequences of the two theories may also be tested with regard to extinct animals. Clearly, the theory proposed here predicts that the same body size relations must also obtain in this case, whereas the neo-Darwinian theory presumes that all the intermediate steps on the road to the optimal body sizes have become extinct. We have already seen that the prediction is corroborated by various members of Pleistocene Cervidae (Figure 5.2), and the same holds for the size distribution of extinct and extant horses. Thus, the outcome of a cluster analysis of data compiled by Robb (1935) shows that the agreement between the values predicted by the above postulate and the observed ones is quite good (Table 5.4), although the significance of the groups is inferior to that recorded in Tables 5.2 and 5.3. This might be expected, since the data in Table 5.4 represent specimens, not means of measurements on several individuals. Another set of data of interest in the present context is the estimated weights of various extinct camelids published by Jerison (1971). Being too few to be subjected to cluster analysis, they are represented in Figure 5.17. Even here the various values are seen to be close to those predicted by the postulate.

The quoted observations may be accounted for by either one of two distinctly different explanations. First, if the evolution from *Hyracotherium* to *Equus* has occurred in 300 Simpsonian steps (Simpson, 1944), then evidently the fossil record is very incomplete. Yet it seems that anyone who supports this interpretation must also explain the fact that, against all odds, the recorded finds happen to fit closely to a geometric scale of the power

Table 5.4
Cluster analysis on extinct and living members of Equidae

Specimen group	Skull length (mm)				Significance of group	Relative body size
	Observed mean value	Standard deviation	Expected value	Ratio E/O		
1–2	125	14·1	[125]	[1·00]	0·50–0·75	1
—	—	—	157	—	—	2
3–5	185	6·4	198	1·07	0·75–0·90	4
6–7	225	7·1	250	1·11	0·75–0·90	8
8–9	325	5·7	315	0·96	0·50–0·75	16
10–25	498	78·4	500	1·00	0·95–0·99	32

Source: R. C. Robb (1935). Questionable measurements not included.

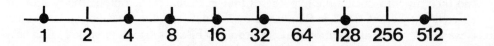

Estimated relative weight

Figure 5.17. Estimated relative body weight of various extinct camelids. 1, *Protylopus*; 4, *Eotylopus*; 8, *Poëbrotherium wilsoni*; 16, *Poëbrotherium labiatum*; 32, *Protolabris*; 128, *Procamelus*; 512, *Camelops*. Based on data from H. Jerison, *Amer. Nat.*, **105** (1971)

of two. Whatever the probability of this happening, it must be so small that it requires faith rather than sense to uphold the neo-Darwinian view. The other possibility is that the theory proposed above is corroborated by palaeozoological findings.

It was known to Darwin that the fossil record does not corroborate the mechanism of evolution at all, and he explained this on the assumption that the fossil record is incomplete. If the conclusion just arrived at is correct, then it follows as a corollary that the fossil record, without being complete, of course, is far less incomplete than asserted by Darwin and his followers. Should this be granted, then it appears to follow that the palaeontological finds suffice to demonstrate that evolution has proceeded in large steps even on the qualitative level, a view expounded by various palaeontologists, for instance Schindewolf (1950). In this way, many observations which have been embarrassing to morphologists and palaeozoologists facing the decrees of Darwinism may find a simple solution.

As regards the consequences of this proposition for Darwinism, I will let Darwin himself pronounce the verdict: 'He who rejects this view of the imperfection of the geological record, will rightly reject the whole theory. For he may ask in vain where are the numberless transitional links which must formerly have connected the closely allied or representative species, found in the successive stages of the same great formation?' (1885, p. 313).

If thus there are serious grounds for doubting the validity of Fisher's axiom with reference to body size variations, there are some other fields where it has also been disproved. One of these is so-called 'quantum' evolution (Simpson, 1944), referring to the number of digits, segments, etc., parameters which must obviously vary by discrete units. A further example is found in the growth of molluscan shells. Thompson (1942) showed in outline how the formation of these can be described mathematically. In a recent paper (Løvtrup and von Sydow, 1974) it was reported that when the shape of the generating curve, i.e. the shape of the body of the animal, or perhaps rather of the mantle edge, and a parameter allowing for accretional growth at the mantle edge are given, the parameters shown in Figure 5.18 suffice to describe the formation of the shells of all living molluscs (cf. however, Løvtrup and von Sydow, 1976).

A number of different gastropod shells are shown in Figure 5.19, with the values of the parameters indicated. The range covered by the figures does not represent all existing shells in which the generating curve is a circle or part of a circle, but it unquestionably encompasses a considerable fraction of them. In view of the relatively slight variations in the three parameters required to generate these shells, it must be concluded that small mutations with respect to the values of the parameters may have large effects with respect to shell morphology.

At this point I shall return to another objection raised by Darwin against the occurrence of large-scale mutations: 'But against the belief in such abrupt changes, embryology enters a strong protest. It is notorious that the wings of birds and bats, and the legs of horses and other quadrupeds, are undistinguishable at an early embryonic period, and that they become differentiated by insensibly fine steps. He who believes that some ancient form was transformed suddenly . . . will further be compelled to believe that many structures beautifully adapted to all other parts of the same creature and to the surrounding condition, have been suddenly produced; and of such complex and wonderful co-adaptation, he will not be able to assign a shadow of an explanation' (1885, pp. 203–4).

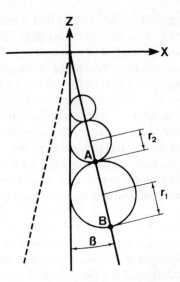

Figure 5.18. For the mathematical description of the formation of the molluscan shell the following form parameters are required over and above the shape of the generating curve, i.e. of the mantle edge: β, half the apical angle of the cone on which the centres of successive whorls are located; $w = r_1/r_2$, the rate of expansion, and $k = v_A/v_B$, where v stands for the rate of growth at the point in question

Thus, (1) ontogenesis proceeds in 'insensibly fine steps' and (2) major changes cannot ensure harmony. It appears to me that in his example Darwin proves what he wants to disprove. The very fact that the cartilaginous primordia of essentially all quadruped limbs are indistinguishable shows that very different limbs may arise from the same starting point. That the processes through which this is accomplished occur by 'insensibly fine steps' is clearly irrelevant from a phylogenetic point of view, and in an evolutionary perspective the changes are surely 'abrupt'. The crucial lesson to be learnt from Darwin's example is that several different forms can result from the same primordium under the influence of growth regulators, presumably in the form of hormones.

And so far as harmony is concerned, it may suffice to dwell on the quantitative aspect and point out that most organisms increase in size substantially without upsetting the harmony of the body. I believe that the processes involved in the epigenetic creation of an animal with the body weight $2W$ from ancestors weighing W are the same as those responsible for the growth of the former from W to $2W$.

This brings me back to a point which I have argued above, that ontogenesis is not the realization of a blueprint encoded in the genome, but a creative process, resulting from a succession of causally related epigenetic events, each of which is the outcome of interaction between agents which may be called the 'ontogenetic substrate' and the 'ontogenetic mechanisms' (Løvtrup, 1974). From this it follows that even slight modifications in those agents may entail far-reaching changes in the ontogenetic end-product, the adult body. That such 'mutations' will most frequently be deleterious cannot be questioned, but that they always must be so, as is implied by Fisher's axiom, is a postulate which, if enforced without empirical support, becomes a dogma. The essence of the present discussion, then, is that ontogenesis, and hence phylogenesis, is more than genetics; it is also, among other things, epigenetics, and therefore it is downright impossible to use observations made in genetic experiments as the basis for statements about what is and what is not possible in evolution.

As will appear from Vorzimmer (1970), Provine (1971) and Hull (1973), opposition to Darwin's notion about the accumulation of small variations was already expressed

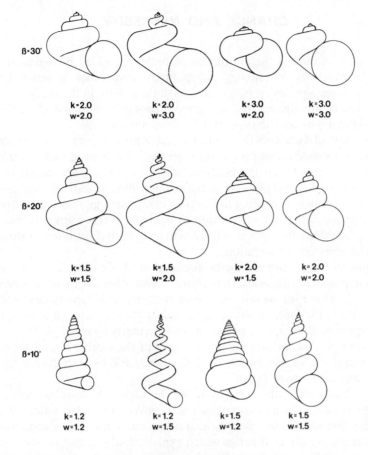

β-30°

k=2.0
w=2.0

k=2.0
w=3.0

k=3.0
w=2.0

k=3.0
w=3.0

β=20°

k=1.5
w=1.5

k=1.5
w=2.0

k=2.0
w=1.5

k=2.0
w=2.0

β=10°

k=1.2
w=1.2

k=1.2
w=1.5

k=1.5
w=1.2

k=1.5
w=1.5

Figure 5.19. Computer-drawn shells corresponding to the indicated
values of the form parameters. Most of the shells resemble shells of
actual gastropods. The 'elephant-tusk' shells of Scaphopoda require
$\beta = 90°$ and $k \gg w$. Likewise, the shells of Bivalvia will be obtained
for $\beta = 90°$ and a very large value for $k = w$

before his book (1859) was published, and elsewhere (1974; 1976) I have listed a number
of outstanding biologists who maintained this view even after the rise of neo-Darwinism.
Most unfortunate of all is that, as pointed out by T. H. Huxley in the letter to Darwin
quoted earlier in this chapter, stubborn adherence to the evolutionary significance of
intraspecific variation was a completely unnecessary burden on his theory. As so many
have pointed out, the acceptance of mutations with large-scale effects does not impair
Darwin's notions about the implications of evolution for the 'origin of species', which
were not his own, by the way, but only his theory about the mechanism of evolution.
To be sure, the process of phylogenesis may become more difficult to visualize if evolutionary innovations are accepted, but is it not better to be unable to visualize the
correct course of this event than to be able to imagine a wrong one?

CHANCE AND NECESSITY

Phylogenetic evolution is the apotheosis of biology, and all biological disciplines, biophysics, biochemistry, physiology, morphology, epigenetics, genetics, taxonomy, palaeontology, biogeography, ecology, etc., have their share in the study of evolution. It may therefore be anticipated that in a good theory of evolution each of the various disciplines will make its contribution of definitions and axioms.

From this point of view neo-Darwinism is not a good theory, since its conceptual basis is confined to genetics and population genetics. The notion that it is a theory of evolution is 'an extrapolation the boldness of which is made acceptable by the impressiveness of its basic conception' (von Bertalanffy, 1952, p. 86). I do not question the virtue of boldness which, according to Popper (1972), distinguishes a good theory, but it must be emphasized that the implication of boldness in this context is that it makes many daring predictions, thus increasing the chances of falsifying the theory through confrontation with empirical observations.

Still, in spite of the narrow axiomatic foundation of neo-Darwinism, its adherents have always imposed a 'categorical imperative' on all other disciplines concerned with evolution, viz., that the results obtained must conform with their theory. For reasons which some time in the future must be explained by the sociology of science, this yoke has been accepted by the majority of biologists, reluctantly by some, gladly, it seems, by most, although it has meant that the practitioners of these other disciplines have been forced to corrupt the interpretation of their own work by introducing teleological explanations and *ad hoc* hypotheses.

Against this background the survival of the Darwinian theories becomes a very interesting phenomenon, which has been dealt with by Woodfield. He writes (1973, p. 36): 'If natural selection weeded out theories as it weeds out plants, Darwinism would yield a second-order theory about theories which explained why it had survived as a first-order theory. It would assert that a theory survives only if it is fit, and that therefore first-order Darwinism must be fit. However, theories are not made true by longevity. Being a fit theory in the Darwinian sense does not necessarily involve being a true theory.'

I believe I can answer the question raised here in two different ways. The first is based on the following quotation: '*Darwinism is much more than a theory, it is a frame of mind which dominates thought*' (Wigand, quoted in Ràdl, 1930, p. 62). But 'frames of mind', like faiths of various kinds, are 'proved', not falsified. As shown on several occasions in the present chapter, Darwinists have been more willing to accept *ad hoc* hypotheses than falsifications of their theory.

The second explanation is based on notions developed above. I believe that Darwinism has a very low PF (potential of falsification) because 'a considerable part of Darwinism is not of the nature of an empirical theory, but is a logical truism' (Popper, 1972, p. 69). I have tried to show that it makes more predictions than is generally assumed, but it is still a rather vague theory, and so it has become randomly isolated in a compartment where few attempts at falsification are made. Neo-Darwinism has a higher PF than Darwinism, but it is a specialized theory, focusing on population genetics, and this has permitted its survival as a non-randomly isolated theory of evolution. Whether the present theory, evidently the secondary twin to neo-Darwinism, will establish itself as the dominant theory can be decided only after vigorous attempts have been made to falsify it.*

However this may be, it is important to note that the present theory in a general, but admittedly superficial, way makes predictions that are relevant to several disciplines. This implies that not only evolution, but also the theory of evolution, becomes part of their domain and not a concern monopolized by geneticists, and it becomes falsifiable through reference to their findings. It must be emphatically stressed that the theory is not intended to replace neo-Darwinism, but only to incorporate it as a part of a more general theory of evolution, a part which, however, is much less important than is currently supposed.

By rejecting Fisher's axiom, and thus allowing for the occurrence of large mutations, the present theory becomes a general theory of evolution and not just one concerned with bisexually reproducing animals, because the accumulation of mutational effects through the process of sexual reproduction is no longer necessary for the establishment of evolutionary innovations.

Furthermore, it allows the elimination of teleology from evolutionary discourse. After the introduction of Darwin's theory, thousands of tons of paper were wasted upon futile discussions of this phenomenon, and thousands of bizarre mechanisms have been proposed to explain the origin of new animal forms. Sober-minded biologists and epistemologists have tried to circumvent the teleological trap, but I fail to see how this can be done. If every existing form and function, pattern and behaviour, have come into existence in minute instalments, then each and every one of the latter must have been advantageous, otherwise they would not have become established.

This difficulty is completely circumvented if large-step mutations are allowed for. Under these circumstances it is possible to postulate that no modifications ever arose *because* of anything, but by chance, and therefore we shall not ask for what purpose a structure was acquired, but in which way it could contribute to the progression and survival of the animals possessing it. The answer may, or must, be that this contribution consists of its use for some beneficial purpose. Even if the purposes invoked in the two cases undoubtedly coincide, it will be envisaged, I hope, that there is an enormous difference between these outlooks. However, the point should not be driven too far; there are, after all, properties that defy all attempts to ascribe a meaningful function to them. Such features are allowed for in the present theory; they may be epigenetic consequences of the acquisition of some advantageous character, or they may have been preserved by chance.

From the preceding discussion it appears that three different kinds of evolutionary

---

*When the present text was written in 1974, I did not realize that 'ecological selection', the most important element of Darwin's natural selection, has been abolished in the synthetic theory. This discovery, of course, supports my view that neo-Darwinism is of no particular importance as a theory of evolution.

My position has received unexpected, but welcome, support through the book recently published by Lewontin (1974), supposedly one of the leading population geneticists of our time. The principal thesis of Lewontin may be represented by the following quotation: 'If one simply cannot measure the state variables or parameters with which [a] theory is constructed, or if their measurement is so laden with error that no discrimination between alternative hypotheses is possible, the theory becomes a vacuous exercise in formal logic that has no points of contact with the contingent world. The theory explains nothing because it explains everything. It is my contention that a good deal of the structure of evolutionary genetics comes perilously close to being of this sort' (pp. 11–12).

If this is correct then it is indeed questionable whether neo-Darwinism is a serious alternative to either Darwin's or the comprehensive theory.

event may be distinguished. The first of these is creative, but undirected, and involves the origination of innovations through fortuitous and unpredictable mutations. This implies that the creativity ascribed to natural selection (cf. Simpson, 1944) can no longer be upheld. This agent can ensure progress through selection among existing mutations, but unless it is assumed that all mutations existed at the beginning of phylogenetic evolution, then even natural selection may, at times, have had to wait for the proper mutations to arise, and if this is true, then evidently the latter must represent the primary creative agent in evolution.

The second type of evolutionary event, involving intertaxonomic competition and extinction, is the destructive, but progressively directed, phase of evolution. In this phase necessity is involved, for it is an unavoidable necessity that all living organisms must succumb through extinction if they cannot, through non-random or random isolation, keep their PP at the level dictated by the dominant forms in their environmental compartment. And if we have been obliged to deprive natural selection of its role as Creator, it appears to me that the role which we can ascribe to it is no less grandiose. In fact, on the jumble of variation created by mutation and recombination, resulting in disorder much more often than in order, natural selection has in many phylogenetic lines imposed a direction towards steady perfection by eliminating everything but the best available at every station. If Darwin had been content to attribute this function to his natural selection, his theory would have survived indefinitely I believe. But the failure of the theory does not detract from the greatness of Darwin and Wallace in invoking natural selection as an evolutionary agent.

The third type of evolutionary event, resulting in taxonomic divergence, is associated with the phenomenon of isolation. In this case chance is also involved, either in the origination of the innovation leading to isolation, or in the process of isolation itself. In non-random isolation a certain measure of necessity is implied since, clearly, only innovations allowing for isolation under the given external conditions have any possibility of becoming established.

One implication of the present theory is that fitness acquires a significance different from the one it has in Darwinian theories. Fitness no longer represents the best possible, but only the best that ever happened to be available, or, in the words of Fraser (1967, p. 107): 'a major feature in understanding the evolution of adaptive form and function is the realization that this is essentially based on the survival of the least inadequate. A successful species is not one which is adapted to a particular environment but rather one which is least badly adapted to that environment. Evolution is based, not on superlatives, but on adequacies'.

We may finish the present discussion with a reference to the Aristotelian antithesis between chance and necessity. As appears from the quotation at the head of the present chapter, the first alternative, advocated by Empedocles and vehemently rejected by Aristotle, corresponds rather closely to the substance of the theory advocated here. To be sure, there are elements of necessity in evolution, but yet this marvellous accomplishment is primarily the fabric of chance.

6

CONCLUSION

On various occasions I have experienced the criticism that my first book, and even the manuscript to the present one, are difficult to read. No doubt this fact may in part be due to my personal shortcomings. Yet surely some of these critics would experience similar frustration with mathematical and philosophical texts, and in that case it is possible to suggest a further reason for the difficulties encountered *vis-à-vis* my work, namely, that it is argumentative rather than descriptive. And arguments are much more difficult to follow than descriptions. The best corroboration of this postulate is given by *On the Origin of Species*. Darwin correctly claimed that it was 'one long argument from the beginning to the end', and this may be one reason why we today, more than one century after the publication of the book, can go on discussing the implications of Darwin's theory.

There are two ways to avoid such a situation. One is the positivist's stratagem, to stick to the description of facts and shun speculation. But this road does not lead to understanding, for that presupposes explanations, and hence theories, and all theories are conjectures. If, therefore, argument from unprovable premises is a necessary pre-requisite for the advance of biology, the other alternative ought to be adopted, namely, to state explicitly the elements, definitions, premises and conclusions on which the reasoning is based. This is the approach I have adopted in my books as far as it has proved possible. If biologists in general would comply with this procedure, then it would be much easier to establish whether prevailing disagreement is factual (i.e. due to different premises) or fictional (i.e. due to faulty reasoning). Also, it would be much more difficult to reject a conclusion should it happen to run counter to prevailing thought.

It is not common to use axiomatization in biological treatises, and when it occurs it is usually of the 'naive' type. My endeavours unquestionably belong to this category. I have revealed my naivety for two reasons.

The first I have given above, the second is that I would like to defy epistemologists and logicians to take a greater interest in biology than they have hitherto done. I hope that in the present work I have been able to show that it is possible to elaborate fairly comprehensive biological theories, to which it may be worthwhile to apply their special skills. Should they do so, I am convinced that they will find more than one *non sequitur*, but that, of course, would only serve to improve the theories.

Popper (1969) has shown that the progress of science consists of continually attempting to falsify existing theories and, if this is successfully achieved, inaugurating new ones to take their place. I have followed this advice in the present book but, 'if an expert or anybody else should take the trouble to refute my criticism I shall be pleased and honoured' (Popper, 1969, p. 137). After all, the principal objective of a scientist is not to be in the right, but to contribute to the promotion of science.

REFERENCES

R. Acher, J. Chauvet and M.-T. Chauvet, 1972, 'Phylogeny of the neurohypophysial hormones. Two new active peptides isolated from a cartilaginous fish, *Squalus acanthias*', *Eur. J. Biochem.*, **29**, 12–19.

R. T. Acton, P. F. Weinheimer, W. H. Hildemann and E. E. Evans, 1969, 'Induced bactericidal response in the hagfish', *J. Bact.*, **99**, 626–628.

H. Adam, 1963a, 'Structure and histochemistry of the alimentary canal', in *The Biology of Myxine* (ed. A. Brodal and R. Fänge), Universitetsforlaget, Oslo, pp. 256–288.

——, 1963b, 'The pituitary gland', in *The Biology of Myxine* (ed. A. Brodal and R. Fänge), Universitetsforlaget, Oslo, pp. 459–476.

J. W. Adelson, 1971, 'Enterosecretory proteins', *Nature*, **229**, 321–325.

G. M. Air, E. O. P. Thompson, B. J. Richardson and G. B. Sharman, 1971, 'Amino-acid sequences of kangaroo myoglobin and haemoglobin and the date of marsupial–eutherian divergence', *Nature*, **229**, 391–394.

G. Aler, G. Båge and B. Fernholm, 1971, 'On the existence of prolactin in cyclostomes', *Gener. Comp. Endocrinol.*, **16**, 498–503.

J. S. Alexandrowicz, 1960a, 'A muscle receptor organ in *Eledone cirrhosa*', *J. Mar. Biol. Ass. U.K.*, **42**, 405–418.

——, 1960b, 'Innervation of the hearts of *Sepia officinalis*', *Acta Zool.*, **41**, 65–100.

A. C. Allison, R. Cecil, P. A. Charlwood, W. B. Gratzer, S. Jacobs and N. S. Snow, 1960, 'Haemoglobin of the lamprey, *Lampetra fluviatilis*', *Biochim. Biophys. Acta*, **42**, 43–48.

D. T. Anderson, 1973, *Embryology and Phylogeny in Annelids and Arthropods*, Pergamon, Oxford.

J. M. Anderson, 1966, 'Aspects of nutritional physiology', in *Physiology of Echinodermata* (ed. R. A. Boolootian), Wiley, New York, pp. 328–357.

S. M. Andrews, 1973, 'Interrelationships of crossopterygians', in *Interrelationships of Fishes* (ed. P. H. Greenwood, R. S. Miles and C. Patterson), Academic Press, London, pp. 137–177.

K. Anno, Y. Kawai and N. Seno, 1964, 'Isolation of chondroitin from squid skin', *Biochim. Biophys. Acta*, **83**, 348–349.

K. Anno, N. Seno and M. Kawaguchi, 1962, 'A comparison of glucosamine and galactosamine content of cartilage from various sources', *Biochim. Biophys. Acta*, **58**, 87–91.

K. Anno, N. Seno, M. B. Mathews, T. Yamagata and S. Suzuki, 1971, 'A new dermatan polysulfate, chondroitin sulfate H, from hagfish notochord', *Biochim. Biophys. Acta*, **237**, 173–177.

A. Anseth and T. C. Laurent, 1961, 'Studies on corneal polysaccharides. I. Separation', *Exptl. Eye Res.*, **1**, 25–38.

J. Anthony and J. Millot, 1972, 'Première capture d'une femelle de Coelacanthe en état de maturité sexuelle', *Compt. Rend. Acad. Sci., Paris* [D], **274**, 1925–1926.

Aristotle, 1862, *Physique d'Aristote ou Leçons sur les Principes Généraux de la Nature* (transl. J. Barthélemy Saint-Hilaire), Vol. 2, Durand, Paris.

L. Arvy, 1960, 'Contribution à l'histo-enzymologie du tube digestif chez *Octopus vulgaris* Lamarck (Céphalopodes)', *Arch. Anat. Microsc. Morphol. Exptl.*, **49**, 229–239.

D. E. Ashhurst and N. M. Costin, 1971a, 'Insect mucosubstances: I. The mucosubstances of developing connective tissue in the locust, *Locusta migratoria*', *Histochem. J.*, **3**, 279–295.

——, 1971b, 'Insect mucosubstances: II. The mucosubstances of the central nervous system', *Histochem. J.*, **3**, 297–310.

P. Ax, 1964, 'Der Begriff Polyphylie ist aus der Terminologie der natürlichen phylogenetischen Systematik zu eliminieren', *Zool. Anz.*, **173**, 52–56.

——, 1966, 'Das chordoide Gewebe als histologisches Lebensformmerkmal der Sandlückenfauna des Meeres', *Naturwiss. Rundschau*, **19**, 282–289.

F. J. Ayala, 1970, 'Competition, coexistence and evolution', in *Essays in Evolution and Genetics in Honor of Theodosius Dobzhansky* (ed. M. K. Hecht and W. C. Steeve), North-Holland, Amsterdam, pp. 121–158.

——, 1971, 'Darwinian *versus* non-Darwinian evolution in natural populations of *Drosophila*', *Proc. 6th Berkeley Symp. Math. Statis. Probab.*, Vol. 5 (ed. L. M. LeCam, J. Neyman and E. L. Scott), University of California Press, Berkeley, pp. 211–236.

Z. M. Bacq and F. Ghiretti, 1953, 'Physiologie des glandes salivaires postérieures dés Céphalopodes Octopodes isolées et perfusées *in vitro*', *Pubbl. Staz. Zool. Napoli*, **24**, 267–277.

A. J. Bailey, 1968, 'The nature of collagen', in *Comprehensive Biochemistry*, Vol. 26B (ed. M. Florkin and E. H. Stotz), Elsevier, Amsterdam, pp. 297–423.

I. L. Baird, 1970, 'The anatomy of the reptilian ear', in *Biology of the Reptilia*, Vol. 2 (ed. C. Gans and T. S. Parsons), Academic Press, London, pp. 193–275.

R. T. Bakker, 1971, 'Dinosaur physiology and the origin of mammals', *Evolution*, **25**, 638–658.

E. A. Balazs, 1965, 'Amino sugar-containing macromolecules in the tissues of the eye and the ear', in *The Amino Sugars*, Vol. 2A (ed. E. A. Balazs and R. W. Jeanloz), Academic Press, New York, pp. 401–460.

E. Baldwin and D. M. Needham, 1937, 'A contribution to the comparative biochemistry of muscular and electrical tissues', *Proc. Roy. Soc. London* [B], **122**, 197–219.

E. Baldwin and J. Needham, 1934, 'Problems of nitrogen catabolism in invertebrates. I. The snail *Helix pomatia*', *Biochem. J.*, **28**, 1372–1392.

V. C. Barber, E. Evans and M. Land, 1967, 'The fine structure of the eye of the mollusc, *Pecten maximus*', *Z. Zellforsch.*, **76**, 295–312.

E. J. W. Barrington, 1938, 'The digestive system of Amphioxus (Branchiostoma) lanceolatus', *Phil. Trans. Roy. Soc. London* [B], **228**, 269–312.

——, 1963, *An Introduction to General and Comparative Endocrinology*, Clarendon, Oxford.

——, 1965, *The Biology of Hemichordata and Protochordata*, Oliver & Boyd, Edinburgh.

——, 1968, 'Phylogenetic perspectives in vertebrate endocrinology', in *Perspectives in Endocrinology Hormones in the Lives of Lower Vertebrates* (ed. E. J. W. Barrington and C. B. Jørgensen), Academic Press, London, pp. 1–46.

R. J. Barrnett, 1953, 'The histochemical distribution of protein-bound sulfhydryl groups', *J. Nat. Cancer Inst.*, **13**, 905–925.

M. Bates, 1960, 'Ecology and evolution', in *The Evolution of Life* (ed. S. Tax), University of Chicago Press, Chicago, pp. 547–568.

W. Bateson, 1886, 'The ancestry of the Chordata', *Quart. J. Micros. Sci.*, **26**, 535–571.

C. J. Bayne, 1966, 'Observations on the composition of the layers of the egg of *Agriolimax reticulatus*, the grey field slug (Pulmonata, Stylomatophora)', *Comp. Biochem. Physiol.*, **19**, 317–338.

R. A. Bayne and F. E. Friedl, 1968, 'The production of externally measurable ammonia and urea in the snail, *Lymnaea stagnalis jugularis* Say', *Comp. Biochem. Physiol.*, **25**, 711–717.

R. E. Bechhofer, S. Elmaghraby and N. Morse, 1959, 'A single-sample multiple-decision procedure for selecting the multinomial event which has the highest probability', *Ann. Math. Statist.*, **30**, 102–119.

M. Beckner, 1959, *The Biological Way of Thought*, Columbia University Press, New York.

G. R. de Beer, 1928, *Vertebrate Zoology. An Introduction to the Comparative Anatomy, Embryology, and Evolution of Chordate Animals*, Sidgwick & Jackson, London.

——, 1937, *The Development of the Vertebrate Skull*, Clarendon, Oxford.

——, 1951, *Embryos and Ancestors* (rev. ed), Clarendon, Oxford.

D. J. Bell and E. Baldwin, 1941, 'The chemistry of galactogen from *Helix pomatia* L-galactose as a component of a polysaccharide of animal origin', *J. Chem. Soc.*, **1941**, 125–132.

A. d'A. Bellairs, 1970, *Reptiles* (3rd ed.), Hutchinson University Library, London.

D. Bellamy and I. C. Jones, 1961, 'Studies on *Myxine glutinosa*—I. The chemical composition of the tissues', *Comp. Biochem. Physiol.*, **3**, 175–183.

L. S. Berg, 1947, *Classification of Fishes, Both Recent and Fossil*, Edwards, Ann Arbor, Mich.

O. Berg, A. Gorbman and H. Kobayashi, 1959, 'The thyroid hormones in invertebrates and lower vertebrates', in *Comparative Endocrinology* (ed. A. Gorbman), Wiley, New York, pp. 302–319.

A. Bernhauser, 1961, 'Zur Knochen- und Zahnhistologie von Latimeria chalumnae Smith und einiger Fossilformen', *Sitz.ber. Österreich. Akad. Wiss.* [1], **170**, 119–137.

N. J. Berrill, 1929, 'Digestion in ascidians and the influence of temperature', *Brit. J. Exptl. Biol.*, **6**, 275–292.

——, 1955, *The Origin of Vertebrates*, Oxford University Press, Oxford.

L. von Bertalanffy, 1952, *Problems of Life. An Evaluation of Modern Biological Thought*, Watts, London.

G. Bertmar, 1968, 'Lungfish phylogeny', in *Current Problems of Lower Vertebrate Phylogeny* (ed. T. Ørvig), Almqvist & Wiksell, Stockholm, pp. 259–283.

——, 1969, 'The vertebrate nose, remarks on its structural and functional adaptation and evolution', *Evolution*, **23**, 131–152.

A. M. Bidder, 1950, 'The digestive mechanisms of the European squids *Loligo vulgaris*, *Loligo forbesi*, *Allotheutis media* and *Allotheutis subulata*', *Quart. J. Micros. Sci.*, **91**, 1–43.

——, 1966, 'Feeding and digestion in cephalopods', in *Physiology of Mollusca*, Vol. 2 (ed. K. M. Wilbur and C. M. Yonge), Academic Press, New York, pp. 96–124.

J. Binyon, 1966, 'Salinity tolerance and ionic regulation', in *Physiology of Echinodermata* (ed. R. A. Boolootian), Wiley, New York, pp. 359–377.

T. W. Blackstad, 1963, 'The skin and the slime glands', in *The Biology of Myxine* (ed. A. Brodal and R. Fänge), Universitetsforlaget, Oslo, pp. 195–230.

G. Bloom, E. Östlund and R. Fänge, 1963, 'Functional aspects of cyclostome hearts in relation to recent structural findings', in *The Biology of Myxine* (ed. A. Brodal and R. Fänge), Universitetsforlaget, Oslo, pp. 317–339.

J. E. V. Boas, 1931, 'Schuppen der Reptilien, Vögel und Säugetiere', in *Handbuch der vergleichenden Anatomie der Wirbeltiere*, Vol. 1 (ed. L. Bolk, E. Göppert, E. Kallius and W. Lubosch), Urban & Schwarzenberg, Berlin, pp. 559–564.

W. J. Bock and G. von Wahlert, 1965, 'Adaptation and the form–function complex', *Evolution*, **19**, 269–299.

H. H. Boer, E. Douma and J. M. A. Koksma, 1968, 'Electron microscope study of neuro-secretory cells and neurohaemal organs in the pond snail *Lymnaea stagnalis*', *Symp. Zool. Soc. London*, **22**, 237–256.

L. Bolognani, A. M. Bolognani Fantin, R. Lusignani and L. Zonta, 1966, 'Presence of sialopolysaccharidic components in egg gelatinous mantle of *Rana latastei* and *Bufo vulgaris*', *Experientia*, **22**, 601–603.

Q. Bone, 1960a, 'The origin of the chordates', *J. Linn. Soc. London*, **44**, 252–269.

——, 1960b, 'The central nervous system in Amphioxus', *J. Comp. Neurol.*, **115**, 27–64.

——, 1963, 'The central nervous system', in *The Biology of Myxine* (ed. A. Brodal and R. Fänge), Universitetsforlaget, Oslo, pp. 50–91.

R. A. Boolootian, 1961, 'Physical properties and chemical composition of perivisceral fluid: Echinodermata', in *Biological Handbook: Blood and Other Body Fluids, Fed. Am. Soc. Exptl. Biol.*, Washington, D.C., pp. 339–344.

C. R. Botticelli, F. R. Hisaw and H. H. Wotiz, 1961, 'Estrogens and progesterone in the sea urchin (*Strongylocentrotus franciscanus*) and Pecten (*Pecten hericius*)', *Proc. Soc. Exptl. Biol. Med.*, **106**, 887–889.

B. B. Boycott, 1960, 'The functioning of the statocysts of *Octopus vulgaris*', *Proc. Roy. Soc. London* [B], **152**, 78–87.

A. Brachet, 1921, *Traité d'embryologie des Vertébrés*. Masson, Paris.

G. Braunitzer and H. Fujiki, 1969, 'Zur Evolution der Vertebraten Die Konstitution und Tertiärstruktur des Hämoglobins der Flussneunauges', *Naturwiss.*, **56**, 322–323.

S. Bricteux-Grégoire, G. Duchateau-Bosson, C. Jeuniaux and M. Florkin, 1964, 'Constituants osmotiquement actifs des muscles adducteurs de *Mytilus edulis* adaptée à l'eau de mer ou à l'eau saumâtre', *Arch. Internat. Physiol. Biochim.*, **72**, 116–123.

S. Bricteux-Grégoire, M. Florkin and C. Grégoire, 1968, 'Prism conchiolin of modern or fossil molluscan shells. An example of protein paleization', *Comp. Biochem. Physiol.*, **24**, 567–572.

P. Brien, 1948, 'Embranchement des Tuniciers', in *Traité de Zoologie Anatomie, Systématique, Biologie*, Vol. 11 (ed. P.-P. Grassé), Masson, Paris, pp. 553–930.

R. J. Britten and E. H. Davidson, 1971, 'Repetitive and non-repetitive DNA sequences and a speculation on the origins of evolutionary novelty', *Quart. Rev. Biol.*, **46**, 111–138.

C. H. Brown, 1952, 'Some structural proteins of *Mytilus edulis*', *Quart. J. Microsc. Sci.*, **93**, 487–502.

L. Brundin, 1966, 'Transantarctic relationships and their significance, as evidenced by chironomid midges. With a monograph of the subfamilies Podonominae and Aphroteniinae and the austral Heptagyiae', *Kungl. Svenska Vetenskapsakad. Handl.* [4], **11**, **1**, 3–472.

L. Brundin, 1968, 'Application of phylogenetic principles in systematics and evolutionary theory', in *Current Problems of Lower Vertebrate Phylogeny* (ed. T. Ørvig), Almqvist & Wiksell, Stockholm, pp. 473–495.

W. von Buddenbrock, 1967, *Vergleichende Physiologie*, Vol. 4, Birkhäuser, Basel.

T. H. Bullock and G. A. Horridge, 1965, *Structure and Function in the Nervous Systems of Invertebrates*, Freeman, San Francisco.

G. L. Bush, 1975, 'Models of animal speciation', *Ann. Rev. Ecol. Syst.*, **6**, 339–364.

S. Butler, 1887, *Luck or Cunning*, Trübner, London.

J. H. Camin and R. R. Sokal, 1965, 'A method for deducing branching sequences in phylogeny', *Evolution*, **19**, 311–326.

J. W. Campbell and S. H. Bishop, 1970, 'Nitrogen metabolism in molluscs', in *Comparative Biochemistry of Nitrogen Metabolism*, Vol. 1 (ed. J. W. Campbell), Academic Press, London, pp. 103–206.

D. Carlström, 1963, 'A crystallographic study of vertebrate otoliths', *Biol. Bull.*, **125**, 441–463.

P. Cerfontaine, 1906, 'Recherches sur le développement de l'Amphioxus', *Arch. Biol.*, **22**, 229–418.

L. Chabry, 1887, 'Contribution à l'embryologie normale et tératologique des Ascidies simples', *J. Anat. Physiol.*, **23**, 167–319.

A.B.Chaet,1964a,'A mechanism for obtaining mature gametes from starfish', *Biol. Bull.*, **126**, 8–13.

——, 1964b, 'Shedding substance and "shedhibin" from the nerves of the starfish, *Patiria miniata*', *Am. Zool.*, **4**, 407.

A. B. Chaet and R. A. McConnaughy, 1959, 'Physiologic activity of nerve extracts', *Biol. Bull.*, **117**, 407–408.

H. Charniaux-Cotton, 1960, 'Sex determination', in *The Physiology of Crustacea*, Vol. 1 (ed. T. H. Waterman), Academic Press, New York, pp. 411–447.

J. J. Christian, 1971, 'Population density and reproductive efficiency', *Biol. Reprod.*, **4**, 248–294.

J. L. Cisne, 1974, 'Trilobites and the origin of arthropods', *Science*, **186**, 13–18.

A. W. Clark, 1963, 'Fine structure of two invertebrate photoreceptor cells', *J. Cell. Biol.*, **19**, 14 A.

H. Clark, 1953, 'Metabolism of the black snake embryo. I. Nitrogen excretion', *J. Exptl. Biol.*, **30**, 492–501.

H. Clark and D. Fischer, 1957, 'A reconsideration of nitrogen excretion by the chick embryo', *J. Exptl. Zool.*, **136**, 1–15.

H. Clark and B. F. Sisken, 1956, 'Nitrogenous excretion by embryos of the viviparous snake *Thamnophis s. sirtalis* (L.)', *J. Exptl. Biol.*, **33**, 384–393.

H. Clark, B. Sisken and J. E. Shannon, 1957, 'Excretion of nitrogen by the alligator embryo', *J. Cell. Comp. Physiol.*, **50**, 129–134.

J. J. Cohen, M. A. Krupp and C. A. Chidsey, 1958, 'Renal conservation of trimethylamine oxide by the spiny dogfish, *Squalus acanthias*', *Am. J. Physiol.*, **194**, 229–235.

M. J. Cohen and S. Dijkgraaf, 1961, 'Mechanoreception', in *The Physiology of Crustacea*, Vol. 2 (ed. T. H. Waterman), Academic Press, New York, pp. 65–108.

P. P. Cohen and G. W. Brown, 1960, 'Ammonia metabolism and urea biosynthesis', in *Comparative Biochemistry. A Comprehensive Treatise*, Vol. 2 (ed. M. Florkin and H. S. Mason), Academic Press, New York, pp. 161–244.

O. Cohnheim, 1901, 'Versuche über Resorption, Verdauung und Stoffwechsel von Echinodermen', *Z. Physiol. Chem.*, **33**, 9–55.

E. H. Colbert, 1969, *Evolution of the Vertebrates. A History of the Backboned Animals Through Time* (2nd ed.), Wiley, New York.

L. J. V. Compagno, 1973, 'Interrelationships of living elasmobranchs', in *Interrelationships of Fishes* (ed. P. H. Greenwood, R. S. Miles and C. Patterson), Academic Press, London.

E. G. Conklin, 1903, 'The cause of inverse symmetry', *Anat. Anz.*, **23**, 577–588.

——, 1906, 'Does half of an ascidian egg give rise to a whole larva?', *Arch. Entw.mech.*, **21**, 727–753.

——, 1911, 'The organization of the egg and the development of single blastomeres of Phallusia mamillata', *J. Exptl. Zool.*, **10**, 393–407.

F. P. Conte, 1969, 'Salt secretion', in *Fish Physiology*, Vol. 1 (ed. W. S. Hoar and D. J. Randall), Academic Press, New York, pp. 241–292.

H. B. S. Cooke, 1972, 'The fossil mammal fauna of Africa', in *Evolution, Mammals and Southern Continents* (ed. A. Keast, F. C. Erk and B. Glass), State Univ. New York Press, Albany, pp. 89–139.

E. L. Cooper, 1968, 'Transplantation immunity in annelids I. Rejection of xenografts exchanged between *Lumbricus terrestris* and *Eisenia foetida*', *Transplantation*, **6**, 322–337.

E. D. Cope, 1896, *The Primary Factors of Organic Evolution*, Open Court, Chicago.

J. B. S. Corrêa, A. Dmytraczenko and J. H. Duarte, 1967, 'Structure of a galactan found in the albumen gland of *Biomphalaria glabrata*', *Carbohydrate Res.*, **3**, 445–452.

C. B. Cox, 1970, 'Migrating marsupials and drifting continents', *Nature*, **226**, 767–770.

J. Cracraft, 1973, 'Vertebrate evolution and biogeography in the Old World tropics: Implications of continental drift and palaeoclimatology', in *Implications of Continental Drift to the Earth Sciences*, Vol. 1 (ed. D. H. Tarling and S. K. Runcorn), Academic Press, London, pp. 373–393.

L. Croizat, 1962 [1964], *Space, Time, Form: The Biological Synthesis*, Caracas.

L. Croizat, G. Nelson and D. E. Rosen, 1974, 'Centers of origin and related concepts', *Syst. Zool.*, **23**, 265–287.

A. W. Crompton and F. A. Jenkins, 1968, 'Molar occlusion in late Triassic mammals', *Biol. Rev.*, **43**, 427–458.

J. F. Crow, 1972, 'Darwinian and non-Darwinian evolution', *Proc. 6th Berkeley Symp. Math. Statis. Probab.*, Vol. 5 (ed. L. M. LeCam, J. Neyman and E. L. Scott), University of California Press, Berkeley, pp. 1–22.

J. F. Crow and M. Kimura, 1970, *An Introduction to Population Genetics*, Harper & Row, New York.

R. A. Crowson, 1970, *Classification and Biology*, Heinemann, London.

J. Daget, 1950, 'Revision des affinités phylogénétiques des Polyptérides', *Mem. Inst. Français Afr. Noire*, **11**, 7–178.

——, 1958, 'Sous-classe des Brachioptérygiens (Brachiopterygii)', in *Traité de Zoologie Anatomie, Systématique, Biologie*, Vol. 13 (ed. P.-P. Grassé), Masson, Paris, pp. 2501–2521.

J. Daget, M.-L. Bauchot, R. Bauchot and J. Arnoult, 1964, 'Développement du chondrocrâne et des arcs aortiques chez *Polypterus senegalis* Cuvier', *Acta Zool.*, **46**, 201–244.

E. Dahl, B. Falck, M. Lindquist and C. von Mecklenburg, 1962, 'Monamines in mollusc neurons', *Kgl. Fysiograf. Sällskap. Lund, Förh.*, **32**, 89–91.

P. J. Darlington, 1957, *Zoogeography: The Geographical Distribution of Animals*, Wiley, New York.

C. Darwin, 1859, *The Origin of Species by Means of Natural Selection. A Facsimile of the First Edition with an Introduction by Ernst Mayr*, Harvard University Press, Cambridge, Mass., 1964.

——, 1885, *The Origin of Species by Means of Natural Selection* (6th rev. ed.), Murray, London.

L. Daudet, 1929, *Écrivains et Artistes*, Vol. 7, Le Capitole, Paris.

J. A. Dawson, 1963, 'The oral cavity, the "jaws" and the horny teeth of Myxine glutinosa', in *The Biology of Myxine* (ed. A. Brodal and R. Fänge), Universitetsforlaget, Oslo, pp. 231–255.

C. Dawydoff, 1959, 'Ontogenèse des Annélides', in *Traité de Zoologie Anatomie, Systematique, Biologie*, Vol. 5.1 (ed. P.-P. Grassé), Masson, Paris, pp. 594–686.

M. O. Dayhoff and C. M. Park, 1969, 'Cytochrome c: building a phylogenetic tree', *Atlas Prot. Seq. Struct.*, **4**, 7–16.

B. Dean, 1899, 'On the embryology of Bdellostoma stouti. A general account of myxinoid development from the egg and segmentation to hatching', *Festschr. Siebenzigsten Geburtstag C. von Kupffer*, Fischer, Jena, pp. 221–276.

E. T. Degens, D. W. Spencer and R. H. Parker, 1967, 'Paleobiochemistry of molluscan shell proteins', *Comp. Biochem. Physiol.*, **20**, 553–579.

H. Delaunay, 1931, 'L'excrétion azotée des Invertébrés', *Biol. Rev.*, **6**, 265–301.

H. Delaunay, 1934, 'Le métabolisme de l'ammoniaque d'après les recherches relatives aux Invertébrés', *Ann. Physiol. Physicochim. Biol.*, **10**, 695–729.

H. C. Delsman, 1913, 'Der Ursprung der Vertebraten Eine neue Theorie', *Mitth. Zool. Stat. Neapel*, **20**, 647–710.

H. C. Delsman, 1921, 'The ancestry of vertebrates as a means of understanding the principal features of their structure and development', *Natuurk. Tijdschr. Nederl.-Indie*, **81**, 187–286.

R. H. Denison, 1974, 'The structure and evolution of teeth in lungfishes', *Fieldiana Geology*, **33**, 31–58.

H. C. Dessauer, 1970, 'Blood chemistry of reptiles: Physiological and evolutionary aspects', in *Biology of the Reptilia*, Vol. 3 (ed. C. Gans and T. S. Parsons), Academic Press, London, pp. 1–72.

C. Devillers, 1973, *Introduction à l'Etude Systématique des Vertébrés*, Doin, Paris.

C. Devillers and J. Corsin, 1968, 'Les os dermiques craniens des Poissons et des Amphibien; points de vue embryologique sur les "territoires osseux" et les "fusions"', in *Current Problems of Lower Vertebrate Phylogeny* (ed. T. Ørvig), Almqvist & Wiksell, Stockholm, pp. 413–428.

R. E. Dickerson, 1971, 'The structure of cytochrome c and the rates of molecular evolution', *J. Mol. Evol.*, **1**, 26–45.

R. S. Dietz and J. C. Holden, 1973, 'The breakup of Pangaea', in *Continents Adrift*, Freeman, San Francisco, pp. 102–113.

N. Dilly, 1964, 'Studies on the receptors in the cerebral vesicle of the ascidian tadpole. 2. The ocellus', *Quart. J. Micros. Sci.*, **105**, 13–20.

T. Dobzhansky, 1970, *Genetics of the Evolutionary Process*, Columbia University Press, New York.

——, 1974, 'Chance and creativity in evolution', in *Studies in Philosophy of Biology Reduction and Related Problems* (ed. F. J. Ayala and T. Dobzhansky), Macmillan, London, pp. 307–338.

J. M. Dodd, 1955, 'The hormones of sex and reproduction and their effects in fish and lower chordates', *Mem. Soc. Endocrin.*, **4**, 166–184.

J. M. Dodd and M. H. I. Dodd, 1966, 'An experimental investigation of the supposed pituitary affinities of the ascidian neural complex', in *Some Contemporary Studies in Marine Science* (ed. H. Barnes), Allen & Unwin, London, pp. 233–252.

A. Dohrn, 1876, *Der Ursprung der Wirbeltiere und das Princip des Funktionswechsel*, Engelmann, Leipzig.

L. Dollo, 1893, 'Les lois de l'evolution (Resumé)', *Bull. Soc. Belge Géol.*, 164–166.

J. Doyle, 1967, 'The mantle mucin of *Pecten maximus* (L)', *Biochem. J.*, **103**, 41 P.

P. Drach, 1948, 'La notion de Procordés et les embranchements des Cordés', in *Traité de Zoologie Anatomie, Systématique, Biologie*, Vol. 11 (ed. P.-P. Grassé), Masson, Paris, pp. 545–551.

F. G. Duerr, 1967, 'The uric acid content of several species of prosobranch and pulmonate snails as related to nitrogen excretion', *Comp. Biochem. Physiol.*, **22**, 333–340.

——, 1968, 'Excretion of ammonia and urea in seven species of marine prosobranch snails', *Comp. Biochem. Physiol.*, **26**, 1051–1059.

M. Durchon, 1967, *L'Endocrinologie des Vers et des Mollusques*, Masson, Paris.

R. M. Eakin, 1965, 'Evolution of photoreceptors', *Cold Spring Harbor Symp. Quant. Biol.*, **30**, 363–370.

——, 1968, 'Evolution of photoreceptors', in *Evolutionary Biology*, Vol. 2 (ed. T. Dobzhansky, M. K. Hecht and W. C. Steere), North-Holland, Amsterdam, pp. 194–242.

T. H. Eaton, 1953, 'Pedomorphosis: an approach to the Chordate–Echinoderm problem', *System. Zool.*, **2**, 1–6.

——, 1959, 'The ancestry of modern Amphibia: a review of the evidence', *Univ. Kansas Publ., Mus. Nat. Hist.*, **12**, 157–180.

N. Eldredge and S. J. Gould, 1972, 'Punctuated equilibria: an alternative to phyletic gradualism', in *Models in Paleobiology* (ed. T. J. M. Schopf), Freeman, Cooper, San Francisco, pp. 82–115.

C. F. Emmanuel, 1956, 'The composition of Octopus renal fluid. II. A chromatographic examination of the constituents', *Z. Vergl. Physiol.*, **39**, 477–482.

——, 1957, 'The composition of Octopus renal fluid. IV. Isolation and identification of the methanol soluble substances', *Z. Vergl. Physiol.*, **40**, 1–7.

C. F. Emmanuel and A. W. Martin, 1956, 'The composition of Octopus renal fluid. I. Inorganic constituents', *Z. Vergl. Physiol.*, **39**, 226–234.

L. Engelman and J. A. Hartigan, 1969, 'Percentage points of a test for clusters', *J. Am. Statist. Ass.*, **64**, 1642–1648.

D. H. Enlow and S. O. Brown, 1958, 'A comparative histological study of fossil and recent bone tissues. Part III', *Texas J. Sci.*, **10**, 187–230.

A. H. Ennor and J. F. Morrison, 1958, 'Biochemistry of phosphagens', *Physiol. Rev.*, **38**, 631–674.

A. Epple, 1969, 'The endocrine pancreas', in *Fish Physiology*, Vol. 2 (ed. W. S. Hoar and D. J. Randall), Academic Press, New York, pp. 275–319.

E. E. Evans, B. Painter, M. L. Evans, P. Weinheimer and R. T. Acton, 1968, 'An induced bactericidin in the spiny lobster, *Panulirus argus*', *Proc. Soc. Exp. Biol. Med.*, **128**, 394–398.

L. Fage, 1949, 'Classe des Mérostomacés (Merostomata, Woodward (1866))', in *Traité de Zoologie Anatomie, Systematique, Biologie*, Vol. 6 (ed. P.-P. Grassé), Masson, Paris, pp. 218–262.

S. Falkmer, 1972, 'Insulin production in vertebrates and invertebrates', *Gen. Comp. Endocrin.*, Suppl. 3, 184–191.

S. Falkmer, J. F. Cutfield, S. M. Cutfield, G. G. Dodson, J. Gliemann, S. Gammeltoft, M. Marques, J. D. Peterson, D. F. Steiner, F. Sundby, S. Emdin, N. Havu, Y. Östberg and L. Windbladh, 1975, 'Comparative endocrinology of insulin and glucagon production', *Amer. Zool.*, Suppl. **1**, 255–270.

S. Falkmer, S. Emdin, N. Havu, G. Lundgren, M. Marques, Y. Östberg, D. F. Steiner and N. W. Thomas, 1973, 'Insulin in invertebrates and cyclostomes', *Amer. Zool.*, **13**, 625–638.

S. Falkmer and L. Winbladh, 1964, 'Some aspects on the blood sugar regulation of the hagfish, *Myxine glutinosa*', in *The Structure and Metabolism of the Pancreatic Islets* (ed. S. E. Brolin, B. Hellman and H. Knutson), Pergamon, Oxford, pp. 33–43.

R. Fänge, 1963, 'Structure and function of the excretory organs of myxinoids', in *The Biology of Myxine* (ed. A. Brodal and R. Fänge), Universitetsforlaget, Oslo, pp. 516–529.

R. Fänge, G. Bloom and E. Östlund, 1963, 'The portal vein heart of myxinoids', in *The Biology of Myxine* (ed. A. Brodal and R. Fänge), Universitetsforlaget, Oslo, pp. 340–351.

A. Farmanfarmaian, 1966, 'The respiratory physiology of echinoderms', in *Physiology of Echinodermata* (ed. R. A. Boolootian), Wiley, New York, pp. 245–265.

J. S. Farris, 1966, 'Estimation of conservatism of characters by constancy within biological populations', *Evolution*, **20**, 587–591.

J. S. Farris, A. G. Kluge and M. J. Eckardt, 1970, 'A numerical approach to phylogenetic systematics', *Syst. Zool.*, **19**, 172–189.

H. B. Fell, 1948, 'Echinoderm embryology and the origin of chordates', *Biol. Rev.*, **23**, 81–107.

M. L. Fernald, 1926, 'The antiquity and dispersal of vascular plants', *Quart. Rev. Biol.*, **1**, 212–245.

B. Fernholm, 1972, 'Neurohypophysial–adenohypophysial relations in hagfish (*Myxinoidea, Cyclostomata*)', *Gener. Comp. Endocrinol.*, Suppl. 3, 1–10.

B. Fernholm and K. Holmberg, 1975, 'The eyes in three genera of hagfish (*Eptatretus, Paramyxine* and *Myxine*)—A case of degenerative evolution', *Vision Res.*, **15**, 253–259.

B. Fernholm and R. Olsson, 1969, 'A cytopharmacological study of the Myxine adenohypophysis', *Gener. Comp. Endocrinol.*, **13**, 336–356.

R. A. Fisher, 1958, *The Genetical Theory of Natural Selection*, Dover, New York.

W. M. Fitch and E. Margoliash, 1969, 'The construction of phylogenetic trees. II. How well do they reflect past history?', *Brookhaven Symp. Biol.*, **21**, 217–242.

P. R. Flood, D. M. Guthrie and J. R. Banks, 1969, 'Paramyosin muscle in the notochord of amphioxus', *Nature*, **222**, 87–88.

E. Florey, 1951, 'Reizphysiologische Untersuchungen an der Ascidie *Ciona intestinalis* L.', *Biol. Zbl.*, **70**, 523–530.

M. Florkin, 1947, *L'Evolution Biochimique* (2nd ed.), Masson, Paris.

——, 1960a, *Unity and Diversity in Biochemistry. An Introduction to Chemical Biology*, Pergamon, Oxford.

——, 1960b, 'Blood chemistry', in *The Physiology of Crustacea*, Vol. 1 (ed. T. H. Waterman), Academic Press, New York, pp. 141–159.

——, 1962, 'La régulation isosmotique intracellulaire chez les Invertébrés marins euryhalins', *Bull. Acad. Belg.* [5], **48**, 687–694.

——, 1966a, *A Molecular Approach to Phylogeny*, Elsevier, Amsterdam.

——, 1966b, 'Nitrogen metabolism', in *Physiology of Mollusca*, Vol. 2 (ed. K. M. Wilbur and C. M. Yonge), Academic Press, New York, pp. 309–351.

M. Florkin and G. Duchateau, 1943, 'Les formes du système enzymatique de l'uricolyse et l'évolution du catabolisme purique chez les animaux', *Arch. Internat. Physiol.*, **53**, 267–307.

B. F. Folkes, R. A. Grant and J. K. M. Jones, 1950, 'Frog spawn mucin', *J. Chem. Soc.*, **8**, 2136–2140.

A. R. Fontaine, 1962, 'Neurosecretion in the ophiurid *Ophiopholis aculeata*', *Science*, **138**, 908–909.

J. Fooden, 1972, 'Breakup of Pangaea and isolation of relict mammals in Australia, South America and Madagascar', *Science*, **175**, 894–898.

R. P. Forster, F. Berglund and B. R. Rennick, 1958, 'Tubular secretion of creatine, trimethylamine oxide, and other organic bases by the aglomerular kidney of Lophius americanus', *J. Gener. Physiol.*, **42**, 319–327.

R. P. Forster and L. Goldstein, 1969, 'Formation of excretory products', in *Fish Physiology*, Vol. 1 (ed. W. S. Hoar and D. J. Randall), Academic Press, New York, pp. 313–350.

D. L. Fox, 1953, *Animal Biochromes and Structural Colours*, Cambridge University Press, Cambridge.

R. M. Frank, R. F. Sognnaes and R. Kern, 1960, 'Calcification of dental tissues with special reference to enamel ultrastructure', in *Calcification in Biological Systems* (ed. R. F. Sognnaes), *Am. Ass. Adv. Sci.*, Washington D.C., pp. 162–202.

V. Franz, 1927, 'Morphologie der Akranier', *Ergebn. Anat. Entw.gesch.*, **27**, 682–692.

A. Fraser, 1967, 'Comments on mathematical challenges to the neo-Darwinian concept of evolution', in *Mathematical Challenges to the neo-Darwinian Interpretation of Evolution* (ed. P. S. Moorhead and M. M. Kaplan), Wistar, Philadelphia, p. 107.

C. Fulton, 1971, 'Centrioles', in *Origin and Continuity of Cell Organelles* (ed. J. Reinert and H. Ursprung), Springer, Berlin, pp. 170–221.

M. Gabe, 1966, *Neurosecretion*, Pergamon, Oxford.

——, 1970, 'The adrenal', in *Biology of the Reptilia*, Vol. 3 (ed. C. Gans and T. S. Parsons), Academic Press, London, pp. 263–318.

P. M. Galton, 1970, 'Ornitischian dinosaurs and the origin of birds', *Evolution*, **24**, 448–462.

B. G. Gardiner, 1973, 'Interrelationships of teleostomes', in *Interrelationships of Fishes* (ed. P. H. Greenwood, R. S. Miles and C. Patterson), Academic Press, London, pp. 105–135.

W. Garstang, 1894, 'Preliminary note on a new theory of the phylogeny of the Chordata', *Zool. Anz.*, **17**, 122–125.

——, 1922, 'The theory of recapitulation: A critical re-statement of the biogenetic law', *J. Linn. Soc. London*, **35**, 81–101.

——, 1928, 'The morphology of the Tunicata, and its bearing on the phylogeny of the Chordata', *Quart. J. Micros. Sci.*, **72**, 51–187.

W. H. Gaskell, 1908, *The Origin of Vertebrates*, Longmans Green, London.

——, 1910, 'Origin of the vertebrates', *Proc. Linn. Soc. London*, **122**, 9–50.

G. F. Gause, 1934, *The Struggle for Existence*, Williams and Wilkins, Baltimore.

E. Geoffroy Saint-Hilaire, 1830, *Principes de Philosophie Zoologique*, Pichon, Didier, Rousseau, Paris.

U. Gerhardt, 1933, 'Kloake und Begattungsorgane', in *Handbuch der vergleichende Anatomie der Wirbeltiere*, Vol. 6 (ed. L. Bolk, E. Göppert, E. Kallius and W. Lubosch), Urban & Schwarzenberger, Berlin, pp. 267–350.

F. Ghiretti, 1966, 'Respiration', in *Physiology of Mollusca*, Vol. 2 (ed. K. M. Wilbur and C. M. Yonge), Academic Press, New York, pp. 175–208.

M. T. Ghiselin, E. T. Degens, D. W. Spencer and R. H. Parker, 1967, 'A phylogenetic survey of molluscan shell matrix proteins', *Mus. Comp. Zool. Harvard, Breviora*, **262**, 1–35.

M. J. Glimcher, 1960, 'Specificity of the molecular structure of organic matrices in mineralization', in *Calcification in Biological Systems* (ed. R. F. Sognnaes), Am. Ass. Adv. Sci., Washington, D.C., pp. 421–487.

O. B. Goin and C. J. Goin, 1967, 'The nuclear DNA of a caecilian', *Copeia*, **1967**, 233.

O. B. Goin, C. J. Goin and K. Bachmann, 1968, 'DNA and amphibian life history', *Copeia*, **1968**, 532–540.

R. Goldschmidt, 1940, *The Material Basis of Evolution*, Yale University Press, New Haven.

R. A. Good, 1969, 'General discussion *Myxina glutinosa*', in *Cellular Recognition* (ed. R. T. Smith and R. A. Good), North-Holland, Amsterdam, pp. 99–102.

I. Goodbody, 1957, 'Nitrogen excretion in Ascidiacea I. Excretion of ammonia and total non-protein nitrogen', *J. Exptl. Biol.*, **34**, 297–305.

——, 1965, 'Nitrogen excretion in Ascidiacea II. Storage excretion and the uricolytic enzyme system', *J. Exptl. Biol.*, **42**, 299–305.

M. Goodman, G. W. Moore and G. Matsuda, 1975, 'Darwinian evolution in the genealogy of haemoglobin', *Nature*, **253**, 603–608.

E. S. Goodrich, 1907, 'On the scales of fish, living and extinct, and their importance in classification', *Proc. Zool. Soc. London*, 751–774.

——, 1909, *A Treatise on Zoology*, Vol. 9 (ed. R. Lankester), Black, London.

——, 1916, 'On the classification of the Reptilia', *Proc. Roy. Soc. London* [B], **89**, 261–276.

——, 1928, 'Polypterus a palaeoniscid?', *Palaeobiologica*, **1**, 87–92.

——, 1958, *Studies on the Structure and Development of Vertebrates*, Dover, New York.

T. W. Goodwin, 1960, 'Biochemistry of pigments', in *The Physiology of Crustacea*, Vol. 1 (ed. T. H. Waterman), Academic Press, New York, pp. 101–140.

A. Gorbman, 1963, 'The myxinoid thyroid gland', in *The Biology of Myxine* (ed. A. Brodal and R. Fänge), Universitetsforlaget, Oslo, pp. 477–480.

S. J. Gould, 1966, 'Allometry and size in ontogeny and phylogeny', *Biol. Rev.*, **41**, 587–640.

——, 1970, 'Dollo on Dollo's law: Irreversibility and the status of evolutionary laws', *J. Hist. Biol.*, **3**, 189–212.

——, 1974, 'The origin and function of "bizarre" structures: Antler size and skull size in the "Irish elk", *Megaloceros giganteus*', *Evolution*, **28**, 191–220.

A. L. Grafflin and R. G. Gould, 1936, 'Renal function in marine teleosts II. The nitrogenous constituents in the urine of sculpin and flounder, with particular reference to trimethylamine oxide', *Biol. Bull.*, **70**, 16–27.

P.-P. Grassé, 1948, *Traité de Zoologie Anatomie, Systematique, Biologie*, Vol. 11, Masson, Paris.

——, 1973, *L'évolution du Vivant*, Michel, Paris.

J. R. Gregg, 1954, *The Language of Taxonomy. An Application of Symbolic Logic to the Study of Classificatory Systems*, Columbia University Press, New York.

C. Grégoire and H. J. Tagnon, 1962, 'Blood coagulation', in *Comparative Biochemistry. A Comprehensive Treatise*, Vol. 4 (ed. M. Florkin and H. S. Mason), Academic Press, New York, pp. 435–482.

W. K. Gregory, 1951, *Evolution Emerging. A Survey of Changing Patterns from Primeval Life to Man*, Macmillan, New York.

M. Grene, 1969, 'Bohm's metaphysics and biology'. 'Notes on Maynard Smith's "Status of neo-Darwinism"', in *Towards a Theoretical Biology*, Vol. 2 (ed. C. H. Waddington), Edinburgh University Press, Edinburgh, pp. 61–69; 97–98.

K. Grobben, 1908, 'Die systematische Einteilung des Tierreiches', *Verh. Zool. Bot. Ges. Wien.*, **58**, 491–511.

J. Gross, 1963, 'Comparative biochemistry of collagen', in *Comparative Biochemistry. A Comprehensive Treatise*, Vol. 5 (ed. M. Florkin and H. S. Mason), Academic Press, New York, pp. 307–346.

J. Gross, B. Dumsha and N. Glazer, 1958, 'Comparative biochemistry of collagen. Some amino acids and carbohydrates', *Biochim. Biophys. Acta*, **30**, 293–297.

J. Gross, A. G. Matoltsy and C. Cohen, 1955, 'Vitrosin: a member of the collagen class', *J. Biophys. Biochem. Cytol.*, **1**, 215–220.

W. Gross, 1933, 'Die phylogenetische Bedeutung der altpaläozoischen Agnathen und Fische', *Palaeont. Z.*, **15**, 102–137.

——, 1950, 'Die paläontologische und stratigraphische Bedeutung der Wirbeltierfaunen des

Old Reds und der marinen altpaläozoischen Schichten', *Abh. dtsch. Akad. Wiss. Berlin*, **1949** [1], 1–130.

——, 1968, 'Fragliche Actinopterygier-Schuppen aus dem Silur Gotlands', *Lethaia*, **1**, 184–218.

W. Haacke, 1893, *Gestaltung und Vererbung*, Weigel, Leipzig.

E. Haeckel, 1857, 'Ueber die Gewebe des Flusskrebses', *Arch. Anat. Physiol. Wiss. Med.*, 469–568.

——, 1866, *Generelle Morphologie der Organismen*, Reiner, Berlin.

D. D. Hagerman, F. M. Wellington and C A. Villee, 1957, 'Estrogen in marine invertebrates', *Biol. Bull.*, **112**, 1–7.

G. Haggag and Y. Fouad, 1968, 'Comparative study of nitrogenous excretion in terrestrial and fresh-water gastropods', *Z. Vergl. Physiol.*, **57**, 428–431.

J. B. S. Haldane, 1932, *The Causes of Evolution*, Longmans Green, London.

——, 1957, 'The cost of natural selection', *J. Genet.*, **55**, 511–524.

W. D. Halliburton, 1885, 'On the occurrence of chitin as a constituent of the cartilages of Limulus and Sepia', *Quart. J. Micros. Sci.*, **25**, 173–181.

C. S. Hammen, H. F. Miller and W. F. Geer, 1966, 'Nitrogen excretion of *Crassostrea virginia*', *Comp. Biochem. Physiol.*, **17**, 1199–1200.

B. Hanström, 1928, *Vergleichende Anatomie des Nervensystems der wirbellosen Tiere unter Berücksichtigung seiner Funktion*, Springer, Berlin.

P. E. Hare and P. H. Abelson, 1965, 'Amino acid composition of some calcified proteins', *Carnegie Inst. Wash. Year Book*, **64**, 223–231.

G. A. D. Haslewood, 1967, *Bile Salts*, Methuen, London.

——, 1968, 'Bile salt differences in relation to taxonomy and systematics', in *Chemotaxonomy and Serotaxonomy* (ed. J. G. Hawkes), Academic Press, London, pp. 159–172.

R. R. Hathaway, 1965, 'Conversion of estradiol-17β by sperm preparations of sea urchins and oysters', *Gener. Comp. Endocrinol.*, **5**, 504–508.

J. D. Hays and W. C. Pitman, 1973, 'Lithospheric plate motion, sea level changes and climatic and ecological consequences', *Nature*, **246**, 18–22.

E. G. Healey, 1948, 'The colour change of the minnow (*Phoxinus laevis* Ag.)', *Bull. Anim. Behav.*, **6**, 5–15.

W. Hennig, 1966, *Phylogenetic Systematics*, University of Illinois Press, Urbana.

K. L. Henriksen, 1931, 'The manner of moulting in Arthropoda', *Notulae Entomol.*, **11**, 103–127.

O. Hertwig and R. Hertwig, 1881, *Die Coelomtheorie Versuch einer Erklärung des mittleren Keimblattes*, Fischer, Jena.

C. P. Hickman and B. J. Trump, 1969, 'The kidney', in *Fish Physiology*, Vol. 1 (ed. W. S. Hoar and D. J. Randall), Academic Press, New York, pp. 91–239.

K. C. Highnam and L. Hill, 1969, *The Comparative Endocrinology of the Invertebrates*, Arnold, London.

W. H. Hildemann and G. H. Thoenes, 1969, 'Immunological responses of pacific hagfish I. Skin transplantation immunity', *Transplantation*, **7**, 506–521.

R. B. Hill and J. H. Welsh, 1966, 'Heart, circulation, and blood cells', in *Physiology of Mollusca* (ed. K. M. Wilbur and C. M. Yonge), Academic Press, New York, pp. 125–174.

R. Hinegardner and D. F. Rosen, 1968, 'Cellular DNA content and the evolution of teleostean fishes', *Amer. Natur.*, **106**, 621–644.

K. Hirose, B.-I. Tamaoki, B. Fernholm and H. Kobayashi, 1975, '*In vitro* bioconversions of steroids in the mature ovary of the hagfish, *Eptatretus Burgeri*', *Comp. Biochem. Physiol.*, **51B**, 403–408.

F. L. Hisaw Jr., C. R. Botticelli and F. L. Hisaw, 1962, 'The relation of the cerebral ganglion–subneural gland complex to reproduction in the ascidian, *Chelyosoma productum*', *Am. Zool.*, **2**, 415.

H. Hiyami, 1949, 'Biochemical studies on carbohydrates. CII. On the jelly of toad and frog eggs', *Tohoku J. Exptl. Med.*, **50**, 373–378. 'CIII. Hexosamine in toad mucin', *Ibid.*, **50**, 379–383. 'CIV. Hexosamine in frog mucin', *Ibid.*, **50**, 385–387.

W. S. Hoar, 1966, *General and Comparative Physiology*, Prentice-Hall, Englewood Cliffs, N.J.

R. Hoffstetter, 1969, 'Un Primate de l'Oligocène inférieur sud-américain: *Branisella boliviana* gen. et sp. nov', *Compt. Rend. Acad. Sci. Paris* [D], **269**, 434–437.

——, 1970, 'Radiation initiale des Mammifères placentaires et biogéographie', *Compt. Rend. Acad. Sci. Paris* [D], **270**, 3127–3030.

R. Hoffstetter and R. Lavocat, 1970, 'Découverte dans le Déséadien de Bolivie de genres penta-lophodontes appuyant les affinités des Rongeurs caviomorphes', *Compt. Rend. Acad. Sci. Paris* [D], **271**, 172–175.

L. Höglund, 1976, 'The comparative biochemistry of invertebrate mucopolysaccharides—V Insecta (*Calliphora erythrocephala*)', *Comp. Biochem. Physiol.*, **53B**, 9–14.

L. Höglund and S. Løvtrup, 1976, 'Changes in acid mucopolysaccharides during the development of the frog *Rana temporaria*', *Acta Embryol. Exp.*, **1976**, 63–79.

K. Holmberg and P. Öhman, 1976, 'Fine structure of retinal synaptic organelles in lamprey and hagfish photoreceptors', *Vision Res.*, **16**, 237–239.

W. N. Holmes and E. M. Donaldson, 1969, 'The body compartments and the distribution of electrolytes', in *Fish Physiology*, Vol. 1 (ed. W. S. Hoar and D. J. Randall), Academic Press, New York, pp. 1–89.

N. Holmgren, 1922, 'Points of view concerning forebrain morphology in lower vertebrates', *J. Comp. Neurol.*, **34**, 391–459.

——, 1933, 'On the origin of the tetrapod limb', *Acta Zool.*, **14**, 185–295.

——, 1950, 'On the pronephros and the blood in Myxine glutinosa', *Acta Zool.*, **31**, 233–248.

N. Holmgren and E. Stensiö, 1936, 'Kranium und Visceralskelett der Akranier, Cyclostomen und Fische', in *Handbuch der vergleichenden Anatomie der Wirbertiere*, Vol. 4 (ed. L. Bolk, E. Göppert, E. Kallius and W. Lubosch), Urban & Schwarzenberg, Berlin, pp. 233–500.

R. Holmquist, 1972, 'Theoretical foundations of paleogenetics', in *Proc. 6th Berkeley Symp. Math. Statis. Probab.*, Vol. 5 (ed. L. M. LeCam, J. Neyman and E. L. Scott), University of California Press, Berkeley, pp. 315–350.

J. Holtfreter, 1938a, 'Differenzierungspotenzen isolierter Teile der Urodelengastrula', *Arch. Entw.mech.*, **138**, 522–656.

——, 1938b, 'Differenzierungspotenzen isolierter Teile der Anurengastrula', *Arch. Entw.mech.*, **138**, 657–738.

——, 1943, 'Properties and functions of the surface coat in the amphibian embryo', *J. Exptl. Zool.*, **93**, 251–323.

F. Horne and V. Boonkoom, 1970, 'The distribution of the ornithine cycle enzymes in twelve gastropods', *Comp. Biochem. Physiol.*, **32**, 141–153.

S. Hörstadius, 1935, 'Über die Determination im Verlaufe der Eiachse bei Seeigeln', *Pubbl. Staz. Zool. Napoli*, **14**, 251–479.

R. Hubbard and R. C. C. St. George, 1958, 'The rhodopsin system of the squid', *J. Gen. Physiol.*, **41**, 501–528.

R. Hubbard and G. Wald, 1960, 'Visual pigment of the horse shoe crab, Limulus polyphemus', *Nature*, **186**, 212–215.

D. L. Hull, 1973, *Darwin and His Critics*, Harvard University Press, Cambridge, Mass.

A. A. Humphries, 1970, 'Incorporation of ^{35}S-sulfate into the oviducts and egg jelly of the newt, *Notophtalmus viridescens*', *Exptl. Cell Res.*, **59**, 157–161.

S. Hunt, 1970, *Polysaccharide–Protein Complexes in Invertebrates*, Academic Press, London.

S. Hunt and F. R. Jevons, 1966, 'The hypobranchial mucin of the whelk *Buccinum undatum* L. The polysaccharide component', *Biochem. J.*, **98**, 522–529.

G. E. Hutchinson, 1959, 'Homage to Santa Rosalia *or* Why are there so many kinds of animals?', *Am. Natur.*, **93**, 145–159.

G. E. Hutchinson and R. H. MacArthur, 1959, 'A theoretical ecological model of size distributions among species of animals', *Am. Natur.*, **93**, 117–125.

J. S. Huxley, 1932, *Problems of Relative Growth*, Methuen, London.

——, 1940, 'Towards the new systematics', in *The New Systematics* (ed. J. Huxley). Clarendon, Oxford, pp. 1–46.

——, 1959, *Evolution: The Modern Synthesis*, Wiley, New York.

L. Huxley, 1900, *Life and Letters of Thomas Henry Huxley*, Macmillan, London.

L. H. Hyman, 1942, *Comparative Vertebrate Anatomy* (2nd ed.), University of Chicago Press, Chicago.

——, 1951, *The Invertebrates*, Vol. 2, McGraw-Hill, New York.

——, 1955, *The Invertebrates*, Vol. 4, McGraw-Hill, New York.

——, 1959, *The Invertebrates*, Vol. 5, McGraw-Hill, New York.

D. R. Idler, G. B. Sangalang and M. Weisbart, 1971, 'Are corticosteroids present in the blood of all fish?', in *Hormonal Steroids* (ed. V. H. T. James and L. Martini), Excerpta Medica, Amsterdam, pp. 983–989.

M. J. Imlay and A. B. Chaet, 1965, 'Microscopic observations of gamete shedding substance in starfish radial nerves', *Fed. Proc.*, **24**, 129.

S. Inoué, 1965, 'Isolation of new polysaccharide sulfates from *Charonia lampas*', *Biochim. Biophys. Acta*, **101**, 16–25.

F. Jacob and J. Monod, 1961, 'Genetic regulatory mechanisms in the synthesis of proteins', *J. Mol. Biol.*, **3**, 318–356.

——, 1963, 'Genetic repression, allosteric inhibition, and cellular differentiation, in *Cytodifferentiation and Macromolecular Synthesis* (ed. M. Locke), Academic Press, New York, pp. 30–64.

G. Jägersten, 1955, 'On the early phylogeny of the Metazoa. The Bilaterogastrea theory', *Zool. Bidrag Uppsala*, **30**, 321–354.

S. P. James, 1951, 'An examination of frog jelly', *Biochem. J.*, **49**, LIV.

N. Jardine and D. McKenzie, 1972, 'Continental drift and the dispersal and evolution of organisms', *Nature*, **235**, 20–24.

E. Jarvik, 1942, 'On the structure of the snout of crossopterygians and lower gnathostomes in general', *Zool. Bidrag Uppsala*, **21**, 235–675.

——, 1952, 'On the fish-like tail in the ichthyostegid stegocephalians, with descriptions of a new stegocephalian and a new crossopterygian from the Upper Devonian at East Greenland', *Medd. Grønland*, **114**, 1–90.

——, 1959, 'Dermal fin-rays and Holmgren's principle of delamination, *Kungl. Sv. Vetenskaps-akad. Handl.* [4], **6**, 3–51.

——, 1964, 'Specializations in early vertebrates', *Ann. Soc. Roy. Zool. Belg.*, **94**, 11–95.

——, 1965, 'Die Raspelzunge der Cyclostomen und die pentadaktyle Extremität der Tetrapoden als Beweise für monophyletische Herkunft', *Zool. Anz.*, **175**, 101–143.

——, 1968a, 'The systematic position of the Dipnoi', in *Current Problems of Lower Vertebrate Phylogeny* (ed. T. Ørvig), Almqvist & Wiksell, Stockholm, pp. 223–245.

——, 1968b, 'Aspects of vertebrate phylogeny', in *Current Problems of Lower Vertebrate Phylogeny* (ed. T. Ørvig), Almqvist & Wiksell, Stockholm, pp. 497–527.

——, 1972, 'Middle and upper Devonian Porolepiformes from East Greenland with special reference to *Glyptolepis* groenlandica n. sp. and a discussion on the structure of the head in the Porolepiformes', *Medd. Grønland*, **187** [2], 1–307.

R. P. S. Jefferies, 1968, 'Some fossil chordates with echinoderm affinities', *Symp. Zool. Soc. London*, **20**, 163–208.

D. Jensen, 1961, 'Cardioregulation in an aneural heart', *Comp. Biochem. Physiol.*, **2**, 181–201.

G. L. Jepsen, 1944, 'Phylogenetic trees', *Trans. N.Y. Acad. Sci.* [2], **6**, 81–92.

H. J. Jerison, 1971, 'Quantitative analysis of the evolution of the camelid brain', *Am. Natur.*, **105**, 227–239.

C. Jeuniaux, 1963, *Chitine et Chitinolyse. Un Chapitre de la Biologie Moléculaire*, Masson, Paris.

——, 1971, 'Chitinous structures', in *Comprehensive Biochemistry*, Vol. 26C (ed. M. Florkin and E. H. Stotz), Elsevier, Amsterdam, pp. 595–632.

M. M. Jezewska, 1968, 'The presence of uric acid, xanthine and guanine in the haemolymph of the snail *Helix pomatia* (*Gastropoda*)', *Bull. Acad. Pol. Sci.* [2], **16**, 73–76.

M. M. Jezewska, B. Gorzkowski and J. Heller, 1963, 'Seasonal changes in the excretion of nitrogen wastes in *Helix pomatia*', *Acta Biochim. Pol.*, **10**, 309–314.

K. Johansen, 1963, 'The cardiovascular system of Myxine glutinosa L', in *The Biology of Myxine* (ed. A. Brodal and R. Fänge), Universitetsforlaget, Oslo, pp. 289–316.

K. Johansen and A. W. Martin, 1962, 'Circulation in the cephalopod, *Octopus dofleini*', *Comp. Biochem. Physiol.*, **5**, 161–176.

A. G. Johnels, 1950, 'On the dermal connective tissue of the head of Petromyzon', *Acta Zool.*, **31**, 177–185.

——, 1956, 'On the peripheral autonomic nervous system of the trunk region of *Lampetra planeri*', *Acta Zool.*, **37**, 251–286.

I. C. Jones, 1963, 'Adrenocorticosteroids', in *The Biology of Myxine* (ed. A. Brodal and R. Fänge), Universitetsforlaget, Oslo, pp. 488–502.

I. C. Jones and J. G. Phillips, 1960, 'Adrenocorticosteroids in fish', *Symp. Zool. Soc. London*, **1**, 17–32.

T. H. Jukes, 1966, *Molecules and Evolution*, Columbia University Press, New York.

T. H. Jukes and R. Holmquist, 1972, 'Evolutionary clock: Nonconstancy of rate in different species', *Science*, **177**, 530–532.

A. Jullien, J. Cardot and J. Ripplinger, 1957, 'De l'existence de fibres élastiques dans l'appareil circulatoire des Mollusques', *Ann. Sci. Univ. Besançon Zool. et Physiol.*, **9**, 25–33.

H. Kanatani, 1964, 'Spawning of starfish: action of gamete-shedding substance obtained from radial nerves', *Science*, **146**, 1177–1179.

H. Kanatani, H. Shirai, K. Nakamishi and T. Kurokawa, 1969, 'Isolation and identification of meiosis inducing substance in starfish *Asterias amurensis*', *Nature*, **221**, 273–274.

R. L. Katzman, A. K. Bhattacharyya and R. W. Jeanloz, 1969, 'Invertebrate connective tissue I. The amino acid and carbohydrate composition of the collagen from *Thyone briareus*', *Biochim. Biophys. Acta*, **184**, 523–528.

R. L. Katzman and R. W. Jeanloz, 1969, 'Acid polysaccharides from invertebrate connective tissue: Phylogenetic aspects', *Science*, **166**, 758–759.

Y. Kawai, N. Seno and K. Anno, 1966, 'Chondroitin polysulfate of squid cartilage', *J. Biochem.* (Tokyo), **60**, 317–321.

A. Keast, 1971, 'Continental drift and the evolution of the biota on Southern Continents', *Quart. Rev. Biol.*, **46**, 335–378.

G. A. Kerkut, 1960, *The Implications of Evolution*, Pergamon, New York.

T. Kerr, 1955, 'The scales of modern lungfish', *Proc. Zool. Soc. London*, **125**, 335–345.

M. Kimura, 1968, 'Evolutionary rate at the molecular level', *Nature*, **217**, 624–626.

——, 1969, 'The rate of molecular evolution considered from the standpoint of population genetics', *Proc. Nat. Acad. Sci. USA*, **63**, 1181–1188.

M. Kimura and T. Ohta, 1971, *Theoretical Aspects of Population Genetics*, Princeton University Press, Princeton.

——, 1972, 'Population genetics, molecular biometry, and evolution', *Proc. 6th Berkeley Symp. Math. Statis. Probab.*, Vol. 5 (ed. L. M. LeCam, J. Neyman and E. L. Scott), University of California Press, Berkeley, pp. 43–68.

J. L. King, 1972, 'The role of mutation in evolution', *Proc. 6th Berkeley Symp. Math. Statis. Probab.*, Vol. 5 (ed. L. M. LeCam, J. Neyman and E. L. Scott), University of California Press, Berkeley, pp. 69–100.

J. L. King and T. H. Jukes, 1969, 'Non-Darwinian evolution', *Science*, **164**, 788–798.

S. Kinoshita, 1969, 'Periodical release of heparin-like polysaccharides within cytoplasms during cleavage of sea urchin egg', *Exp. Cell Res.*, **56**, 39–43.

L. Klein and J. D. Currey, 1970, 'Echinoid skeleton: Absence of collagenous matrix', *Science*, **169**, 1209–1210.

L. Kleinholz, 1961, 'Pigmentary effectors', in *The Physiology of Crustacea*, Vol. 2 (ed. T. H. Waterman), Academic Press, New York, pp. 133–169.

H. Kobayashi, 1964, 'On the photo-perceptive function in the eye of the hagfish, *Myxine garmani* Jordan et Snyder', *J. Shimonoseki Univ. Fish.*, **13**, 67–83.

E. Korschelt and K. Heider, 1936, *Vergleichende Entwicklungsgeschichte der Tiere* (rev. ed., ed. E. Korschelt), Fischer, Jena.

A. Kowalevsky, 1867, 'Entwicklungsgeschichte der einfachen Ascidien', *Mém. Acad. Imp. Sci. St.-Petersb.* [7], **10**, 1–19.

——, 1871, 'Weitere Studien über die Entwicklung der einfachen Ascidien', *Arch. Mikros. Anat.*, **7**, 101–130.

F. B. Krasne and P. A. Lawrence, 1966, 'Structure of the photoreceptors in the compound eyespots of *Branchiomma vesiculosum*', *J. Cell Sci.*, **1**, 239–248.

B. J. Krijgsman, 1956, 'Contractile and pacemaker mechanisms of the heart of tunicates', *Biol. Rev.*, **31**, 288–312.

B. J. Krijgsman and G. A. Divaris, 1955, 'Contractile and pacemaker mechanisms of the heart of molluscs', *Biol. Rev.*, **30**, 1–39.

A. Krogh, 1939, *Osmotic Regulation in Aquatic Animals*, Cambridge University Press, Cambridge.

H. Kuhlenbeck, 1967, *The Central Nervous System*, Vol. 1, Karger, Basel.

A. Lang, 1900, *Lehrbuch der vergleichenden Anatomie der wirbellosen Tiere* (2nd rev. ed.), Vol. 3, 1. Mollusca (rev. K. Hescheler), Fischer, Jena.

O. Larsell, 1967, *The Comparative Anatomy and Histology of the Cerebellum from Myxinoids through Birds*, University of Minnesota Press, Minneapolis.

L. O. Larsen, 1973, *Development in Adult, Freshwater River Lampreys and its Hormonal Control. Starvation, Sexual Maturation and Natural Death*, University of Copenhagen, Copenhagen.

R. L. Larson and W. C. Pitman, 1972, 'World-wide correlation of Mesozoic magnetic anomalies, and its implications', *Geol. Soc. Am. Bull.*, **83**, 3645–3662.

J. W. Lash and M. W. Whitehouse, 1960a, 'An unusual polysaccharide in chondroid tissue of the snail *Busycon*: polyglucose sulphate', *Biochem. J.*, **74**, 351–355.

——, 1960b, 'Variation in the polysaccharide composition of cartilage with age', *Arch. Biochem. Biophys.*, **90**, 159–160.

P. Laviolette, 1954, 'Role de la gonade dans le déterminisme humoral de la maturité glandulaire du tractus génital chez quelques Gastéropodes *Arionidae* et *Limacidae*', *Bull. Biol.*, **38**, 310–332.

R. Lavocat, 1969, 'La systématique des Rongeurs hystricomorphes et la dérive des continents', *Compt. Rend. Acad. Sci. Paris* [D], **269**, 1496–1497.

P. A. Lee, 1967, 'Studies of frog oviducal jelly secretion I. Chemical analysis of secretory product', *J. Exptl. Zool.*, **166**, 99–106.

J.-P. Lehman, 1966a, 'Actinopterygii', in *Traité de Paléontologie*, Vol. 4,3 (ed. J. Piveteau), Masson, Paris, pp. 1–242.

——, 1966b, 'Dipnoi et Crossopterygii', in *Traité de Paléontologie*, Vol. 4,3 (ed. J. Piveteau), Masson, Paris, pp. 243–412.

——, 1966c, 'Brachiopterygii', in *Traité de Paléontologie*, Vol. 4,3 (ed. J. Piveteau), Masson, Paris, pp. 413–420.

C. A. Leone, 1954, 'Serological studies of some arachnids, other arthropods, and molluscs', *Physiol. Zoöl.*, **27**, 317–325.

A. B. Lerner, J. D. Case and R. V. Heinzelman, 1959, 'Structure of melatonin', *J. Am. Chem. Soc.*, **81**, 6064–6085.

J. Lever, J. Jansen and T. A. De Vlieger, 1961, 'Pleural ganglia and water balance in the fresh water pulmonate *Limnaea stagnalis* L.', *Proc. Kon. Nederl. Akad. Wetensch.* [C], **64**, 531–542.

P. T. Levine, M. J. Glimcher, J. M. Seyer, J. I. Huddleston and J. W. Hein, 1966, 'Noncollagenous nature of the proteins of shark enamel', *Science*, **154**, 1192–1194.

E. B. Lewis, 1951, 'Pseudoallelism and gene evolution', *Cold Spring Harbor Symp. Quant. Biol.*, **16**, 159–174.

J. B. Lewis, 1967, 'Nitrogenous excretion in the tropical sea urchin Diadema antillarum Phillippi', *Biol. Bull.*, **132**, 34–37.

R. C. Lewontin, 1974, *The Genetic Basis of Evolutionary Change*, Columbia University Press, New York.

J. A. Lillegraven, 1969, 'Latest Cretaceous mammals of upper part of Edmonton formation of Alberta, Canada, and review of marsupial–placental dichotomy in mammalian evolution', *Paleont. Contrib. Univ. Kansas*, **50**, 1–122.

T. J. Linna, J. Finstad, K. E. Fichtelius and R. A. Good, 1970, 'Proliferative responses of cyclostomes to stimulation with antigen in adjuvant', *Acta Path. Microbiol. Scand.* [A], **78**, 169–178.

B. Lofts, 1968, 'Patterns of testicular activity', in *Perspectives in Endocrinology. Hormones in the Lives of Lower Vertebrates* (ed. E. J. W. Barrington and C. B. Jørgensen), Academic Press, London, pp. 239–304.

S. Løvtrup, 1973, 'Classification, convention and logic', *Zool. Scripta*, **2**, 49–61.

——, 1974, *Epigenetics—A Treatise on Theoretical Biology*, Wiley, London.

——, 1975a, 'Validity of the Protostomia–Deuterostomia theory', *Syst. Zool.*, **24**, 96–108.

——, 1975b, 'A reexamination of the Arachnid theory on the origin of Vertebrata', *Zool. Scr.*, **4**, 125–131.

——, 1975c, 'On phylogenetic classification', *Acta Zool. Cracov.*, **20**, 499–524.

S. Løvtrup, F. Rahemtulla and N.-G. Höglund, 1974, 'Fisher's axiom and the body size of animals', *Zool. Scripta*, **3**, 53–58.

S. Løvtrup and B. von Sydow, 1974, 'D'Arcy Thompson's theorems and the shape of the molluscan shell', *Bull. Math. Biol.*, **36**, 567–575.

——, 1976, 'An addendum to D'Arcy Thompson's theorems and the shape of the molluscan shell', *Bull. Math. Biol.*, **38**, 321–322.

P. Lubet and J. P. Pujol, 1963, 'Sur l'évolution du système neurosécréteur de *Mytilus gallopro-vincialis* Lmk. (Mollusque Lamellibranche) lors de variations de la salinité', *Compt. rend. Acad. Sci. Paris* [D], **257**, 4032–4034.

S. C. Lum and C. S. Hammen, 1964, 'Ammonia excretion of Lingula', *Comp. Biochem. Physiol.*, **12**, 185–190.

G. D'A. Lunetta, 1971, 'An analysis of the ovular mucin of *Physa acuta* (Pulmonate Gastropod)', *Acta Embryol. Exp.*, 103–108.

N. Macbeth, 1971, *Darwin Retried An Appeal to Reason*, Gambit, Boston.

W. N. McFarland and F. W. Munz, 1958, 'A re-examination of the osmotic properties of the Pacific hagfish, Poliostrema stouti', *Biol. Bull.*, **114**, 348–356.

P. J. McLaughlin and M. O. Dayhoff, 1973, 'Eukaryote evolution: a view based on cytochrome c sequence data', *J. Molec. Evol.*, **2**, 99–116.

W. Makarewicz, 1963, 'AMP-aminohydrolase and glutaminase activities in the kidneys and gills of some freshwater vertebrates', *Acta Biochim. Pol.*, **10**, 363–369.

W. Manski, S. P. Halbert, T. Auerbach-Pascal and P. Javier, 1967a, 'On the use of antigenic relationships among species for the study of molecular evolution I. The lens proteins of Agnatha and Chondrichthyes', *Internat. Arch. Allergy*, **31**, 38–56.

W. Manski, S. P. Halbert and P. Javier, 1967b, 'On the use of antigenic relationships among species for the study of molecular evolution II. The lens proteins of the Choanichthyes and early Actino-pterygii', *Internat. Arch. Allergy*, **31**, 475–489.

W. Manski, S. P. Halbert, P. Javier and T. Auerbach-Pascal, 1967c, 'On the use of antigenic rela-tionships among species for the study of molecular evolution III. The lens proteins of the late Actinopterygii', *Internat. Arch. Allergy*, **31**, 529–545.

S. M. Manton, 1973, 'Arthropod phylogeny—a modern synthesis', *J. Zool. London*, **171**, 111–130.

C. Manwell, 1963, 'The blood proteins of cyclostomes. A study in phylogenetic and ontogenetic biochemistry', in *The Biology of Myxine* (ed. A. Brodal and R. Fänge), Universitetsforlaget, Oslo, pp. 372–455.

C. De Marco and E. Antonini, 1958, 'Amino-acid composition of haemoglobin from *Thunnus thynnus*', *Nature*, **181**, 1128.

E. Margoliash, 1973, 'The molecular variation of cytochrome c as a function of the evolution of species', in *From Theoretical Physics to Biology* (ed. M. Marois), Karger, Basel, pp. 175–239.

C. L. Markert (ed.), 1975, *Isozymes III Developmental Biology*, Academic Press, New York.

A. W. Martin and F. M. Harrison, 1966, 'Excretion', in *The Physiology of Mollusca*, Vol. 2 (ed. K. M. Wilbur and C. M. Yonge), Academic Press, New York, pp. 353–386.

M. B. Mathews, 1967, 'Macromolecular evolution of connective tissue', *Biol. Rev.*, **42**, 499–551.

——, 1975, *Connective Tissue Macromolecular Structure and Evolution*, Springer, Berlin.

M. B. Mathews, J. Duh and P. Person, 1962, 'Acid mucopolysaccharides of invertebrate cartilage', *Nature*, **193**, 378–379.

A. G. Matoltsy, 1962, 'Mechanisms of keratinization', in *Fundamentals of Keratinization* (ed. E. O. Butcher and R. F. Sognnaes), *Am. Ass. Adv. Sci.*, Washington, D.C., pp. 1–25.

W. D. Matthew, 1915, 'Climate and evolution', *Ann. N.Y. Acad. Sci.*, **24**, 171–318.

R. Matthey, 1949, *Les Chromosomes des Vertébrés*, Rouge, Lausanne.

——, 1954, 'Les chromosomes des Vertébrés, Généralités', in *Traité de Zoologie Anatomie, Systématique, Biologie*, Vol. 12 (ed. P.-P. Grassé), Masson, Paris, pp. 1044–1063.

D. M. Maynard, 1960, 'Circulation and heart function', in *The Physiology of Mollusca*, Vol. 1 (ed. T. H. Waterman), Academic Press, New York, pp. 161–226.

E. Mayr, 1960, in *Issues in Evolution*, Vol. 3 (ed. S. Tax and C. Callender), University of Chicago Press, Chicago, p. 141.

——, 1969, *Principles of Systematic Zoology*, McGraw-Hill, New York.

E. Mayr and D. Amadon, 1951, 'A classification of recent birds', *Am. Mus. Novitates*, No. 1496.

V. R. Meenakshi, 1954, 'Galactogen in some common South Indian gastropods with special reference to *Pila*', *Curr. Sci.*, **23**, 301–302.

V. R. Meenakshi and B. T. Scheer, 1968, 'Studies on the carbohydrates of the slug *Ariolimax columbianus* with special reference to their distribution in the reproductive system', *Comp. Biochem. Physiol.*, **26**, 1091–1097.

V. R. Meenakshi and B. T. Scheer, 1969, 'Regulation of galactogen synthesis in the slug *Ariolimax columbianus*', *Comp. Biochem. Physiol.*, **29**, 841–845.

E. H. Mercer, 1961, *Keratin and Keratinization. An Essay in Molecular Biology*, Pergamon, New York.

R. S. Miles, 1967, 'Observations on the ptyctodont fish, *Rhamphodopsis* Watson', *J. Linn. Soc. London (Zool.)*, **47**, 99–120.

——, 1968, 'Jaw articulation and suspension in *Acanthodes* and their significance', in *Current Problems of Lower Vertebrate Phylogeny* (ed. T. Ørvig), Almqvist & Wiksell, Stockholm, pp. 109–127.

——, 1973, 'Relationships of acanthodians', in *Interrelationships of Fishes* (ed. P. H. Greenwood, R. S. Miles and C. Patterson), Academic Press, London, pp. 63–104.

W. H. Miller, 1958, 'Derivatives of cilia in the distal sense cells of the retina of *Pecten*', *J. Biophys. Biochem. Cytol.*, **4**, 227–228.

J. Millot and J. Anthony, 1956, 'Note préliminaire sur le thymus et la glande thyroide de *Latimeria chalumnae* (Crossoptérygien coelacanthidé)', *Compt. Rend. Acad. Sci. Paris* [D], **242**, 560–561.

——, 1958a, 'Crossoptérygiens actuels: *Latimeria chalumnae* derniers de Crossoptérygiens', in *Traité de Zoologie Anatomie, Systématique, Biologie*, Vol. 13 (ed. P.-P. Grassé), Masson, Paris, pp. 2553–2597.

J. Millot and J. Anthony, 1958b, *Anatomie de Latimeria chalumnae*, Vol. 1, Centre Nat. Rech. Sci., Paris.

J. Millot and N. Carasso, 1955, 'Note préliminaire sur l'oeil de *Latimeria chalumnae* (Crossoptérygien coelacanthidé)', *Compt. Rend. Acad. Sci. Paris* [D], **241**, 576–577.

N. Millott, 1954, 'Sensitivity to light and the reactions to changes in light intensity of the echinoid *Diadema antillarum* Phillippi', *Phil. Trans. Roy. Soc. London* [B], **238**, 187–220.

N. Millott and H. G. Vevers, 1955, 'Carotenoid pigment in the optic cushion of *Marthasterias glacialis* (L.)', *J. Mar. Biol. Ass. U.K.*, **34**, 279–287.

A. Minganti, 1955, 'Chemical investigations on amphibian egg jellies', *Exptl. Cell Res.*, Suppl. 3, 248–251.

A. Minganti and T. D'Anna, 1958, 'Sulla composizione della mucina ovulare di *Discoglossus pictus*', *Ric. Sci.*, **28**, 2090–2094.

C. S. Minot, 1897, 'Cephalic homologies: a contribution to the determination of the ancestry of vertebrates', *Am. Natur.*, **31**, 927–943.

S. G. Mivart, 1871, *On the Genesis of Species* (2nd ed.), Macmillan, London.

R. Morris, 1960, 'General problems of osmoregulation with special reference to cyclostomes', *Symp. Zool. Soc. London*, **1**, 1–16.

M. L. Moss, 1968, 'Comparative anatomy of vertebrate dermal bone and teeth I. The epidermal co-participation hypothesis', *Acta Anat.*, **71**, 178–208.

M. L. Moss and M. M. Meehan, 1967, 'Sutural connective tissue in the test of an echinoid *Arbacia punctulata*', *Acta Anat.*, **66**, 279–304.

F. W. Munz and W. N. McFarland, 1964, 'Regulative function of a primitive vertebrate kidney', *Comp. Biochem. Physiol.*, **13**, 381–400.

R. G. Myers, 1920, 'A chemical study of the blood of several invertebrate animals', *J. Biol. Chem.*, **41**, 119–143.

R. Nagabushanam, 1963, 'Neurosecretory changes in the nervous system of the oyster, *Crassostrea virginia*, induced by various experimental conditions', *Ind. J. Exptl. Biol.*, **2**, 1–4.

F. Nansen, 1887, 'The structure and combination of the histological elements of the central nervous system', *Bergens Mus. Aarsb. 1886*, 29–214.

H. V. Neal and H. W. Rand, 1936, *Comparative Anatomy*, Lewis, London.

D. M. Needham, J. Needham, E. Baldwin and J. Yudkin, 1931, 'A comparative study of the phosphagens, with some remarks on the origin of vertebrates', *Proc. Roy. Soc. London* [B], **110**, 260–294.

J. Needham, 1931, *Chemical Embryology*, Cambridge University Press, Cambridge.

——, 1938, 'Contributions of chemical physiology to the problem of reversibility in evolution', *Biol. Rev.*, **13**, 225–251.

G. J. Nelson, 1972, 'Comments on Hennig's "Phylogenetic systematics" and its influence on ichthyology', *Syst. Zool.*, **21**, 364–374.

J. A. C. Nicol, 1964, 'Special effectors: luminous organs, chromatophores, pigments and poison glands', in *Physiology of Mollusca*, Vol. 1 (ed. K. M. Wilbur and C. M. Yonge), Academic Press, New York, pp. 353–381.

R. Nieuwenhuys, 1963, 'The comparative anatomy of the actinopterygian forebrain', *J. Hirnforsch.*, **6**, 171–192.

——, 1964, 'Comparative anatomy of the spinal cord', in *Organization of the Spinal Cord* (ed. J. C. Eccles and J. P. Schadé), Elsevier, Amsterdam, pp. 1–57.

R. Nieuwenhuys and M. Hickey, 1965, 'A survey of the forebrain of the Australian lungfish Neoceratodus forsteri', *J. Hirnforsch.*, **7**, 433–452.

G. K. Noble, 1954, *The Biology of the Amphibia*, Dover, New York.

E. R. Norris and G. J. Benoit, 1945, 'Studies on trimethylamine oxide I. Occurrence of trimethylamine oxide in marine organisms', *J. Biol. Chem.*, **158**, 433–438.

M. M. Novikoff, 1952, 'Regularity of form in organisms', *System. Zool.*, **1**, 57–62.

S. Ohno, 1970, *Evolution by Gene Duplication*, Springer, Berlin.

T. Ohta, 1974, 'Mutational pressure as the main cause of molecular evolution and polymorphism', *Nature*, **252**, 351–354.

E. C. Olson, 1971, *Vertebrate Paleozoology*, Wiley, New York.

R. Olsson, 1961, 'The skin of Amphioxus', *Z. Zellforsch.*, **54**, 90–104.

——, 1965, 'Comparative morphology and physiology of the *Oikopleura* notochord', *Israel J. Zool.*, **14**, 213–220.

——, 1972, 'Reissner's fiber in ascidian tadpole larvae', *Acta Zool.*, **53**, 17–21.

J. H. Orton, 1914, 'On ciliary mechanisms in brachiopods and some polychaetes, with a comparison of the ciliary mechanisms on the gills of molluscs, Protochordata, brachiopods, and cryptocephalous polychaetes, and an account of the endostyle of Crepidula and its allies', *J. Marine Biol. Ass. U.K.*, **10**, 283–311.

T. Ørvig, 1967, 'Phylogeny of tooth tissues: evolution of some calcified tissues in early vertebrates', in *Structural and Chemical Organization of Teeth*, Vol. 1 (ed. A. E. W. Miles), Academic Press, New York, pp. 45–110.

D. T. Osuga and R. E. Feeney, 1968, 'Biochemistry of the egg-white proteins of ratites', *Arch. Biochem. Biophys.*, **124**, 560–574.

G. Owen, 1966, 'Digestion', in *Physiology of Mollusca*, Vol. 2 (ed. K. M. Wilbur and C. M. Yonge), Academic Press, New York, pp. 53–96.

A. Packard, 1972, 'Cephalopods and fishes: the limits of convergence,' *Biol. Rev.*, **47**, 241–307.

S. Paléus and G. Liljequist, 1972, 'The hemoglobins of *Myxine glutinosa* L.—II. Amino acid analyses, end group determinations and further investigations', *Comp. Biochem. Physiol.*, **42B**, 611–617.

C. F. A. Pantin, 1951, 'Organic design', *Adv. Sci.*, **8**, 138–150.

B. C. Parker and B. L. Turner, 1961, '"Operational niches" and "community-interaction" as determined from *in vitro* studies of some soil algae', *Evolution*, **15**, 228–238.

G. H. Parker, 1948, *Animal Colour Changes and Their Neurohumours*, Cambridge University Press, Cambridge.

H. W. Parker, 1956, 'Viviparous caecilians and amphibian phylogeny', *Nature*, **178**, 250–252.

G. Parry, 1960, 'Excretion', in *The Physiology of Crustacea*, Vol. 1 (ed. T. H. Waterman), Academic Press, New York, pp. 341–393.

T. S. Parsons, 1970, 'The nose and Jacobson's organ', in *Biology of Reptilia*, Vol. 2 (ed. C. Gans and T. S. Parsons), Academic Press, London, pp. 99–191.

T. S. Parsons and E. E. Williams, 1962, 'The teeth of Amphibia and their relation to amphibian phylogeny', *J. Morphol.*, **110**, 375–389.

——, 1963, 'The relationships of the modern Amphibia: a re-examination', *Quart. Rev. Biol.*, **38**, 26–53.

L. M. Passano, 1960, 'Molting and its control', in *The Physiology of Crustacea*, Vol. 1 (ed. T. H. Waterman), Academic Press, New York, pp. 473–536.

W. Patten, 1912, *The Evolution of the Vertebrates and Their Kin*, Churchill, London.

B. Patterson and R. Pascual, 1972, 'The fossil mammal fauna of South America', in *Evolution, Mammals and Southern Continents* (ed. A. Keast, F. C. Erk and B. Glass), State University of New York Press, Albany, pp. 247–309.

C. Patterson, 1967, 'Are the teleosts a monophyletic group?', in *Problèmes Actuels de Palé-ontologie. Evolution des Vertébrés* (ed. J.-P. Lehman), Centre Nat. Rech. Sci., Paris, pp. 93–109.

——, 1973, 'Interrelationships of holosteans', in *Interrelationships of Fishes* (ed. P. H. Greenwood, R. S. Miles and C. Patterson), Academic Press, London, pp. 233–305.

D. Penny, 1974, 'Evolutionary clock: the rate of evolution of rattlesnake cytochrome c', *J. Mol. Evol.*, **3**, 179–188.

J. Pereda, 1970, 'Etude histochimique des mucopolysaccharides de l'oviducte et des gangues muqueuses des ovocytes de *Rana pipiens*: incorporation du $^{35}SO_4^{2-}$', *Developm. Biol.*, **21**, 318–330.

A. M. Perks, 1969, 'The neurohypophysis', in *Fish Physiology*, Vol. 2 (ed. W. S. Hoar and D. J. Randall), Academic Press, New York, pp. 111–205.

P. Perlmann, H. Boström and A. Vestermark, 1959, 'Sialic acid in the gametes of the sea urchin', *Exptl. Cell Res.*, **17**, 439–446.

P. Person, 1969, 'Cartilaginous scales in cephalopods', *Science*, **164**, 1404–1405.

P. Person and M. B. Mathews, 1967, 'Endoskeletal cartilage in a marine polychaete, Eudistylia polymorpha', *Biol. Bull.*, **132**, 244–252.

P. Person and D. E. Philpott, 1969, 'The nature and significance of invertebrate cartilages', *Biol. Rev.*, **44**, 1–16.

W. Peters, 1966, 'Chitin in Tunicata', *Experientia*, **22**, 820–821.

B. Peyer, 1968, *Comparative Odontology* (transl. and ed. R. Zangerl), University of Chicago Press, Chicago.

G. E. Pickford and F. B. Grant, 1967, 'Serum osmolatity in the coelacanth, *Latimeria chalumnae*: urea retention and ion regulation', *Science*, **155**, 568–570.

K. A. Piez, 1962, 'Chemistry of the protein matrix of enamel', in *Fundamentals of Keratinization* (ed. E. O. Butcher and R. F. Sognnaes), *Am. Ass. Adv. Sci.*, Washington, D.C., pp. 173–184.

K. A. Piez and J. Gross, 1959, 'The amino acid composition and morphology of some invertebrate and vertebrate collagens', *Biochim. Biophys. Acta*, **34**, 24–39.

J. Pikkarainen, 1968, 'The molecular structures of vertebrate skin collagens. A comparative study', *Acta Physiol. Scand.*, Suppl. **309**, 3–72.

J. Pikkarainen and E. Kulonen, 1969, 'Comparative chemistry of collagen', *Nature*, **223**, 839–841.

A. Pisanò, 1949, 'Lo sviluppo dei primi due blastomeri separati dell'uovo di Ascidie', *Pubbl. Staz. Zool. Napoli*, **22**, 11–25.

K. R. Popper, 1968, *The Logic of Scientific Discovery* (rev. ed.), Hutchinson, London.

——, 1969, *Conjectures and Refutations. The Growth of Scientific Knowledge* (3rd rev. ed.), Routledge & Kegan Paul, London.

——, 1972, *Objective Knowledge: An Evolutionary Approach*, Clarendon, Oxford.

W. T. W. Potts, 1965, 'Ammonia excretion in *Octopus dofleini*', *Comp. Biochem. Physiol.*, **14**, 339–355.

——, 1967, 'Excretion in the molluscs', *Biol. Rev.*, **42**, 1–41.

E. M. Prager and A. C. Wilson, 1975, 'Slow evolutionary loss of the potential for interspecific hybridization in birds: a manifestation of slow regulatory evolution', *Proc. Nat. Acad. Sci. USA*, **72**, 200–204.

P. S. Prescott, 1974, 'The great agitator', *Newsweek*, May 6, 52–52A.

C. L. Prosser and F. A. Brown, 1961, *Comparative Animal Physiology* (2nd ed.), Saunders, Phila-delphia.

W. B. Provine, 1971, *The Origins of Theoretical Population Genetics*, University of Chicago Press, Chicago.

M. G. M. Pryor, 1962, 'Sclerotization', in *Comparative Biochemistry. A comprehensive treatise*, Vol. 4 (ed. M. Florkin and H. S. Mason), Academic Press, New York, pp. 371–396.

S. J. Przylecki, 1926, 'La répartition et la dégradation de l'acide urique chez les invertébrés', *Arch. Internat. Physiol.*, **27**, 157–202.

I. Pucci-Minafra, C. Casano and C. La Rosa, 1972, 'Collagen synthesis and spicule formation in sea urchin embryos', *Cell Diff.*, **1**, 157–165.

E. Ràdl, 1930, *The History of Biological Theories*, Oxford University Press, London.

F. Rahemtulla, N.-G. Höglund and S. Løvtrup, 1976, 'Acid mucopolysaccharides in the skin of some lower vertebrates (Hagfish, lamprey and *Chimaera*)', *Comp. Biochem. Physiol.*, **53B**, 295–298.

F. Rahemtulla and S. Løvtrup, 1975, 'The comparative biochemistry of invertebrate mucopoly-saccharides—IV. Bivalvia. Phylogenetic implications', *Comp. Biochem. Physiol.*, **50B**, 631–635.

F. Rahemtulla and S. Løvtrup, 1976, 'The comparative biochemistry of invertebrate mucopoly-saccharides—VI. Crustacea', *Comp. Biochem. Physiol.*, **53B**, 15–18.

D. P. Rall and J. W. Burger, 1967, 'Some aspects of hepatic and renal excretion in Myxine', *Am. J. Physiol.*, **212**, 354–356.

G. N. Ramachandran, 1963, 'Molecular structure of collagen', *Intern. Rev. Connective Tissue Res.*, **1**, 127–182.

A. Raynaud, 1969, 'Les organes génitaux des Mammifères', in *Traité de Zoologie Anatomie, Systématique, Biologie*, Vol. 16, 6 (ed. P.-P. Grassé), Masson, Paris, pp. 149–636.

K. R. H. Read, 1966, 'Molluscan hemoglobin and myoglobin', in *Physiology of Mollusca*, Vol. 2 (ed. K. M. Wilbur and C. M. Yonge), Academic Press, New York, pp. 209–232.

——, 1968, 'The myoglobins of the gastropod molluscs *Busycon contrarium* Conrad, *Lunatia heros* Say, *Littorina littorea* L. and *Siphonaria gigas* Sowerby', *Comp. Biochem. Physiol.*, **25**, 81–94.

C. A. Reed, 1960, 'Polyphyletic or monophyletic ancestry of mammals, or: What is a class?', *Evolution*, **14**, 314–322.

A. Remane, 1932, 'Archiannelida', in *Tierwelt der Nord- und Ostsee* Lief. 22, Teil VIa (ed. G. Grimpe), Akademische Verlagsgesellschaft, Leipzig, pp. 1–36.

——, 1936, 'Wirbelsaüle und ihre Abkömmlinge', in *Hanbuch der vergleichenden Anatomie der Wirbeltiere*, Vol. 4 (ed. L. Bolk, E. Göppert, E. Kallius and W. Lubosch), Urban & Schwarzenberg, Berlin, pp. 1–206.

——, 1956, *Die Grundlagen des natürlichen Systems, der vergleichenden Anatomie und der Phylogenetik. Theoretische Morphologie und Systematik*, Vol. 1 (2nd ed.), Akademisches Verlagsgesellschaft, Leipzig.

G. Retzius, 1890, 'Ueber die Ganglienzellen der Cerebrospinalganglien und über subcutane Ganglienzellen bei *Myxine glutinosa*', *Biol. Untersuch. N.F.*, **1**, 97–99.

G. Reverberi, 1961, 'The embryology of ascidians', *Adv. Morphogen.*, **1**, 55–101.

G. Reverberi and G. Ortolani, 1962, 'Twin larvae from halves of the same egg in ascidians', *Developm. Biol.*, **5**, 84–100.

A. G. Richards, 1951, *The Integument of Arthropods*, University of Minnesota Press, Minneapolis.

A. de Ricqlès, 1975, 'Quelques remarques paléo-histologiques sur le problème de la néoténie chez les Stégocéphales', in *Problèmes Actuels de Paléontologie (Évolution des Vertébrés)*, Centre Nat. Rech. Sci., Paris, pp. 351–363.

R. C. Robb, 1935, 'A study of mutations in evolution. I. Evolution in the equine skull', *J. Genet.*, **31**, 39–46.

C. S. Robbins, B. Bruun and H. S. Zim, 1968, *Birds of North America*, Golden Press, New York.

J. D. Robertson, 1954, 'The chemical composition of the blood of some aquatic chordates, including members of the Tunicata, Cyclostomata and Osteichtyes', *J. Exptl. Biol.*, **31**, 424–442.

——, 1957, 'The habitat of the early vertebrates', *Biol. Rev.*, **32**, 156–187.

——, 1960, 'Osmotic and ionic regulation', in *The Physiology of Crustacea*, Vol. 1 (ed. T. H. Waterman), Academic Press, New York, pp. 317–339.

——, 1963, 'Osmoregulation and ionic composition of cells and tissues', in *The Biology of Myxine* (ed. A. Brodal and R. Fänge), Universitetsforlaget, Oslo, pp. 503–515.

——, 1965, 'Studies on the chemical composition of muscle tissue III. The mantle muscle of cephalopod molluscs', *J. Exptl. Biol.*, **42**, 153–175.

——, 1966a, 'Osmotic constituents of the blood plasma and parietal muscle of *Myxine glutinosa* L', in *Some Contemporary Studies in Marine Sciences* (ed. H. Barnes), Allen & Unwin, London, pp. 631–644.

——, 1966b, 'Osmotic and ionic regulation', in *Physiology of Mollusca*, Vol. 1 (ed. K. M. Wilbur and C. M. Yonge), Academic Press, New York, pp. 283–311.

P. L. Robinson, 1975, 'The functions of the hooked fifth metatarsal in lepidosaurian reptiles', in *Problèmes Actuels de Paléontologie (Évolution des Vertébrés)*, Centre Nat. Rech. Sci., Paris, pp. 461–483.

J. Roche and M. Fontaine, 1940, 'Le pigment réspiratoire de la Lamproie marine et la répartition zoologiques des érythrocruorines et des hemoglobines', *Ann. Inst. Océanograph.*, **20**, 77–86.

J. Roche, M. Fontaine and J. Leloup, 1963, 'Halides', in *Comparative Biochemistry. A Comprehensive Treatise*, Vol. 5 (ed. M. Florkin and H. S. Mason), Academic Press, New York, pp. 493–547.

J. Roche, Y. Robin, N.-V. Thoai and L. A. Pradel, 1960, 'Sur les dérivés guanidiques et le phosphagène de quelques Annélides polychètes de la baie de Naples et du Mollusque *Arca noae* L.', *Comp. Biochem. Physiol.*, **1**, 44–55.

J. Roche, G. Salvatore and G. Rametta, 1962, 'Sur la présence et la biosynthèse d'hormones thyroidiennes chez un Tunicier, *Ciona intestinalis* L.', *Biochim. Biophys. Acta*, **63**, 154–165.

J. Roche, N.-V. Thoai and Y. Robin, 1957, 'Sur la présence de la créatine chez les Invertébrés et sa signification biologique', *Biochim. Biophys. Acta*, **24**, 514–519.

M. Rockstein, 1971, 'The distribution of phosphoarginine and phosphocreatine in marine invertebrates', *Biol. Bull.*, **141**, 167–175.

A. S. Romer, 1937, 'The brain case of the Carboniferous crossopterygian *Megalichthys nitidus*', *Bull. Mus. Comp. Zool.*, **82**, 1–73.

——, 1955a, 'Fish origins—fresh or salt water?', *Deep-Sea Res.*, **3**, 261–280.

——, 1955b, 'Herpeichthyes, Amphibioidei, Choanichthyes or Sarcopterygii?', *Nature*, **176**, 126–127.

——, 1962, *The Vertebrate Body* (3rd ed.), Saunders, Philadelphia.

——, 1966, *Vertebrate Paleontology* (3rd ed.), University of Chicago Press, Chicago.

A. E. Romero-Herrera, H. Lehmann, K. A. Joysey and A. E. Friday, 1973, 'Molecular evolution of myoglobin and the fossil record: a phylogenetic synthesis', *Nature*, **246**, 389–395.

D. M. Ross, 1963, 'The sense organs of *Myxine glutinosa* L.', in *The Biology of Myxine* (ed. A. Brodal and R. Fänge), Universitetsforlaget, Oslo, pp. 150–160.

[N. M. V.] Rothschild, 1965, *A Classification of Living Animals* (2nd ed.), Longmans Green, London.

G. H. Roux, 1942, 'The microscopic anatomy of the *Latimeria* scale', *South African J. Med. Sci.*, **7** (Biol. Suppl.), 1–18.

K. M. Rudall, 1952, 'The proteins of the mammalian epidermis', *Adv. Protein Chem.*, **7**, 253–290.

——, 1955, 'The distribution of collagen and chitin', *Symp. Soc. Exptl. Biol.*, **9**, 49–71.

——, 1968, 'Intracellular fibrous proteins and the keratins', in *Comprehensive Biochemistry*, Vol. 26B (ed. M. Florkin and E. H. Stotz), Elsevier, Amsterdam, pp. 559–591.

K. M. Rudall and W. Kenchington, 1973, 'The chitin system', *Biol. Rev.*, **49**, 597–636.

L. S. Russell, 1965, 'Body temperature of dinosaurs, and its relationship to their extinction', *J. Paleont.*, **39**, 497–501.

G. Ruud, 1925, 'Die Entwicklung isolierter Keimfragmente frühester Stadien von Triton taeniatus', *Arch. Entw.mech.*, **105**, 209–293.

P. de Saint-Seine, C. Devillers and J. Blot, 1969, 'Holocéphales et Élasmobranches', in *Traité de Paléontologie*, Vol. 4,2 (ed. J. Piveteau), Masson, Paris, pp. 693–776.

S. N. Salthe and N. O. Kaplan, 1966, 'Immunology and rates of enzyme evolution in the Amphibia in relation to the origins of certain taxa', *Evolution*, **20**, 603–616.

A. Sannasi and H. R. Hermann, 1970, 'Chitin in the Cephalochordata, *Branchisotoma floridae*', *Experientia*, **26**, 351–352.

L. Sanzo, 1907, 'Stickstoff-Stoffwechsel bei marinen wirbellosen Tieren', *Biol. Zentralbl.*, **27**, 479–491.

V. M. Sarich, 1970, 'Primate systematics with special reference to Old World monkeys', in *Old World Monkeys, Evolution, Systematics and Behavior* (ed. J. R. Napier and P. H. Napier), Academic Press, New York, pp. 175–226.

G. Säve-Söderberg, 1934, 'Some points of view concerning the evolution of the vertebrates and the classification of this group', *Arkiv Zool.*, **26A**, 17, 1–20.

B. Schaeffer, 1968, 'The origin and basic radiation of the Osteichthyes', in *Current Problems of Lower Vertebrate Phylogeny* (ed. T. Ørvig), Almqvist & Wiksell, pp. 207–222.

——, 1973, 'Interrelationships of chondrosteans', in *Interrelationships of Fishes* (ed. P. H. Greenwood, R. S. Miles and C. Patterson), Academic Press, London, pp. 207–226.

E. Schiffman, G. R. Martin and E. J. Miller, 1970, 'Matrices that calcify', in *Biological Calcification: Cellular and Molecular Aspects* (ed. H. Schraer), North Holland, Amsterdam, pp. 27–67.

O. H. Schindewolf, 1950, *Grundfragen der Paläontologie*, Schweizerbart, Stuttgart.

H. Schirner, 1963, 'The pancreas', in *The Biology of Myxine* (ed. A. Brodal and R. Fänge), Universitetsforlaget, Oslo, pp. 481–487.

I. I. Schmalhausen, 1949, *Factors of Evolution: The Theory of Stabilizing Selection*, Blakiston, Philadelphia.

K. C. Schneider, 1912, 'Zur Theorie des Systems', *Zool. Jahrb.* Suppl. XV., **3**, 135–154.

G. D. Schnell, 1970, 'A phenetic study of the suborder Lari (Aves). II. Phenograms, discussion, and conclusions', *Syst. Zool.*, **19**, 264–302.

E. Schoffeniels and R. Gilles, 1970, 'Nitrogenous constituents and nitrogen metabolism in arthropods', in *Chemical Zoology*, Vol. 5A (ed. M. Florkin and B. T. Scheer), Academic Press, New York, pp. 199–227.

A. W. Schuetz, 1969, 'Chemical properties and physiological actions of a starfish radial nerve factor and ovarian factor', *Gen. Comp. Endocrin.*, **12**, 209–221.

H.-P. Schultze, 1970, 'Die Histologie der Wirbelkörper der Dipnoer', *N. Jb. Geol. Paläont. Abh.*, **135**, 311–336.

——, 1973, 'Crossopterygier mit heterozerker Schwanzflosse aus dem Oberdevon Kanadas, nebst einer Beschreibung von Onychodontida-Resten aus dem Mitteldevon Spaniens und aus dem Karbon der USA', *Palaeontographica* **143** [A], 188–208.

G. R. Seaman and N. L. Robert, 1968, 'Immunological response of male cockroaches to injection of Tetrahymena pyriformis', *Science*, **161**, 1359–1361.

C. Semper, 1875, 'Die Verwandtschaftsbeziehungen der gegliederten Thiere', *Arb. Zool. -Zoot. Inst. Würzburg*.

N. Seno, F. Akiyama and K. Anno, 1972, 'A novel dermatan polysulphate from hagfish skin, containing trisulphated disaccharide residues', *Biochim. Biophys. Acta*, **264**, 229–233.

E. Sereni, 1930, 'The chromatophores of the cephalopods', *Biol. Bull.*, **59**, 247–268.

R. V. Seshaya, P. Ambujabai and M. Kalyani, 1963, 'Amino acid composition of ichtylepidin from fish scales', in *Aspects of Protein Structure* (ed. G. N. Ramachandran), Academic Press, London, pp. 343–349.

L. R. Sillman, 1960, 'The origin of the vertebrates', *J. Paleontol.*, **34**, 540–544.

G. G. Simpson, 1940, 'Mammals and land bridges', *J. Washington Acad. Sci.*, **30**, 137–163.

——, 1944, *Tempo and Mode in Evolution*, Columbia University Press, New York.

——, 1953, *The Major Features of Evolution*, Colombia University Press, New York.

——, 1959, 'Mesozoic mammals and the polyphyletic origin of mammals', *Evolution*, **13**, 405–414.

——, 1961, *Principles of Animal Taxonomy*, Columbia University Press, New York.

——, 1967, *The Meaning of Evolution* (rev. ed.), Yale University Press, New Haven.

B. H. Slaughter, 1968, 'Earliest known marsupials', *Science*, **162**, 254–255.

W. C. Sloan, 1954, 'The accumulation of nitrogenous compounds in terrestrial and aquatic eggs of prosobranch snails', *Biol. Bull.*, **126**, 302–306.

C. L. Smith, C. S. Rand, B. Schaeffer and J. W. Atz, 1975, '*Latimeria*, the living coelacanth, is ovoviviparous', *Science*, **190**, 249–250.

H. W. Smith, 1932, 'Water regulation and its evolution in fishes', *Quart. Rev. Biol.*, **7**, 1–26.

——, 1953, *From Fish to Philosopher*, Little Brown, Boston.

J. E. Smith, 1965, 'Echinodermata', in T. H. Bullock and G. A. Horridge, *Structure and Function in the Nervous System of Invertebrates*, Freeman, San Francisco, pp. 1519–1558.

——, 1966, 'The form and function of the nervous system', in *Physiology of Echinodermata* (ed. R. A. Boolootian), Wiley, New York, pp. 503–511.

J. L. B. Smith, 1940, 'A living coelacanthid fish from South Africa', *Trans. Roy. Soc. South Africa*, **28**, 1–106.

M. M. Smith, M. H. Hobdell and W. A. Miller, 1972, 'The structure of the scales of *Latimeria chalumnae*', *J. Zool. London*, **167**, 501–509.

P. H. A. Sneath and R. R. Sokal, 1973, *Numerical Taxonomy. The Principles and Practice of Numerical Classification*, Freeman, San Francisco.

G. Sommerhoff, 1950, *Analytical Biology*, Oxford University Press, Oxford.

T. H. Sonneborn, 1965, 'Degeneracy of the genetic code: extent, nature and genetic implications', in *Evolving Genes and Proteins* (ed. V. Bryson and H. J. Vogel), Academic Press, New York, pp. 377–397.

K. V. Speeg and J. W. Campbell, 1968, 'Formation and volatilization of ammonia gas by terrestrial snails', *Am. J. Physiol.*, **214**, 1392–1402.

N. Spjeldnaes, 1965, 'The palaeozoology of the Ordovician vertebrates of the Harding formation', in *Problèmes Actuels de Paléontologie* (*Evolution des Vértebrés*) (ed. J.-P. Lehman), Centre Nat. Rech. Sci., Paris, pp. 11–20.

S. R. Srinivasan, B. Radhakrishnamurthy, E. R. Dalferes and G. S. Berenson, 1969, 'Glycosaminoglycans from squid skin', *Comp. Biochem. Physiol.*, **28**, 169–176.

S. M. Stanley, 1973, 'An explanation for Cope's rule', *Evolution*, **27**, 1–26.

E. A. Stensiö, 1958, 'Les Cyclostomes fossiles ou Ostracoderms', in *Traité de Zoologie Anatomie, Systématique, Biologie*, Vol. 13,1 (ed. P.-P. Grassé), Masson, Paris, pp. 173–425.

——, 1968, 'The cyclostomes with special reference to the diphyletic origin of the Petromyzontida and Myxinoidea', in *Current Problems of Lower Vertebrate Phylogeny* (ed. T. Ørvig), Almqvist & Wiksell, Stockholm, pp. 13–71.

G. C. Stephens, J. F. Van Pilsum and D. Taylor, 1965, 'Phylogeny and the distribution of creatine in invertebrates', *Biol. Bull.*, **129**, 573–581.

L. Størmer, 1949, 'Classe de Trilobites. Classes de Merostomoidea Marellomorpha et Pseudocrustacea', in *Traité de Zoologie Anatomie, Systématique, Biologie*, Vol. 6 (ed. P.-P. Grassé), Masson, Paris, pp. 160–210.

H. Szarski, 1962, 'The origin of the Amphibia', *Quart. Rev. Biol.*, **37**, 189–241.

K. Takahashi, 1966, 'Muscle physiology', in *Physiology of Echinodermata* (ed. R. A. Boolootian), Wiley, New York, pp. 513–527.

L. Tauc, 1966, 'Physiology of the nervous system', in *Physiology of Mollusca*, Vol. 2 (ed. K. M. Wilbur and C. M. Yonge), Academic Press, New York, pp. 387–454.

L. Thaler, 1973, 'Nanisme et gigantisme insulaires', *La Recherche*, **4**, 741–750.

G. H. Thoenes and W. H. Hildemann, 1970, 'Immunological responses of pacific hagfish II. Serum antibody production to soluble antigen', in *Developmental Aspects of Antibody Formation and Structure*, Vol. 11 (ed. J. Šterzl and J. Riha), Academic Press, New York, pp. 711–726.

D'Arcy W. Thompson, 1942, *On Growth and Form* (2nd ed.), Cambridge University Press, Cambridge.

K. S. Thomson, 1969, 'The biology of the lobe-finned fishes', *Biol. Rev.*, **44**, 91–154.

——, 1971, 'The adaptation and evolution of early fishes', *Quart. Rev. Biol.*, **46**, 139–166.

R. A. Thulborn, 1971, 'Origins and evolution of ornitischian dinosaurs', *Nature*, **234**, 75–78.

H. Timofeeff-Ressovsky, 1935, 'Auslösung von Vitalitätsmutationen durch Röntgenbestrahlung bei *Drosophilia melanogaster*', *Nachr. Ges. Wiss. Göttingen. Biol. N.F.*., **1**, 163–180.

B. P. Toole, G. Jackson and J. Gross, 1972, 'Hyaluronate in morphogenesis: inhibition of chondrogenesis *in vitro*', *Proc. Nat. Acad. Sci. USA*, **69**, 1384–1386.

D. F. Travis, 1970, 'The comparative ultrastructure and organization of five calcified tissues', in *Biological Calcification: Cellular and Molecular Aspects* (ed. H. Schraer), North-Holland, Amsterdam, pp. 203–311.

J. L. Travis, 1971, 'A criticism of the use of the concept of "dominant group" in arguments for evolutionary progressivism', *Philos. Sci.*, **38**, 369–375.

Y. Tsuchiya and Y. Suzuki, 1962, 'Biochemical studies of the ascidian, *Cynthia roretzi* v. Drasche VI. The presence of pseudokeratin in the test', *Bull. Jap. Soc. Sci. Fish*, **28**, 222–234.

T. C. Tung, S. C. Wu and Y. F. Y. Tung, 1958, 'The development of isolated blastomeres of Amphioxus', *Scientia Sinica*, **7**, 1280–1320.

R. Tuomikoski, 1967, 'Notes on some principles of phylogenetic systematics', *Ann. Ent. Fenn.*, **33**, 137–147.

H. Unger, 1962, 'Experimentelle und histologische Untersuchungen über Wirkfaktoren aus dem Nervensystem von *Asterias* (*Marthasterias*) *glacialis*, (Asteroidea: Echinodermata)', *Zool. Jahrb.* [3], **69**, 481–536.

J. W. Valentine, 1969, 'Patterns of taxonomic and ecological structure of the shelf benthos during Phanerozoic time', *Palaeontology*, **12**, 684–709.

J. A. Valverde, 1964, 'Remarques sur la structure et l'évolution des communautés des Vertébrés terrestres. I. Structure d'une communauté. II. Rapports entre prédateurs et proies', *La Terre et la Vie*, **111**, 121–154.

A. Vandel, 1958, *L'Homme et l'Evolution* (2nd ed.), Gallimard, Paris.

——, 1965, *Biospeology. The Biology of Cavernicolous Animals*, Pergamon, Oxford.

A. J. P. Van den Broek, 1933, 'Gonaden und Ausführungsgänge', in *Handbuch der vergleichende Anatomie der Wirbeltiere*, Vol. 6 (ed. L. Bolk, E. Göppert, E. Kallius and W. Lubosch), Urban & Schwarzenberg, Berlin, pp. 1–154.

H. C. Van der Heyde, 1923, 'Petites contributions à la physiologie comparée IV. Sur l'excretion chez les Echinodermes', *Arch. Neérl. Physiol.*, **8**, 151–160.

L. Van Valen, 1976, 'Energy and evolution', *Evolut. Theory*, **1**, 179–229.

E. Vasseur, 1952, *The Chemistry and Physiology of the Jelly Coat of the Sea-urchin Egg*, Stockholm.

N. Vicente, 1963, 'Ablation des ganglions nerveux et osmorégulation chez *Aplysia rosea* Rathke (Gastérop. Opistobr.)', *Compt. Rend. Acad. Sci. Paris* [D], **256**, 2928–2930.

W. Vogt, 1929, 'Gestaltungsanalyse am Amphibienkeim mit örtlicher Vitalfärbung. II. Teil. Gastrulation und Mesodermbildung bei Urodelen und Anuren', *Arch. Entw. mech.*, **120**, 384–706.

H. J. Vonk, 1960, 'Digestion and metabolism', in *The Physiology of Crustacea*, Vol. 1 (ed. T. H. Waterman), Academic Press, New York, pp. 291–316.

P. J. Vorzimmer, 1970, *Charles Darwin: The Years of Controversy*, Temple University Press, Philadelphia.

M. Wagner, 1868, Die Darwin'sche Theorie und das Migrationsgesetz der Organismen, Duncker & Humblot, Leipzig.

G. von Wahlert, 1968, *Latimeria und die Geschichte der Wirbeltiere Eine evolutionsbiologische Untersuchung*, Fischer, Stuttgart.

G. Wald, 1960, 'The distribution and evolution of visual systems', in *Comparative Biochemistry. A Comprehensive Treatise*, Vol. 1 (ed. M. Florkin and H. S. Mason), Academic Press, New York, pp. 311–345.

E. P. Walker, 1968, *Mammals of the World* (2nd ed.), Johns Hopkins, Baltimore.

B. Wallace, 1968, 'Polymorphism, population size, and genetic load', in *Population Biology and Evolution* (ed. R. C. Lewontin), Syracuse University Press, Syracuse, pp. 87–108.

——, 1970, *Genetic Load. Its Biological and Conceptual Aspects*, Prentice-Hall, Englewood Cliffs.

L. Warren, 1963, 'The distribution of sialic acids in nature', *Comp. Biochem. Physiol.*, **10**, 153–171.

T. H. Waterman, 1961, 'Comparative physiology', in *The Physiology of Crustacea*, Vol. 2 (ed. T. H. Waterman), Academic Press, New York, pp. 521–593.

D. M. S. Watson, 1937, 'The acanthodian fishes', *Phil. Trans. Roy. Soc. London* [B], **228**, 49–146.

R. L. Watts and D. C. Watts, 1968, 'Gene duplication and the evolution of enzymes', *Nature*, **217**, 1125–1130.

C. K. Weichert, 1965, *Anatomy of the Chordates* (3rd ed.), McGraw-Hill, New York.

M. J. Wells, 1960, 'Optic glands and the ovary of *Octopus*', *Symp. Zool. Soc. London*, **2**, 87–107.

M. J. Wells and J. Wells, 1969, 'Pituitary analogue in the octopus', *Nature*, **222**, 293–294.

J. H. Welsh, 1961, 'Neurohumors and neurosecretion', in *The Physiology of Crustacea*, Vol. 2 (ed. T. H. Waterman), Academic Press, New York, pp. 281–311.

E. I. White, 1946, '*Jamoytius kerwoodi*, a new chordate from the Silurian of Lanarkshire', *Geol. Mag.*, **88**, 89–97.

——, 1958, 'Original environment of the craniates', in *Studies of Fossil Vertebrates* (ed. T. S. Westoll), Athlone, London, pp. 213–234.

M. Whitear, 1959, 'Some remarks on the ascidian affinities of vertebrates', *Ann. Mag. Nat. Hist.* [12], **10**, 338–348.

C. O. Whitman, 1895, 'Evolution and epigenesis', *Biol. Lectures Wood's Holl*, **1894**, 205–224.

A. Wierzejski, 1905, 'Embryologie von *Physa fontinalis* L.', *Z. Wiss. Zool.*, **83**, 502–705.

K. M. Wilbur and K. Simkiss, 1968, 'Calcified shells', in *Comprehensive Biochemistry*, Vol. 26A (ed. M. Florkin and E. H. Stotz), Elsevier, Amsterdam, pp. 229–295.

E. O. Wiley, 1975, 'Karl R. Popper, systematics, and classification: a reply to Walter Bock and other evolutionary taxonomists', *Syst. Zool.*, **24**, 233–243.

A. Willey, 1894, *Amphioxus and the Ancestry of the Vertebrates*, Macmillan, New York.

M. B. Williams, 1970, 'Deducing the consequences of evolution: a mathematical model', *J. Theor. Biol.*, **29**, 343–385.

J. C. Willis, 1922, *Age and Area. A Study in Geographical Distribution and Origin of Species*, Cambridge University Press, Cambridge.

——, 1940, *The Course of Evolution by Differentiation or Divergent Mutation rather than by Selection*, Cambridge University Press, Cambridge.

E. N. Willmer, 1974, 'Nemertines as possible ancestors of the vertebrates', *Biol. Rev.*, **49**, 321–363.

A. C. Wilson, G. L. Bush, S. M. Case and M.-C. King, 1975, 'Social structuring of mammalian populations and rate of chromosomal evolution', *Proc. Nat. Acad. Sci. USA*, **72**, 5061–5065.

A. C. Wilson, L. R. Maxson and V. M. Sarich, 1974, 'Two types of molecular evolution: evidence from studies of interspecific hybridization', *Proc. Nat. Acad. Sci. USA*, **71**, 2843–2847.

A. C. Wilson, V. M. Sarich and L. R. Maxson, 1974, 'The importance of gene arrangement in evolution: evidence from studies on rates of chromosomal, protein and anatomical evolution', *Proc. Nat. Acad. Sci. USA*, **71**, 3028–3030.

E. B. Wilson, 1904, 'Experimental studies in germinal localization. II. Experiments on the cleavage-mosaic in Patella and Dentalium', *J. Exptl. Zool.*, **1**, 197–268.

J. T. Wilson, 1973, 'Continental drift', in *Continents Adrift*, Freeman, San Francisco, pp. 41–55.

S. Wilson and S. Falkmer, 1965, 'Starfish insulin', *Can. J. Biochem.*, **43**, 1615–1624.

H. P. Wolvekamp and T. H. Waterman, 1960, 'Respiration', in *The Physiology of Crustacea*, Vol. 1 (ed. T. H. Waterman), Academic Press, New York, pp. 35–100.

S. C. Wood, K. Johansen and R. E. Weber, 1972, 'Haemoglobin of the coelacanth', *Nature*, **239**, 283–285.

A. Woodfield, 1973, 'Darwin, teleology and taxonomy', *Philosophy*, **48**, 35–49.

J. H. Woodger, 1929, *Biological Principles*, Routledge & Kegan Paul, London.

J. H. Woodger, 1952, *Biology and Language, An Introduction to the Methodology of the Biological Sciences including Medicine*, Cambridge University Press, Cambridge.

S. Wright, 1931, 'Evolution in Mendelian populations', *Genetics*, **16**, 97–159.

T. Yamagata and K. Okazaki, 1974, 'Occurrence of a dermatan sulfate isomer in sea urchin larvae', *Biochim. Biophys. Acta*, **372**, 469–473.

G. Yasuzumi, H. Tanaka and O. Tezuka, 1960, 'Spermatogenesis in animals as revealed by electron microscopy. VIII. Relation between the nutritive cells and the developing spermatids in a pond snail *Cipangopalludina malleata* Reeve', *J. Biophys. Biochem. Cytol.*, **7**, 499–504.

C. M. Yonge, 1925, 'Studies on the comparative physiology of digestion. III. Secretion, digestion, and assimilation in the gut of *Ciona intestinalis*', *Brit. J. Exptl. Biol.*, **2**, 373–388.

——, 1937, 'Evolution and adaptation in the digestive system of the Metazoa', *Biol. Rev.*, **12**, 87–115.

M. Yoshida, 1966, 'Photosensitivity', in *Physiology of Echinodermata* (ed. R. A. Boolootian), Wiley, New York, pp. 435–464.

M. Yoshida and N. Millott, 1960, 'The shadow reaction of *Diadema antillarum* Philippi. III. Re-examination of the spectral sensitivity', *J. Exptl. Biol.*, **37**, 390–397.

J. Z. Young, 1936, 'The photoreceptors of lampreys II. The functions of the pineal complex', *J. Exptl. Biol.*, **12**, 254–270.

——, 1960, 'The statocysts of *Octopus vulgaris*', *Proc. Roy. Soc. London* [B], **152**, 3–29.

——, 1962, *The Life of Vertebrates* (2nd ed.), Clarendon, Oxford.

W. H. Yudkin, 1954, 'Transphosphorylation in echinoderms', *J. Cell. Comp. Physiol.*, **44**, 507–518.

G. U. Yule, 1924, 'A mathematical theory of evolution, based on the conclusions of Dr. J. C. Willis, F.R.S.', *Phil. Trans. Roy. Soc. London* [B], **213**, 21–87.

E. Zuckerkandl, 1975, 'The appearance of new structures and functions in proteins during evolution', *J. Mol. Evol.*, **7**, 1–57.

E. Zuckerkandl and L. Pauling, 1965, 'Evolutionary divergence and convergence in proteins', in *Evolving Genes and Proteins* (ed. V. Bryson and H. J. Vogel), Academic Press, New York, pp. 97–166.

AUTHOR INDEX

SUBJECT INDEX

The text printed in small print on the pages 84–133 is a survey of non-morphological characters used for testing the Protostomia-Deuterostomia theory. A wealth of different subjects are dealt with on these pages, most of them peripheral to the main theme of the book. However, the text is subdivided in quite small sections, and it should therefore be easy to use the list of contents for information retrieval. For these reasons the text in question has not been included in the subject index.